One-Dimensional Turbulence and the Stochastic Burgers Equation

Mathematical
Surveys
and
Monographs

Volume 255

One-Dimensional Turbulence and the Stochastic Burgers Equation

Alexandre Boritchev
Sergei Kuksin

AMERICAN MATHEMATICAL SOCIETY
Providence, Rhode Island

2020 *Mathematics Subject Classification.* Primary 60H15;
Secondary 35L65, 35Q35, 37A25, 76F02, 76F25.

For additional information and updates on this book, visit
www.ams.org/bookpages/surv-255

Library of Congress Cataloging-in-Publication Data

Names: Boritchev, Alexandre, 1986- author. | Kuksin, Sergej B., 1955- author.
Title: One-dimensional turbulence and the stochastic Burgers equation / Alexandre Boritchev, Sergei Kuksin.
Description: Providence, Rhode Island: American Mathematical Society, [2021] | Series: Mathematical surveys and monographs, 0076-5376; volume 255 | Includes bibliographical references and index.
Identifiers: LCCN 2021004451 | ISBN 9781470464363 (paperback) | (ebook)
Subjects: LCSH: Stochastic partial differential equations. | Burgers equation. | Turbulence–Mathematical models. | AMS: Probability theory and stochastic processes – Stochastic analysis – Stochastic partial differentia | Partial differential equations – Hyperbolic equations and systems – Conservation laws. | Partial differential equations – Equations of mathematical physics and other areas of application – PDEs in connection with fluid mechanics. | Dynamical systems and ergodic theory – Ergodic theory – Ergodicity, mixing, rates of mixing. | Fluid mechanics – Turbulence – Fundamentals. | Fluid mechanics – Turbulence – Turbulent transport, mixing.
Classification: LCC QA274.25 .B67 2021 | DDC 515/.353–dc23
LC record available at https://lccn.loc.gov/2021004451

∞ The paper used in this book is acid-free and falls within the guidelines
established to ensure permanence and durability.
Visit the AMS home page at `https://www.ams.org/`

10 9 8 7 6 5 4 3 2 1 26 25 24 23 22 21

Table of contents

Introduction

The goal of this book is to present in a systematic form the results obtained in the last 20 years for the qualitative study of solutions to the small-viscosity stochastic 1d Burgers equation under periodic boundary conditions and applications of these results to certain physical problems. Due to some remarkable properties of the equation, it proved to be possible to get a description of the solutions which becomes asymptotically sharp as the viscosity goes to zero. The corresponding results may be reformulated in terms of turbulence in the Burgers equation, and allow to prove rigorously a number of predictions made in the 1940's–1990's and established in [**1**, **10**] on the physical level of rigour. For the turbulence in the Burgers equation Uriel Frisch coined the name "burgulence", which we use below.

The suggested rigorous justification of burgulence relies on a basic theory of the stochastic Burgers equation. The latter could be derived from the general theory of SPDEs (i.e. stochastic PDEs), developed in the last decades, see the books [**43**, **44**, **60**, **97**, **121**, **132**, **134**, **148**]. But SPDEs are a rather technical topic, and their general theory (practically) relies on the reader's experience both in PDEs and in stochastic analysis, which makes the use of these books rather complicated and time-consuming. On the other hand, since the Burgers equation is one of the easiest nonlinear PDEs, then its stochastic version may be examined directly, without evoking the general theory. The corresponding self-consistent study makes up the content of the first chapters of our book. The simplicity of the model allows us to present there in a light form many of the main ideas from the theory of SPDEs. Therefore this part of the text may be regarded as a basic introduction to stochastic PDEs and may serve as a foundation for further reading on that topic.

The book is concerned with the stochastic Burgers equation

(0.0.1)

$$u_t + uu_x - \nu u_{xx} = \eta(t,x), \quad u = u(t,x),\ t \geq 0,\ x \in S^1 := \mathbb{R}/\mathbb{Z}, \ \int_{S^1} u(t,x)\,dx = 0,$$

where $0 < \nu \leq 1$, $\eta = \partial_t \xi(t,x)$ and ξ a Wiener process in the space of functions of x, of the form

$$\xi^\omega(t,x) = \sum_{s=\pm1,\pm2,\dots} b_s \beta_s^\omega(t) e_s(x), \qquad 0 < B_0 = \sum b_s^2 < \infty. \tag{0.0.2}$$

Here $\{e_s(x), s = \pm1, \pm2, \dots\}$ is the trigonometric basis in the space of periodic functions with zero mean, $\{\beta_s^\omega(t)\}$ are standard independent Brownian processes and $\{b_s\}$ are real numbers. For the purposes of this introduction we assume that b_s goes to zero faster than every negative degree of $|s|$. Below we refer to (0.0.1) with any force η as *the Burgers equation*, and to (0.0.1), (0.0.2) - as *the stochastic Burgers equation.*

Although (0.0.1) is a relatively simple nonlinear partial differential equation of mathematical physics, it is still highly non-trivial. Without entering into considerations of the most recent research, let us say a few words about the equation and its relevance, both in the deterministic and random setting. If one only considers the free Burgers equation (eq. (0.0.1) with $\eta = 0$), one recognises in it a 1D version of the free incompressible Navier-Stokes system with the gradient of the pressure removed. So the equation describes a 1d media where evolution of a particle is governed only by the advection term uu_x and the viscous damping term νu_{xx}. This was the original motivation for the work starting in 1930s of the Dutch physicist who played a capital role in advocating eq. (0.0.1) as a hydrodynamical model, and whose name the equation bears [**35**]. As Johannes Burgers stated it, the equation describes 'infinitely compressible medium without pressure'.

On the purely mathematical level, the equation is integrable in the sense that it can be mapped to the heat equation using the Cole-Hopf transformation ([**39**, **64**, **76**]; see [**19**] for a historical account). However, this transformation does not exist in the limiting inviscid case $\nu = 0$, while for the case of small positive ν, which interests us the most in this book, it is not instrumental since it involves exponentials of a certain quantity divided by ν. So it does not provide information about the behaviour of solutions for small ν, at least in a transparent way. Moreover, the properties of solutions for the Burgers equation are similar to those for a large class of nonlinear PDEs, called *viscous scalar conservation laws,* for which an integrating transformation is not available, while for some equations from this group the approach of our book still applies (e.g. see below equations (0.0.12), (0.0.13)). The theory of the latter class of PDEs has been intensively developed starting from the pioneering works of Lax, Oleinik and Kruzhkov in the 1950s-60s [**95**, **96**, **116**, **129**]. In particular, it was realised that when ν goes to zero, their solutions form shocks of a special type, which burst the process of dissipation of energy. Accordingly, for a fixed positive time, the solutions stay of order one when the viscosity ν decays to zero. The behaviour of shocks in the Burgers equation and in other viscous scalar conservation laws in the limit $\nu \to 0$, and its relevance for the small-viscosity properties of the solutions, make an interesting and complicated theory, both in the free and forced equations, which we do not touch in our book.

In physics and mechanics, among other applications, the Burgers equation has been used as an approximation in a number of settings from the gas dynamics. There it provides natural pressureless models, ranging from acoustics [**151**] to granular media [**12**]. In the late 1940s, V. A. Florin[1] used the equation to describe the consolidation of wet soil, and in the work [**64**], a few years before Hopf and Cole [**39**, **76**], discovered the transformation integrating the equation. In cosmology, a multidimensional inviscid Burgers equation was considered by Zeldovich with collaborators to describe large-scale movements of galaxies [**136**]. Let us also mention that the equation is used as a model for traffic flow, where the shocks correspond to the jamming effects [**38**].

From the physical point of view, when using the Burgers equation, it is natural to add to it a random r.h.s. (and/or random initial data) to represent various influences neglected by the model. So physical works on the Burgers equation with randomness appeared practically simultaneously with those on the deterministic

[1] Victor Anatolyevich Florin (1899-1960) was a prominent Soviet engineer, expert in the mechanics of soil, and the author of the fundamental work [**65**].

version; some references may be found in [**44**, Section 14]. Burgers himself in [**34**] manipulates correlations of solutions $u(t,x)$, thus implicitly assuming their randomness. Starting from the 1980's, when reliable computer numerics of PDEs with random terms became available, conclusions of such works became more convincing. Eventually in the 1990's the problem of statistical properties of solutions for the Burgers equation with stochasticity drew the attention of Sinai [**140, 141**], and after the influential paper [**56**] – of many other mathematicians.

In parallel, different forms of the randomly forced Burgers equation have been considered from the point of view of statistical mechanics and field theory, e.g. see in [**94**]. The stochastic Burgers equation with white in time and singular in space forces $\eta(t,x)$ appears in the study of one-dimensional interfaces, where its solutions correspond to derivatives of the slopes. The evolution of the latter is referred to as the Kardar-Parisi-Zhang (KPZ) model [**81**]. The discrete processes and hydrodynamical limits linked to KPZ-type models have given birth to an important research activity in the past 30 years [**133**].

Now let us describe the structure and content of our book.

In Chapter 1 we construct solutions of the equation by firstly solving the linear heat equation $u_t - \nu u_{xx} = \eta(t,x)$, and then studying the corresponding nonlinear correction. There we also introduce the Markov process defined by solutions of the stochastic equation and discuss the two corresponding Markov semigroups.

In Chapter 2 we obtain upper and lower bounds for moments of Sobolev norms of the solutions, which turn out to be asymptotically sharp as $\nu \to 0$ in a sense which we will soon explain. We start the section with Oleinik's maximum principle in the form given to it by Kruzhkov. This extremely powerful result provides estimates for L_∞-norms in x of a solution $u(t,x)$ and of $u_x^+(t,x) = \max(u_x(t,x), 0)$, uniformly in ν, in the initial data $u(0,x)$ and in $t \geq \theta > 0$, for any fixed $\theta > 0$. Next we use this estimate and basic PDE techniques to bound from above moments of the Sobolev norms $\|u(t,\cdot)\|_m$ with $m \geq 1$; in particular we obtain that $\mathbb{E}\|u(t,\cdot)\|_m^2 \leq C_m \nu^{-(2m-1)}$ for any $t \geq 1$ and $m \geq 1$. Then we evoke the energy balance

$$\frac{1}{2}\mathbb{E}\int_{S^1} |u(T+\sigma,x)|^2 dx - \frac{1}{2}\mathbb{E}\int_{S^1} |u(T,x)|^2 dx + \nu\mathbb{E}\int_T^{T+\sigma}\int_{S^1} |u_x(s,x)|^2 dx ds = \frac{\sigma}{2}B_0,$$

where $T \geq 1$. By Oleinik's maximum principle the sum of the first two terms in the l.h.s is bounded by a constant C, depending only on the process ξ. Choosing $\sigma \geq \sigma_0 = 4C/B_0$, we see that the squared first Sobolev norm of a solution u satisfies $\langle\langle \|u\|_1^2 \rangle\rangle \geq B_0 \nu^{-1}/4$. Here the double brackets denote the averaging in ensemble and local averaging in time given by

$$\text{(0.0.3)} \qquad \langle\langle f^\omega(t) \rangle\rangle = \langle\langle f^\omega(t) \rangle\rangle_{T,\sigma} = (1/\sigma)\int_T^{T+\sigma} \mathbb{E} f(t)\, dt,$$

with $T \geq 1$ and $\sigma \geq \sigma_0$. Interpolating the just found lower bound for $\|u\|_1$ with the already obtained upper estimates for the norms of solutions, we get in Theorem 2.3.4 lower bounds for the averaged Sobolev norms, and thus arrive at two-sided inequalities

$$C_m^{-1} \nu^{-(2m-1)} \leq \langle\langle \|u\|_m^2 \rangle\rangle \leq C_m \nu^{-(2m-1)}, \quad m = 1, 2, \dots,$$

independent from the initial data $u(0,x)$. On the other hand, in Theorem 2.3.15 we show that the averaged squared L_2–norm of a solution is a quantity of order one. This is how it should be since the stochastic Burgers equation models "one-dimensional fluids", and $\frac{1}{2}\langle\langle\|u\|_0^2\rangle\rangle$ is the averaged energy of the fluid which should be of order one.[2]

In the inequalities above and in other similar situations in our book, for relevant quantities depending on the small parameter ν, we obtain two–sided estimates which are *asymptotically sharp* in the sense that the lower and upper bounds involve the same power of ν, but we cannot establish for them asymptotics as $\nu \to 0$. For example, for $m \geq 2$ we cannot show that $\langle\langle\|u\|_m^2\rangle\rangle = C_m^* \nu^{-(2m-1)} + o(\nu^{-(2m-1)})$ for a suitable C_m^*, and do not exclude that such an asymptotic does not exist. We will abbreviate the obtained two–sided estimate for the Sobolev norm $\|u\|_m$ as

$$\langle\langle\|u\|_m^2\rangle\rangle \sim \nu^{-(2m-1)}, \qquad m = 1, 2 \ldots. \tag{0.0.4}$$

In Chapter 3 we turn to the problem of the asymptotic behaviour of solutions for the stochastic equation in the limit $t \to \infty$. Namely, we prove there that the equation is mixing in Sobolev spaces H^m, $m \geq 1$. That is, in H^m there exists a measure μ^ν, such that for any continuous functional $f(u)$ on H^m which does not grow too fast as $\|u\|_m \to \infty$ and for any solution u of the stochastic Burgers equation, we have

$$\mathbb{E} f(u(t,\cdot)) \to \int f(u)\, \mu^\nu(du) \quad \text{as} \quad t \to \infty. \tag{0.0.5}$$

The proof uses another feature of the Burgers equation – the L_1–non-expansion. That is, the L_1–distance between solutions corresponding to the same force η and different initial conditions is non-increasing in time.

In Chapter 4 we prove that the Markov process defined by the equation in Sobolev spaces extends by continuity to the space L_1 and is mixing there at a rate which is at least logarithmic in time and ν-independent. Moreover, we show that the process is mixing uniformly in ν with respect to any L_p-norm with finite p.

In Part II of the book, made of Chapters 6–9, we apply the asymptotic results from Part I to the problem of burgulence. Namely, there we discuss the qualitative behaviour of solutions for the stochastic Burgers equation when $\nu > 0$ is small, mimicking the main laws of the Kolmogorov theory of hydrodynamical turbulence, known as the *K41 theory*. Chapter 6, opening Part II, contains a rather detailed discussion of this theory. Later in that section, motivated by the work of Frisch and his collaborators, we suggest natural 1d versions of the main laws of the K41 theory, and in Chapter 7 use the results of Part I to prove these laws for "Burgers fluid", described by the stochastic Burgers equation. Now we will review the most important of those results, referring the reader to Chapter 6 for a detailed discussion and to Chapter 7 for proofs. When comparing our results with predictions of the K41 theory we always assume that the latter is being applied to a velocity field of fluid (or gas), where the units are chosen in such a way that the space-scale of the velocity field is of order one (e.g. it is one-periodic in all three coordinates), and the fluid's energy per volume also is of order one. Then the main remaining parameter of the fluid's motion is its kinematic viscosity ν.

The basic quantity characterising a solution $u(t,x)$ of the stochastic Burgers equation, regarded as a 1d turbulent flow, is its *dissipation scale* l_d, also known as *Kolmogorov's inner scale*. We define it in the Fourier presentation as follows: the scale is $l_d = \nu^{-c_d}$, where the exponent c_d is the smallest number such that for

[2]Since otherwise for small ν the fluid either boils or freezes.

$|s| \gg l_d$ the squared norm of the s-th Fourier coefficient $|\hat{u}_s(t)|^2$, under suitable averaging $\langle\langle\cdot\rangle\rangle$, decays very fast. Namely, for any $n \in \mathbf{N}$ and $\gamma > 0$ there exists $C_{N,\gamma} > 0$ such that

$$\langle\langle|\hat{u}_s(t)|^2\rangle\rangle \le C_{N,\gamma}|s|^{-N}, \quad \forall\,|s| \ge \nu^{-c_d-\gamma}, \ \forall\, 0 < \nu \le 1. \tag{0.0.6}$$

Here the constants $C_{N,\gamma}$ do not depend on ν (as well as all other constants in the introduction).

Understanding the double brackets as in (0.0.3), in Section 7.1 we derive from the results of Part I that

$$l_d = \nu^{-1}.$$

This is similar to a prediction of the K41 theory, where (under the specified above choice of the units) the hydrodynamical dissipation scale is $l_d = \nu^{-3/4} =: l_d^K$.

Two main characteristics of a turbulent flow, which are in a certain duality to each other, are the structure function (corresponding to the x-presentation) and the distribution of energy along the spectrum (corresponding to the Fourier presentation). The structure function S for a 1d velocity field $u(t,x)$ is defined as an average of powers of its increments:

$$S_{p,r} = S_{p,r}(u) = \langle\langle \int |u(x+r) - u(x)|^p dx\rangle\rangle, \quad |r| \le 1/2,\ p > 0.$$

In Section 7.2 we show that if in this definition $u(t,x)$ is a solution of the stochastic Burgers equation and we understand the double bracket as in (0.0.3), then the asymptotic behaviour of S as a function of r drastically depends on whether $|r|$ is larger or smaller than the inverse dissipation scale $l_d^{-1} = \nu$. Namely, under the assumption that $0 < \nu \le c_*$, where $c_* > 0$ depends only on the force η, we prove for solutions u the law of averaged powers of space-increments :[3]

$$S_{p,r}(u) \sim |r|^p\, \nu^{1-\max(p,1)}, \quad \text{if } |r| \le C\nu, \ \forall p > 0, \tag{0.0.7}$$

$$S_{p,r}(u) \sim |r|^{\min(p,1)}, \quad \text{if } C\nu \le |r| \le c', \ \forall p > 0, \tag{0.0.8}$$

for some constants $C, c' > 0$, depending only on the force. As in (0.0.4), in (0.0.8) the notation $S_{p,r}(u) \sim |r|^{\min(p,1)}$ means that $C_p^{-1}|r|^{\min(p,1)} \le S_{p,r}(u) \le C_p|r|^{\min(p,1)}$ for all $0 < \nu \le c_*$, $C\nu \le |r| \le c'$ and a suitable $C_p \ge 1$, and similarly in (0.0.7). The interval $[Cl_d^{-1}, c'] = [C\nu, c']$ is called the *inertial range,* and its complement $(0, C\nu)$ – the dissipation range of u, in the x-presentation.[4] The behaviour of the structure function in the inertial range is the most interesting since it does not depend on ν and since there S is equivalent to r raised to a degree, named the scaling exponent, which is a non-linear function of p.

In the K41 theory the (longitudinal) structure function $S_{p,r}^{||}$ is defined by the same relation as $S_{p,r}$, but with the vector-increment $u(x+r) - u(x)$ of a 3d velocity field u replaced by its projection to the direction of the vector r. The K41 theory studies $S^{||}$ for the increments r whose length belongs to the hydrodynamical inertial range in the x-presentation, which is an interval $[(l_d^K)^{-1}, (l_e^K)^{-1}]$, where $l_e^K \ll l_d^K$ (see Chapter 6 concerning the definition of l_e). The celebrated *2/3-law of K41* predicts that

$$S_{2,r}^{||} \sim |r|^{2/3} \quad \text{for} \quad r \in [(l_d^K)^{-1}, (l_e^K)^{-1}].$$

[3] It is easy to see that the two relations below follow from (7.2.3)-(7.2.5).

[4] For simplicity, in the introduction we do not discuss the energy range $(c', 1]$ (see in Chapter 6).

Kolmogorov's proof of this relation is based on a heuristic dimension argument, which is also valid for other values of the exponent p and implies that

$$S^{\|}_{p,r} \sim |r|^{p/3} \quad \text{for} \quad p>0, \ r\in[(l_d^K)^{-1},(l_e^K)^{-1}]. \tag{0.0.9}$$

The latter relation was not claimed in the K41 papers, but appeared frequently in later works. The asymptotic (0.0.8), which is obviously similar to the 2/3-law, was first obtained in the work [**1**] by a convincing heuristic argument. Since in (0.0.8) the scaling exponent is $\min(p,1)$ while in (0.0.9) it is $p/3$ for any p, then in [**1**] the obtained behaviour of the structure function for burgulence was referred to as *anomalous scaling.*

At the end of Section 7.2 we explain that the law (0.0.8) implies that the random fields $u(t,x)$, given by solutions of the stochastic Burgers equation with small ν, are essentially non-Gaussian.

The second celebrated law of the Kolmogorov theory deals with the distribution of the kinetic energy $\langle\langle\frac12\int|u|^2dx\rangle\rangle$ along the spectrum. For a solution $u(t,x)$, regarded as a 1d turbulent flow, we consider the averaged energy of its k-th Fourier mode $\frac12\langle\langle|\hat u_k|^2\rangle\rangle$. Next for $k\ge1$ we define $E_k(u)$ as the averaging of the latter quantity along the layer $\Sigma_k=\{n\in\mathbb{Z}^*: M^{-1}k\le|n|\le Mk\}$, $M>1$. That is,

$$E_k(u)=\langle\langle e_k(u)\rangle\rangle, \quad e_k(u)=\frac{1}{|\Sigma_k|}\sum_{n\in\Sigma_k}\tfrac12|\hat u_n|^2.$$

The function $k\mapsto E_k$ is called the *energy spectrum.* It is immediate that in view of (0.0.6) for $k\gg l_d$ it decays faster than any negative degree of k (uniformly in ν). But for $k\le l_d$ the behaviour of E_k is quite different: in Section 7.3 we derive from (0.0.8) and estimates in Part I the following spectral power law :

$$E_k\sim k^{-2} \quad \text{for} \quad 1\le k\le (c_1'\nu)^{-1}=l_d/c_1', \tag{0.0.10}$$

for a suitable $c_1'>0$. The interval $[1,l_d/c_1']$ (resp., its complement $(l_d/c_1',\infty)$) is called the inertial range (resp. the dissipation range) of u in the Fourier presentation. Relation (0.0.10) is an analogue of the celebrated Kolmogorov–Obukhov law which states that for the hydrodynamical turbulence, under the same stipulations concerning the units as before, the energy spectrum satisfies

$$E_k\sim k^{-5/3} \quad \text{for} \quad k\in[l_e^K,l_d^K].$$

For burgulence the spectral power law (0.0.10) was predicted by Burgers in [**34**] and then by other physicists, including Frisch with collaborators [**1**].

In Section 7.4 we consider the energy spectrum of the stationary measure μ^ν (see (0.0.5)) which is the function $\mathbf{N}\ni k\mapsto\langle e_k,\mu^\nu\rangle$, and the instant energy spectrum of a solution u at time t – the function $k\mapsto\mathbb{E}e_k(u(t))$. We derive from (0.0.10) and the mixing (0.0.5) the spectral power law for the measure μ^ν :

$$\langle e_k,\mu^\nu\rangle\sim k^{-2} \quad \text{for } k\le C\nu^{-1}.$$

That is, if $u^{st}(t)$ is a stationary solution, $\mathcal{D}u^{st}(t)\equiv\mu^\nu$, then

$$\mathbb{E}e_k(u^{st}(t))\sim k^{-2} \quad \text{for any } t \text{ and for } k\le C\nu^{-1}.$$

These relations (and again the mixing) imply that any solution $u(t)$ satisfies

$$\mathbb{E}e_k(u(t))\sim k^{-2} \quad \text{for} \quad k\le C'\nu^{-1} \ \text{ if } t\ge t_k, \tag{0.0.11}$$

for suitable positive constants C' and t_k. Similar results hold for the structure function of the measure μ^ν and for the instant structure function of solutions. This (and the limit (0.0.5)) establish for burgulence what is a general belief in the theory of turbulence: statistical characteristics of turbulent flows converge, as time grows, to a universal equilibrium, see [**7**].

In Section 7.5 we show that the law of averaged powers of increments and the spectral power law hold with high probability if we understand the brackets $\langle\langle\cdot\rangle\rangle$ as the local averaging in time along sufficiently large intervals (then $S_{p,l}$ and E_k are random variables).

In Chapter 8 we re-prove the known fact that as $\nu \to 0$, solutions $u^\nu(t,x)$ of the stochastic Burgers equation converge, for every ω, to non-trivial *inviscid limit solutions* (also known as the *entropy solutions*), denoted $u^e(t,x)$. Next we pass to the limit in the theorems from Sections 7.2, 7.3 to get that

$$S_{p,r}(u^e) \sim |r|^{\min(p,1)} \quad \text{for } |r| \le c',$$

and

$$E_k(u^e) \sim k^{-2} \quad \text{for} \quad k \ge 1.$$

Note that now the two laws hold for all $|r| \le c'$ and all k. So under the inviscid limit the inertial range in the x- and in Fourier-presentations becomes the whole interval $(0, c']$ and the whole set $\mathbf{N}$, and thus the dissipation range vanishes. In Section 8.5 we show that the dynamics defined by the entropy solutions is mixing, and that for the corresponding stationary measure μ^e the energy spectrum satisfies $\langle e_k, \mu^e\rangle \sim k^{-2}$, for all k. We derive from this fact that the inviscid instant energy spectrum $\mathbb{E}e_k(u^e(t))$ of an entropy solution u^e meets (0.0.11) for all k and for $t \ge t_k$. Similar results hold for the structure function of μ^e and for instant structure functions of entropy solutions. The results of this chapter justify *inviscid burgulence*, whose analogue does not exist in K41.

Part III is made of chapter *Miscellanea*, eight appendices and a chapter comprising solutions for selected exercises. In Section 10.1 of Miscellanea we discuss versions of the results in Parts I, II, corresponding to other random forces η in (0.0.1) and to the free Burgers equation, and in Section 10.3 – ergodic properties of solutions for the equation with various random forces. In Section 10.2 we examine solutions for the Burgers equation perturbed by random kick-forces with high time-frequency, appropriately rescaled – such forces naturally appear in the computer modelling of the stochastic Burgers equation. We prove there that stochastic properties of these solutions are close to those of solutions for the equation (0.0.1),(0.0.2), uniformly in time. In Section 10.4 we present analogues of some of the results of our book for other nonlinear PDEs and SPDEs with small linear parts. The appendices contain additional material which may be less familiar to the reader.

The bulk of available mathematical literature on SPDEs and on the Burgers equation, and of physical literature on burgulence, is immense. Our bibliography contains only a small part of the former and just a few works from the latter.

Generalisations. Since in this book we do not use the Hopf–Cole transformation which linearises the Burgers equation, our main results [5] remain true for modified Burgers equations

$$u_t + \frac{\partial}{\partial x} f(u) - \nu u_{xx} = \eta(t,x), \qquad \int_{S^1} u\,dx = 0, \quad 0 < \nu \le 1, \tag{0.0.12}$$

[5] including Theorems 2.1.2, 2.3.4, 3.3.4, 7.1.3, 7.2.3, 7.3.2

where the smooth function $f(u)$ is close to Cu^2 in a suitable sense. In particular – if

$$
\begin{aligned}
&f \in C^\infty(\mathbb{R}), \quad f'' \geq c, \quad f(0) = f'(0) = 0 \quad \text{and} \\
&f(u) = C_1 u^2 \quad \text{for} \quad u \geq C_2, \quad f(u) = C_3 u^2 \quad \text{for} \quad u \leq -C_2,
\end{aligned}
\tag{0.0.13}
$$

for some positive constants c, C_1, C_2, C_3. A corresponding generalisation is straightforward, and proofs of some of its steps are given as exercises, see a discussion at the beginning of Chapter 7. For another class of admissible nonlinearities f, see in [**28**]. Our argument uses specific properties of the Burgers equation and generalizes poorly outside the class of modified Burgers equations (0.0.12) with $f(u)$ close to Cu^2. See Section 10.4 for some related results, and see the recent papers [**13, 14**], where very similar results are obtained for a non-local nonlinear equation. The argument is less sensitive to the form of the random force η, see Section 10.1. See Chapter 9 for a further discussion.

What is not in our book. Since the book is motivated by analysis of turbulence in the 1d Burgers equation, then, firstly, we touch the multidimensional Burgers equation only in the form of a brief discussion in Section 10.4. Secondly, we do not deal with the equation perturbed by rough in space random forces; in particular – with the KPZ equation, where significant progress was achieved recently, see [**71, 74, 110, 133**] (also see the earlier papers [**41, 111**]). Similarly we do not discuss solutions of the 1d Burgers equation under Dirichlet boundary conditions, as well as those of the equation on the whole line with finite energy. Homogeneous in space non-periodic solutions are only briefly mentioned at the end of Chapter 6.

Readers of this book. The book is aimed at mathematicians and physicists with basic knowledge of probability and analysis; in particular it is well-suited for M2 courses at mathematical departments as well as at mathematically minded physical departments (cf. lecture notes [**82**]). The appendices in Part III make the book self-sufficient for most interested readers, and more than 40 problems, spread throughout the text, should help them to understand the material better.

Notation and agreement.

Abbreviations. The expressions l.h.s., r.h.s. stand, respectively, for "left-hand side" and "right-hand side". The abbreviations r.v, a.e. and a.s. stand, respectively, for "random variable", "almost everywhere" and "almost surely". Finally, i.i.d. means "independent identically distributed".

Constants. By $C, C', C_1, \dots$ etc we denote various positive constants. Usually in the same formula different quantities are denoted by different letters, but in different formulas the same letter may stand for different constants.

Functional spaces. For $T \in (0, +\infty]$ and a Banach space B endowed with a norm $\|\cdot\|_B$, $C(0, T; B)$ stands for the space of continuous B-valued curves defined for $t \in [0, T]$. If $T < \infty$, this is a Banach space with the supremum norm $\sup_{t \in [0,T]} \|u(t)\|_B$. If $T = \infty$, then by definition $[0, T] = [0, \infty)$; in this case $C(0, T; B)$ is a metric space with the distance (11.2.1).

Everywhere in the book, except in Chapter 8, we consider functions of the space-variable $x \in S^1$ with zero space mean-value (cf. (1.1.1)).

Let $m \in \mathbb{R}$. For a zero mean function $u(x)$ on S^1, $\|u\|_m$ stands for its homogeneous Sobolev norm of order m (see (1.2.12)). If $m = 0$, we write $\|u\| := \|u\|_0 = |u|_2$. For $p \geq 1$, $|\cdot|_p$ signifies the L_p–norm, and for $m = 0, 1, 2, \dots$, $|\cdot|_{m,p}$ stands for the homogeneous norm in the Sobolev space W_p^m (see Section 1.3). Note that on S^1, the Sobolev norms are totally ordered. Namely, for $u \in W_p^m$ we have:

$$|u|_{n,q} \leq |u|_{m,p}, \text{ for } n < m, \ q \in [1, +\infty] \text{ or } n = m, \ q \in [1, p]. \tag{0.0.14}$$

Moreover, we have the classical interpolation inequality [**119, 144**]

$$\|u\|_n \leq \|u\|_k^{\theta} \|u\|_m^{(1-\theta)}, \text{ for } 0 \leq k \leq n \leq m, \tag{0.0.15}$$

where $\theta = (m - n)/(m - k)$. We use the abbreviations:

$$X_T^m = C(0, T; H^m), \quad X_T^0 = X_T,$$

and denote $\dot{X}_T^m = \{\xi \in X_T^m : \xi(0) = 0\}$, and similarly for $\dot{X}_T$.

We define the spaces C^n, $n \in \mathbf{N} \cup \{0\}$, of functions of x in the standard way, and we denote the corresponding norm

$$\|u\|_{C^n} := \sum_{i=0}^{n} \left|\frac{d^i u}{dx^i}\right|_\infty.$$

We call complete separable metric spaces *Polish spaces.* For a Polish space X we denote by $\|\cdot\|_{L(X)}$ the norm (1.7.2) in the Banach space of bounded Lipschitz functions on X, and by $\|\cdot\|_{L(X)}^*$ – the norm in the dual Banach space.

Functions, mappings and sets. We denote by $\lceil\ \rceil$ the ceiling function

$$\lceil x \rceil = \min\{z \in \mathbb{Z} : \ z \geq x\}, \tag{0.0.16}$$

by $\lfloor x \rfloor$ – the integer part of x,

$$\lfloor x \rfloor = \max\{z \in \mathbb{Z} : \ z \leq x\}, \tag{0.0.17}$$

by $\mathbf{1}$ – the function identically equal to one, and by χ_A – the indicator function of a set A (equal to one on A and vanishing outside this set). We write $f \not\equiv 0$ if the function f does not vanish identically. The smoothness of mappings between Banach spaces is understood in the sense of Fréchet [**36**]. A mapping between metric spaces whose Lipschitz constant is bounded by K is called *K–Lipschitz.* A map between Banach spaces B_1 and B_2 is said to be *locally Lipschitz* if its restriction to any ball in B_1 is Lipschitz, and similarly for locally bounded maps. For a metric space M, $C_b(M)$ indicates the space of bounded real-valued continuous functions on M, given the sup-norm.

We denote by $\mathbf{N}$ the set of natural numbers $\{1, 2, \dots\}$, by $\mathbf{N}_0$ – the set $\mathbf{N} \cup \{0\}$, by $\mathbb{Z}^*$ – the set of non-zero integers and by $\mathbb{R}^+$ – the set of non-negative real numbers. For a set A we denote by $\bar{A}$ its closure and by A^c – its complement. For a metric space M, a positive number R and $v \in M$, $B_M^R(v)$ stands for the open ball:

$$B_M^R(v) := \{u \in M : \ \mathrm{dist}(u, v) < R\}.$$

For a Banach space X we abbreviate $B_X^R := B_X^R(0)$ and $B_m^R(v) := B_{H^m}^R(v)$, $B_m^R := B_{H^m}^R$.

Measurable spaces, measures and integrals. A null-set is a mesurable set of measure zero. The Dirac measure concentrated at a point v is denoted δ_v. For a measure μ on a measurable space and an integrable function f we write

$$\langle \mu, f\rangle = \langle f, \mu\rangle = \int f\, d\mu\,.$$

The counting measure on a discrete set is the one which assigns to each point of the set the measure 1. For a measurable space $(X, \mathcal{F})$ we denote by $\mathcal{P}(X, \mathcal{F})$ the space of probability measures on $(X, \mathcal{F})$, and for a measurable map $F : (X, \mathcal{F}_X) \to (Y, \mathcal{F}_Y)$ we denote by $F\circ$ the induced mapping from $\mathcal{P}(X, \mathcal{F}_X)$ to $\mathcal{P}(Y, \mathcal{F}_Y)$. A metric space M usually is equipped with the Borel sigma–algebra $\mathcal{B} = \mathcal{B}_M$, and measurable maps between metric spaces with Borel sigma–algebras are called *Borel-measurable* or just measurable. We abbreviate $\mathcal{P}(M, \mathcal{B}_M)$ to $\mathcal{P}(M)$.

Random variables and random processes. The law of a r.v. ξ is denoted by $\mathcal{D}(\xi)$. If ξ is a Gaussian r.v. with mean value m and dispersion σ^2, we write $\xi \sim N(m, \sigma^2)$. All random processes which we consider in this book have continuous trajectories. In the notation of random variables and processes we often drop the random parameter ω.

Miscellanea. Let $K > 0$. We use the notation $K \gg 1$ (resp. $K \ll 1$) if K is large enough (resp. small enough). The m-th derivative of a function f of a scalar argument is denoted $f^{(m)}$. The time derivative is denoted by a subindex t. Notation $X \Subset Y$, where X, Y are Banach spaces, means that the former is compactly embedded in the latter.

Exercises. Solutions of selected exercises are given in Chapter 12. In particular, this section contains solutions to all exercises whose results are used in the main text. An asterisk after an exercise's number means that its proof uses some non-standard results, not included in this book.

Part 1

Stochastic Burgers Equation

CHAPTER 1

Basic results

1.1. Introduction.

We begin by introducing the equation playing a central role in this book – the viscous Burgers equation with periodic boundary conditions, perturbed by a random force:

$$\begin{aligned} &u_t(t,x) + u(t,x)u_x(t,x) - \nu u_{xx}(t,x) = \eta(t,x), \\ &t \geq 0, \quad x \in S^1 = \mathbb{R}/\mathbb{Z}\,. \end{aligned} \tag{B}$$

Here the viscosity coefficient satisfies $\nu \in (0,1]$, and all constants in the estimates in this chapter do not depend on ν, unless otherwise stated. The random force $\eta = \eta^\omega$ is defined on a probability space $(\Omega, \mathcal{F}, \mathbf{P})$ (see Appendix 11.5). The force and the solution u are real-valued functions, and we assume that for $t \geq 0$

$$\int_{S^1} \eta^\omega(t,x)dx = \int_{S^1} u(0,x)dx = 0.$$

Integrating (B) in dx, using the equality above, we get that for $t \geq 0$ the solutions of (B) satisfy:

$$\int_{S^1} u(t,x)dx = 0. \tag{1.1.1}$$

We denote by H the Hilbert space defined by:

$$H = \{v \in L_2(S^1) : \int_{S^1} v(x)dx = 0\},$$

endowed with the usual L_2–norm and scalar product, denoted, respectively, by $\|\cdot\|$ and $\langle\cdot,\cdot\rangle$. Based on (1.1.1), we are looking for solutions of (B) in the space H and write them either as functions $u(t,x)$, or as curves

$$u(t) := u(t,\cdot) \in H, \qquad t \geq 0.$$

We endow the space H with the trigonometric Hilbert basis $\{e_s(x), s \in \mathbb{Z}^*\}$, where $\mathbb{Z}^* = \mathbb{Z} \setminus \{0\}$ and

$$\begin{cases} e_k = \sqrt{2}\cos(2\pi kx), \\ e_{-k} = \sqrt{2}\sin(2\pi kx), \end{cases} \quad k \in \mathbf{N}. \tag{1.1.2}$$

If

$$H \ni u(x) = \sum_{s \in \mathbb{Z}^*} u_s e_s(x)\,, \tag{1.1.3}$$

then $u(x)$ may be re-written as a Fourier series:

$$u(x) = \sum_{s \in \mathbb{Z}^*} \hat{u}_s e^{2i\pi sx}, \quad \hat{u}_s = \bar{\hat{u}}_{-s} = \left(\sqrt{2}\right)^{-1} (u_s - iu_{-s}) \text{ for all } s \in \mathbf{N}. \tag{1.1.4}$$

The force $\eta(t,x)$ is a generalised function of time of the form

$$\eta(t,x) = \partial_t \xi(t,x), \tag{1.1.5}$$

where $\xi(t,x) = \xi^\omega(t,x)$ is a regular random field defined by

$$\xi^\omega(t,x) = \sum_{s\in\mathbb{Z}^*} b_s \beta_s^\omega(t) e_s(x)\,. \tag{1.1.6}$$

Here b_s are real numbers and β_s^ω are standard independent Wiener processes (cf. Appendix 11.5), defined on $(\Omega, \mathcal{F}, \mathbb{P})$. By a small abuse of notation, we also call the random field $\xi^\omega(t,x)$ "a random force".

For $m \geq 0$ we denote

$$B_m = \sum_{s\in\mathbb{Z}^*} |2\pi s|^{2m} b_s^2 \leq +\infty\,, \tag{1.1.7}$$

and always assume that $B_0 < \infty$ (so, in particular, $b_s \to 0$ as $|s| \to \infty$). By $\vec{B}$ we denote the sequence

$$\vec{B} = (B_0, B_1, \dots), \tag{1.1.8}$$

and by b_{max} – the number

$$b_{max} = \max_j |b_j| < \infty. \tag{1.1.9}$$

Naturally, $b_{max} \geq 0$ and $b_{max} > 0$ if $B_0 \neq 0$.

In the next Section 1.2 we will show that for almost all ω trajectories of the process $\xi^\omega(t)$ are continuous in H, and the distribution $\mathcal{D}(\xi)$ is a Gaussian measure in the space of continuous curves in H. Accordingly, we refer to $\xi^\omega(t)$ as a *Wiener process in the space* H,[1] and to its derivative $\eta^\omega = \partial_t \xi^\omega$ – as a *white noise* in H .

We will say that u solves eq. (B), (1.1.5), where $\xi(t,\cdot)$ is a continuous curve in H, if the equality

$$u(t,x) - u(0,x) + \int_0^t (u(s,x)u_x(s,x) - \nu u_{xx}(s,x))\,ds = \xi(t,x) - \xi(0,x) \tag{1.1.10}$$

holds for all $t \geq 0$. Our solutions will be continuous curves in some Sobolev space H^m, where $m \geq 1$. If $m \geq 2$, then the equality (1.1.10) with a fixed t holds for a.e. x, while if $m = 1$, then it should be understood in the sense of generalised functions. A random function $u^\omega(t,x)$ solves (B), (1.1.5), (1.1.6) if it defines a random process with continuous trajectories in some Sobolev space H^m, $m \geq 1$, and a.s. relation (1.1.10) holds for all t. If the equality is valid not for all $t \geq 0$, but for $t \in [0,T]$, we say that u solves (B) for $t \in [0,T]$. See below in this chapter (and in Appendix 11.5) for rigorous definitions and results.

Equation (B), (1.1.5), (1.1.6) is the *stochastic Burgers equation*; being supplemented with the initial condition

$$u(0,x) = u_0(x)\,, \tag{1.1.11}$$

it becomes the *Cauchy problem for the stochastic Burgers equation.* For convenience, we re-write in a more compact form the former:

(1.1.12)

$$u_t(t,x) + u(t,x)u_x(t,x) - \nu u_{xx}(t,x) = \partial_t \xi^\omega(t,x), \quad \xi^\omega(t,x) = \sum_{s\in\mathbb{Z}^*} b_s \beta_s^\omega(t) e_s(x)\,,$$

[1] Unlike in the one-dimensional case (see Appendix 11.5), in an infinite-dimensional Hilbert space there are many Wiener processes, and none of them is canonical.

and the latter:

(1.1.13) $\quad u_t(t,x)+u(t,x)u_x(t,x)-\nu u_{xx}(t,x)=\partial_t\xi(t,x), \quad u(0,x)=u_0(x)\,.$

Sometimes below we will consider and discuss problem (1.1.13), where $\xi(t,x)$ is a fixed function. The type of forcing under consideration will always be clear from the context.

The equation (1.1.12) and the problem (1.1.13) are the main objects of study in our book. It is not excluded that $B_0=0$, i.e. that in (1.1.6) all b_s are zero. In this case (B) is the deterministic Burgers equation with zero r.h.s.

1.2. Properties of the random force ξ.

In this section we discuss properties of the random force ξ, see (1.1.6). Since for any $t\geq 0, x\in S^1$ and any s the r.v. $b_s\beta_s^\omega(t)e_s(x)$ is Gaussian with zero mean-value and its dispersion is bounded by $2b_s^2t$, and since $\sum b_s^2=B_0<\infty$, then $\xi^\omega(t,x)$ is a well-defined Gaussian r.v. such that $\mathbb{E}\xi=0$ and $\mathbb{E}\xi^2\leq 2B_0t$, see [**139**, Section II.13]. As $\mathbb{E}\beta_{s_1}(t_1)\beta_{s_2}(t_2)=\delta_{s_1,s_2}\min(t_1,t_2)$, then it is easy to see that the correlation between the r.v.'s $\xi^\omega(t,x)$ may be expressed via the cos-Fourier series with the coefficients $\{b_s^2\}$ and $\{(\operatorname{sgn} s)b_s^2\}$:

$$\mathbb{E}\xi^\omega(t_1,x_1)\xi^\omega(t_2,x_2)=\min(t_1,t_2)\sum_{s\in\mathbb{Z}^*}b_s^2\big(\cos 2\pi s(x_1-x_2)+(\operatorname{sgn} s)\cos 2\pi s(x_1+x_2)\big).$$

These relations completely define joint distributions of the system of Gaussian r.v.'s $\xi^\omega(t,x)$, see [**139**]. But for the analysis in our book we need another kind of information about the random force ξ: we should interpret it as a random process $t\mapsto\xi^\omega(t,\cdot)$, in a space of functions in x and examine its distribution in a space of curves in t. This requires a more detailed analysis. We start with some definitions.

For $T>0$, consider the Banach space $C(0,T;H)=X_T^0$. We abbreviate it to X_T and call its elements curves. By $\dot X_T\subset X_T$ we denote the subspace of curves vanishing at $t=0$. For $T<\infty$ the spaces X_T and $\dot X_T$ are separable Banach spaces, endowed with the supremum–norm $\|\xi\|_{X_T}=\sup_{t\in[0,T]}\|\xi(t)\|$, while the spaces X_∞ and $\dot X_\infty$ are given the distance (11.2.1), which make them *Polish spaces* (i.e. – complete separable metric spaces); see Appendix 11.2.

THEOREM 1.2.1. *There exists a random process $\xi^\omega(t)$ with trajectories in $\dot X_\infty$*[2] *and a null-set Q such that*

1) for any $T<\infty$ and for $\omega\notin Q$ the series in the r.h.s. of (1.1.6) *with $t\in[0,T]$ converge to $\xi\mid_{[0,T]}$ in $\dot X_T$;*

2) for any $T<\infty$,

(1.2.1) $\quad \mathbb{E}\,\exp(\alpha\|\xi\|^2_{X_T})\leq 4\exp(2T\alpha B_0)-3 \quad \textit{if} \quad \alpha\leq\alpha_*=(4Tb_{max}^2)^{-1},$

(1.2.2) $\quad \mathbb{E}\sup_{0\leq t\leq T}\|\xi^\omega(t)\|^p\leq C(p,T,B_0) \quad \textit{if} \quad p\geq 1,$

(1.2.3) $\quad \mathbb{E}\,\|\xi^\omega(t)\|^2=tB_0, \quad \forall\, 0\leq t\leq T$

(b_{max} is defined in (1.1.9)*).*

Due to this theorem the random force (1.1.6) is defined for a.e. ω. It is convenient to extend it to all $\omega\in\Omega$, as we do this in the theorem, but that

[2]So ξ defines a r.v. $\Omega\to\dot X_\infty$, and $\xi\mid_{[0,T]}$ – a r.v. $\Omega\to\dot X_T$, for any $T<\infty$.

extension certainly is not canonical, cf. below Theorem 1.2.3, Corollary 1.2.4 and a discussion after the corollary.

PROOF. *Step 1 : truncation.* Fix $T>0$, and for every $N\in\mathbb{N}$ set

(1.2.4) $H^{(N)}=\text{span}\{e_j,|j|\le N\}\subset H$ and

$$\xi^N(t,x)=\xi^N_T(t,x)=\sum_{|s|\le N} b_s\beta_s^\omega(t)e_s(x),\quad 0\le t\le T.$$

Then $\xi^N(t)$ s a continuous random process in $H^{(N)}$, vanishing at $t=0$. So for every ω the curve $t\longmapsto\xi^N(t)\in H^{(N)}$ is continuous. For $t\ge0$, we have:

(1.2.5)
$$\|\xi^N(t)\|^2=\sum_{|s|\le N} b_s^2\beta_s^\omega(t)^2,\quad \mathbb{E}\|\xi^N(t)\|^2=tB_0^N,\quad\text{where}\quad B_0^N:=\sum_{|s|\le N}b_s^2\le B_0.$$

Now we will estimate the exponential moments for $\sup_{0\le t\le T}\|\xi^N(t)\|$. For $\alpha>0$ and $0\le t\le T$:

$$\mathbb{E}\ \exp(\alpha\|\xi^N(t)\|^2)=\mathbb{E}\ \exp\Big(\alpha\sum_{|s|\le N}b_s^2\beta_s^\omega(t)^2\Big)=\prod_{|s|\le N}\mathbb{E}\ \exp(\alpha b_s^2\beta_s^\omega(t)^2),$$

since the processes $\{\beta_s\}$ are independent. As $\beta_s(t)\sim N(0,t)$ (see Notation), then

$$\mathbb{E}\ \exp(\alpha b_s^2\beta_s^\omega(t)^2)=\frac{1}{\sqrt{2\pi t}}\int_{\mathbb{R}}\exp(\alpha b_s^2x^2)\exp\Big(-\frac{x^2}{2t}\Big)\,dx=\frac{1}{\sqrt{1-2t\alpha b_s^2}}$$
$$\text{for}\quad \alpha<\frac{1}{2tb_s^2}.$$

So

$$\mathbb{E}\ \exp(\alpha\|\xi^N(t)\|^2)=\exp\Big(-\frac12\sum_{|s|\le N}\ln(1-2t\alpha b_s^2)\Big)\quad\text{for}\quad\alpha<\frac{1}{2tb^2_{\max}}.$$

Since for every $a\in[0,\frac12]$ we have $-\ln(1-a)\le 2a$, then the equality above implies that

$$\mathbb{E}\ \exp(\alpha\|\xi^N(t)\|^2)\le\exp\Big(\sum_{|s|\le N}2t\alpha b_s^2\Big)=\exp(2t\alpha B_0^{(N)})\quad\text{for}\quad\alpha\le\frac{1}{4tb^2_{\max}}=\alpha_*.$$

Now by Doob's inequality (11.5.3),

(1.2.6)
$$\mathbb{E}e^{\alpha\|\xi^N\|_{\dot X_T}}=\mathbb{E}e^{\alpha\sup_{0\le t\le T}\|\xi^N(t)\|^2}=\mathbb{E}\sup_{0\le t\le T}\exp(\alpha\|\xi^N(t)\|^2)\le 4\exp(2T\alpha B_0^{(N)})-3$$

if $\alpha\le\alpha_*$, for every N.

Step 2: passing to the limit. For $N<M$, consider

$$\xi_N^M:=\xi^M-\xi^N=\sum_{N<|s|\le M}b_s\beta_s^\omega e_s.$$

Since $\mathbb{E}\|\xi_N^M(T)\|^2=T\sum_{N<|s|\le M}b_s^2$, then Doob's inequality (11.5.2) with $p=2$ yields:

(1.2.7)
$$\mathbb{E}\Big[\|\xi_N^M\|^2_{\dot X_T}\Big]=\mathbb{E}\sup_{t\in[0,T]}\|\xi_N^M(t)\|^2\le 4T\sum_{N<|s|\le M}b_s^2.$$

Consider the Banach space $\mathbb{L}_2^T = L_2(\Omega; \dot{X}_T)$ (see discussion and references in Section 11.2) and note that the l.h.s. of (1.2.7) equals $\|\xi^M - \xi^N\|^2_{\mathbb{L}_2}$. Since $\sum b_s^2 = B_0 < \infty$, then we derive from (1.2.7) that $\{\xi^M\} \subset \mathbb{L}_2^T$ is a Cauchy sequence, so there exists ξ'_T such that

$$\xi^M \to \xi'_T \ \text{ as } M \to \infty \text{ in } \mathbb{L}_2^T . \tag{1.2.8}$$

Let us fix any r.v. $\xi_T : \Omega \to \dot{X}_T$, representing the class of equivalence $\xi'_T \in \mathbb{L}_2^T$. Then $\|\xi^M - \xi_T\|^2_{\dot{X}_T} \to 0$ in $L_1(\Omega)$, and for a suitable null set Q^T and a sequence $M_j \to +\infty$ we have that

$$\|\xi^{M_j} - \xi_T\|^2_{\dot{X}_T} \to 0 \quad \forall \omega \notin Q^T. \tag{1.2.9}$$

Since for $N < M$ and any ω the quantity $\|\xi^M - \xi^N\|^2_{\dot{X}_T} = \|\xi_N^M\|^2_{\dot{X}_T}$ is a decreasing function of N, then for $M_j > N_2 > N_1$ we have

$$\|\xi^{M_j} - \xi^{N_2}\|^2_{\dot{X}_T} \le \|\xi^{M_j} - \xi^{N_1}\|^2_{\dot{X}_T} \quad \forall \omega.$$

Passing in this relation to the limit as $M_j \to \infty$ using (1.2.9) we get that $\|\xi^N - \xi_T\|^2_{\dot{X}_T}$ is a decreasing function of N if $\omega \notin Q^T$. Exploiting (1.2.9) once again we derive from it that

$$\|\xi^M - \xi_T\|_{\dot{X}_T} \to 0 \quad \text{as } M \to \infty, \quad \text{if } \omega \notin Q^T. \tag{1.2.10}$$

Finally, taking $T = 1, 2, \dots$ and putting $Q = Q^1 \cup Q^2 \cup \cdots$ we note that $\xi_K(t) = \xi_{K'}(t)$ for $\omega \notin Q$ if $t \le \min(K, K')$. Thus we have obtained a process $\xi^\omega(t)$, $t \ge 0$, defined for $\omega \in \Omega \setminus Q$, with trajectories in $\dot{X}_\infty$, such that the sum $\sum_{s\in\mathbb{Z}^*} b_s \beta_s^\omega(t) e_s$ converges to $\xi(t)$ for $\omega \in \Omega \setminus Q$ and for all t. Extending $\xi^\omega(t)$ by zero for $\omega \in Q$ we arrive at a process ξ satisfying the first assertion of the theorem.

Using (1.2.6), (1.2.10) and the monotone convergence theorem we obtain the inequality (1.2.1). Similarly (1.2.5) and (1.2.10) yield (1.2.3). Finally, (1.2.2) follows from (1.2.1) since

$$x^p \le C_{p,\alpha} e^{\alpha x^2} \qquad \forall x \ge 0, \tag{1.2.11}$$

for any $p, \alpha > 0$. □

We will denote

$$\pi_T = \mathcal{D}(\xi \mid_{[0,T]}) \in \mathcal{P}(\dot{X}_T) \subset \mathcal{P}(X_T).$$

Take any continuous linear functional $f \in (\dot{X}_T)^*$ and consider the random variable

$$f(\xi^\omega) = f\left(\sum b_s \beta_s^\omega(\cdot) e_s\right) = \sum b_s f\left(\beta_s^\omega(\cdot) e_s\right).$$

By (1.2.8) this series converges in $L_2(\Omega)$. Since the r.v.'s $b_s f\left(\beta_s^\omega(\cdot) e_s\right) =: \xi_s$ are independent and Gaussian, then π_T is a Gaussian measure (see [**139**], Section II.13). As the law of ξ_s depends only on f, s and b_s, then, evoking Definition 11.5.2 we have:

COROLLARY 1.2.2. *For any $0 < T < \infty$ the law π_T of $\xi \mid_{[0,T]}$ is a Gaussian measure in the space $\dot{X}_T$, which depends only on the coefficients $\{b_s, s \in \mathbb{Z}_*\}$.*

Notice that Corollary 1.2.2 and Fernique's Theorem (see (11.5.1)) jointly imply the assertion (1.2.1) of Theorem 1.2.1, without specifying the constant in the r.h.s. of the inequality and the restriction on the exponent α.

For $m \in \mathbf{N}_0$, we denote by H^m the Sobolev space defined by

$$H^m = \{v \in H, v^{(m)} \in H\},$$

where $u^{(m)} := \partial_x^m u$ is the m-th derivative of u in the variable x in the sense of distributions (so $H^0 = H$). We endow H^m with the homogeneous scalar product

$$\langle u, v\rangle_m = \int_{S^1} u^{(m)}(x)v^{(m)}(x)\,dx = \sum_{s\in\mathbb{Z}^*} |2\pi s|^{2m} u_s v_s \tag{1.2.12}$$

(assuming that the functions u and v are written as series (1.1.3)), and we denote by $\|\cdot\|_m$ the norm associated with this product :

$$\|u\|_m^2 = \sum_{s\in\mathbb{Z}^*} |2\pi s|^{2m} |u_s|^2. \tag{1.2.13}$$

Note that $\partial_x : H^{m+1} \to H^m$ is an isomorphism and $\|\partial_x^k u\|_m = \|u\|_{m+k}$, for every $m, k \in \mathbf{N}_0$.

For any $m \in \mathbb{R}$ relation (1.2.13) defines a norm $\|\cdot\|_m$. For a real number $m \geq 0$ we define the space H^m as

$$H^m = \{u \in H : \|u\|_m < \infty\},$$

and for $m < 0$ we denote by H^m the completion of H with respect to the norm $\|\cdot\|_m$. Elements of spaces H^m with negative m are generalised functions, while elements of spaces H^m with $m \geq 0$ are defined for a.e. x.

We recall the Sobolev embedding

$$H^m \hookrightarrow C^k(S^1), \quad m > k + \tfrac{1}{2}, \quad k \in \mathbf{N}_0\,. \tag{1.2.14}$$

According to (1.2.14), for $m > 1/2$ we will regard elements of H^m as continuous functions, well-defined at each $x \in S^1$. By (1.2.14),

$$\bigcap_{m\in\mathbf{N}} H^m = C^\infty(S^1). \tag{1.2.15}$$

For $m > \frac{1}{2}$ the space H^m is a Banach algebra:

$$\|uv\|_m \leq C_m \|u\|_m \|v\|_m\,. \tag{1.2.16}$$

By X_T^m and $\dot{X}_T^m$, $T > 0$, we denote the spaces

$$X_T^m = C(0, T; H^m), \quad \dot{X}_T^m = \{u \in X_T^m : u(0) = 0\},$$

endowed with the Banach norm $\|u\|_{X_T^m} = \sup_{t\in[0,T]} \|u(t)\|_m$ if $T < \infty$, and with the distance (11.2.1) if $T = \infty$.

THEOREM 1.2.3. *Let ξ be the process constructed in Theorem 1.2.1. If $B_m < \infty$ for some real number $m > 0$, then after we redefine $\xi(\cdot)$ on a suitable null set Q^m as zero, it becomes a process in H^m:*

$$\xi^\omega\,|_{[0,T]} \in \dot{X}_T^m \quad \textit{for any } T > 0 \textit{ and every } \omega. \tag{1.2.17}$$

Its law

$$\pi_T^m = \mathcal{D}(\xi\,|_{[0,T]}) \in \mathcal{P}(\dot{X}_T^m)$$

is a Gaussian measure on $\dot{X}_T^m$. Besides,

$$\mathbb{E}\|\xi(t)\|_m^2 = tB_m, \quad t \geq 0 \tag{1.2.18}$$

(see (1.1.7)*), and*
(1.2.19)

$$\mathbb{E}\,\exp(\alpha\|\xi\|^2_{X_T^m}) \le 4\exp(2T\alpha B_m) - 3 \quad \textit{if} \quad \alpha \le \alpha_*^m = \frac{1}{4T\max_s\big((2\pi|s|)^{2m}|b_s|^2\big)}.$$

Note that, in particular, we can choose in (1.2.19) $\alpha = (4TB_m)^{-1}$. As before, in view of (1.2.11), relation (1.2.19) implies that all moments of the r.v. $\sup_{t\in[0,T]}\|\xi(t)\|_m$ are bounded. The proof of the theorem is similar to that of Theorem 1.2.1 and is omitted.

COROLLARY 1.2.4. *If $B_m < \infty$ for all m, then ξ may be modified on a null-set in such a way that it becomes a process in each space H^m, and relations* (1.2.17)-(1.2.19) *hold for every m.*

Agreement. Below, if in some part of the text it is assumed that $B_m < \infty$ for an $m > 0$ (or for all m), then there we modify the process ξ as in Theorem 1.2.3 (or as in Corollary 1.2.4). E.g. see the beginnings of Section 1.4 and of Chapter 7. These modifications give us the advantage that then $\xi^\omega \in \dot X^m_\infty$ for all ω (for all ω and all m); not just for a.e. ω, which makes the presentation a bit lighter.

PROBLEM 1.2.5. Prove that if $B_1 < \infty$, then for every ω the function $(t,x) \mapsto \xi^\omega(t,x)$ is continuous.
Hint: Use that $|\xi^\omega(t+\varepsilon,x) - \xi^\omega(t,x)| = o(1)$ as $\varepsilon \to 0$, uniformly in x.

THEOREM 1.2.6. *If $B_m < \infty$ for some $m \ge 0$, then*

1) The Gaussian measures π_T^m on the spaces $\dot X_T^m$ agree in the sense that for $T_2 > T_1 > 0$ the restriction mapping $\dot X^m_{T_2} \mapsto \dot X^m_{T_1}$ sends $\pi^m_{T_2}$ to $\pi^m_{T_1}$.

2) There exists a Borel measure π^m_∞ on the Polish space $\dot X^m_\infty$ such that for any $T > 0$ the restriction mapping $\dot X^m_\infty \to \dot X^m_T$ sends π^m_∞ to π^m_T.

PROOF. To simplify notation we only consider the case $m = 0$. Next, for $0 < T_1 < T_2 \le \infty$ let $\Pi^{T_2}_{T_1} : \dot X_{T_2} \to \dot X_{T_1}$ be the restriction map, and $\iota^{T_1}_{T_2} : \dot X_{T_1} \to \dot X_{T_2}$ be a mapping which continues any curve $\zeta(t), 0 \le t \le T_1$, as $\zeta(T_1)$ for $T_1 \le t \le T_2$. Then $\Pi^{T_2}_{T_1} \circ \iota^{T_1}_{T_2} = \mathrm{id}$. For the random force $\xi^\omega = \xi^\omega(t,x)$ we denote by ξ^ω_T its restriction to $t \in [0,T]$.

1) By (1.2.8) with $T{=}T_2$, ξ_{T_2} is the $\mathbb{L}_2^{T_2}$–limit of $\xi^N_{T_2}$, where $\xi^N_{T_2}$ stands for the process ξ^N in (1.2.4) with $T = T_2$. Then $\Pi^{T_2}_{T_1}\xi_{T_2} = \Pi^{T_2}_{T_1}(\lim_N \xi^N_{T_2}) = \lim_N \xi^N_{T_1} = \xi_{T_1}$ since obviously $\Pi^{T_2}_{T_1}\xi^N_{T_2} = \xi^N_{T_1}$. As $\pi^m_T = \mathcal{D}(\xi_T) \in \mathcal{P}(H^m)$ for any T, then this proves the first assertion of the theorem.

2) To establish the second assertion let us consider the measures $\pi^\infty_T = \iota^T_\infty \circ \pi_T \in \mathcal{P}(\dot X_\infty)$. We claim that $\{\pi^\infty_T, T > 0\}$ is a Cauchy family with respect to the Lipschitz-dual distance in $\mathcal{P}(\dot X_\infty)$ (see below in Section 1.7). To do this let us take any $f \in L(\dot X_\infty)$, $\|f\|_{L(\dot X_\infty)} \le 1$ (see (1.7.3)), and consider $d := \langle f, \pi^\infty_{T_2} - \pi^\infty_{T_1}\rangle$, where $\infty > T_2 > T_1 > 0$. We have $d = \langle f\circ\iota^{T_2}_\infty, \pi_{T_2}\rangle - \langle f\circ\iota^{T_1}_\infty, \pi_{T_1}\rangle$. Let us write $\langle f\circ\iota^{T_2}_\infty, \pi_{T_2}\rangle$ as

$$\langle f\circ\iota^{T_1}_\infty\circ\Pi^{T_2}_{T_1}, \pi_{T_2}\rangle + \langle (f\circ\iota^{T_2}_\infty - f\circ\iota^{T_1}_\infty\circ\Pi^{T_2}_{T_1}), \pi_{T_2}\rangle.$$

By 1), the first term equals $\langle f\circ\iota^{T_1}_\infty, \Pi^{T_2}_{T_1}\circ\pi_{T_2}\rangle = \langle f\circ\iota^{T_1}_\infty, \pi_{T_1}\rangle$. So

$$d = \langle (f\circ\iota^{T_2}_\infty - f\circ\iota^{T_1}_\infty\circ\Pi^{T_2}_{T_1}), \pi_{T_2}\rangle.$$

Consider the integrand in this integral. For any $\xi \in \dot{X}_{T_2}$, if we denote $N = \lfloor T_1 \rfloor$ (see (0.0.17)), then

$$\operatorname{dist}(\iota_\infty^{T_2}(\xi), \iota_\infty^{T_1} \circ \Pi_{T_1}^{T_2}(\xi)) \leq \sum_{n=N+1}^{\infty} 2^{-n} = 2^{-N},$$

where dist is the distance (11.2.1) in the space $\dot{X}_\infty$. So since $\operatorname{Lip}(f) \leq 1$, then $|d| \leq 2^{-N}$. Accordingly, $\|\pi_{T_2}^\infty - \pi_{T_1}^\infty\|_{L(\dot{X}_\infty)}^* \leq 2^{-N}$, and

$$\pi_T^\infty \rightharpoonup \pi_\infty^0 \in \mathcal{P}(\dot{X}_\infty) \quad \text{if} \quad T \to \infty.$$

for some measure π_∞^0. As for any $\infty > T_1 \geq T > 0$ we have $\Pi_T^\infty \circ \pi_{T_1}^\infty = \Pi_T^{T_1} \circ \pi_{T_1} = \pi_T$ by 1), then due to the convergence above $\Pi_T^\infty \circ \pi_\infty^0 = \pi_T$, and the theorem is proved. □

The measure π_∞^m is the law of the r.v. $\omega \mapsto \dot{X}_\infty^m$. This is a Gaussian measure on the Fréchet space $\dot{X}_\infty^m$ (see (11.2.2)). We avoid the complicated topic of Gaussian measures on linear topological spaces and refer an interested reader to [**23**].

1.3. Deterministic Cauchy problem.

In this section we study the problem (1.1.13), where $u_0(x)$ and $\xi(t,x)$ are fixed non-random functions, sufficiently smooth in x.

Preliminaries.

We begin by considering the heat equation

$$\begin{cases} v_t(t,x) - \nu v_{xx}(t,x) = \xi_t(t,x), \quad \xi \in \dot{X}_T^m, \ m \in \mathbb{R}, \\ v(0,x) = u_0(x). \end{cases} \tag{1.3.1}$$

where $T > 0$ is fixed. We say that $v \in X_T^m$ solves (1.3.1) if for all $t \in [0,T]$,

$$v(t) - \nu \int_0^t v_{xx}(s)ds = u_0 + \xi(t)\,, \tag{1.3.2}$$

where the equality holds in the space H^{m-2}.

LEMMA 1.3.1. *Fix $m \in \mathbb{R}$ and choose any real number $m_* > m$. Then for every $\xi \in \dot{X}_T^{m_*}$ and $u_0 \in H^m$, there exists a unique solution $v \in X_T^m$ of (1.3.1). Moreover, the mapping*

$$H^m \times \dot{X}_T^{m_*} \to X_T^m, \quad (u_0, \xi) \longmapsto v\,, \tag{1.3.3}$$

is linear and continuous.

PROOF. *The uniqueness.* Let v^1 and v^2 be two solutions. Their difference $v = v^1 - v^2$ satisfies (1.3.1) with $u_0 = 0$, $\xi = 0$. So by (1.3.2) any Fourier coefficient $v_s(t)$ of $v(t)$ is a continuous function satisfying the equation

$$v_s(t) - \nu(2\pi s)^2 \int_0^t v_s(\tau)\, d\tau = 0 \quad \forall t\,.$$

From here $v_s = 0$ for all s, so $v = 0$.

The existence. Without loss of generality we may assume that $m_* \in (m, m+2]$. Suppose first that $u_0 \in H^m, \xi \in \dot{X}_T^{m_*}, \dot{\xi} \in X_T^{m_*}$ and consider the curve

(1.3.4)
$$v(t) = e^{\nu\Delta t}u_0 + \int_0^t e^{\nu\Delta(t-\tau)}\xi_\tau(\tau)\,d\tau = e^{\nu\Delta t}u_0 + \xi(t) + \nu\int_0^t \Delta e^{\nu\Delta(t-\tau)}\xi(\tau)\,d\tau,$$

where $\Delta = \frac{\partial^2}{\partial x^2}$ and $\{e^{\nu t\Delta}, t \geq 0\}$, is the semigroup of operators generated by $\nu\Delta$. That is, $\left(e^{\nu\tau\Delta}u_0\right)(x) = v(\tau, x)$, where $v(t,x)$ is the solution of the problem (1.3.1) with $\xi = 0$. Denote

$$\Delta e^{\nu\Delta(t-\tau)}\xi(\tau) =: \Theta(\tau) = \sum \Theta_s(\tau)e_s\,.$$

Then $\Theta_s(\tau) = -|2\pi s|^2 e^{-\nu|2\pi s|^2(t-\tau)}\xi_s(\tau)$, so

$$\begin{aligned}
&|\Theta_s(\tau)|\,|2\pi s|^m \\
&\quad = (\nu(t-\tau))^{-1+\frac{m_*-m}{2}}\Big((\nu|2\pi s|^2(t-\tau))^{\frac{2+m-m_*}{2}}e^{-\nu|2\pi s|^2(t-\tau)}\Big)|2\pi s|^{m_*}|\xi_s(\tau)| \\
&\quad \leq C(t-\tau)^{-1+\frac{m_*-m}{2}}|2\pi s|^{m_*}|\xi_s(\tau)|\,, \quad C = C(\nu, m, m_*),
\end{aligned}$$

since $2 + m - m_* \geq 0$ and $\sup_{r\geq 0} r^a e^{-r} < \infty$ if $a \geq 0$. Therefore we have

$$\|\Theta(\tau)\|_m^2 \leq C(t-\tau)^{-2+m_*-m}\sum |2\pi s|^{2m_*}|\xi_s(\tau)|^2 = C(t-\tau)^{-2+m_*-m}\|\xi(\tau)\|_{m_*}^2\,.$$

Hence,

$$\begin{aligned}
\|v(t)\|_m &\leq \|u_0\|_m + \|\xi(t)\|_m + C_1\int_0^t \|\Theta(\tau)\|_m\,d\tau \\
&\leq \|u_0\|_m + C_2\|\xi\|_{\dot{X}_T^{m_*}}\big(1 + \int_0^t \theta^{-1+\frac{m_*-m}{2}}\,d\theta\big) \leq \|u_0\|_m + C_3\|\xi\|_{\dot{X}_T^{m_*}},
\end{aligned}$$

since $m_* > m$, where C_1, C_2, C_3 depend on ν, m, m_*, T. That is,

(1.3.5)
$$\|v\|_{X_T^m} \leq \|u_0\|_m + C_3\|\xi\|_{\dot{X}_T^{m_*}}\,.$$

Relation (1.3.4) implies that v satisfies eq. (1.3.2) in the sense that each of its Fourier coefficients satisfies the corresponding equation which follows from (1.3.2). Since $v \in X_T^m$, then for $\xi \in \dot{X}_T^{m_*}$, $\xi_t \in X_T^{m_*}$ relation (1.3.2) holds in H^{m-2}, so v is a solution of (1.3.1). By continuity the constructed mapping $(u_0, \xi) \mapsto v$ extends to a bounded linear operator (1.3.3), and $v(u_0, \xi)$ solves (1.3.1) for any $(u_0, \xi) \in H^m \times \dot{X}_T^{m_*}$. □

It is not very natural that to construct a solution $v \in X_T^m$ of (1.3.1) we have to assume that $\xi \in \dot{X}_T^{m_*}$ for some $m_* > m$. This discrepancy may be removed paying the price that the mapping (1.3.3), considered as a function of the second argument, becomes measurable and defined a.e. with respect to the measure $\pi_T^{m_*}$ on the space $\dot{X}_T^{m_*}$ with $m_* = m$:

PROBLEM* 1.3.2. Fix any $m \geq 0$ and assume that $B_m < \infty$. Prove that there exists a null-set Q_m and a measurable mapping $\Sigma : H^m \times \Omega \to X_T^m$ such that $\mathbb{E}|\Sigma(u_0, \cdot)|_{X_T^m}^2 < \infty$ for each u_0, and for any $\omega \notin Q_m$, $u_0 \in H^m$, the function $v = \Sigma(u_0, \omega)$ solves (1.3.1) with $\xi = \xi^\omega$.

The results below where we demand that $B_{m_*} < \infty$ for some $m_* > m$ usually may be proved under the weaker assumption $B_m < \infty$ if we use the result of Problem 1.3.2 instead of Lemma 1.3.1. Since this way to study solutions of (B) is less transparent, we decided to work with the non–optimal pathwise construction which uses Lemma 1.3.1, unless a proof with m_* replaced by m is equally short. Similarly, if in a result concerning the behaviour of solutions for (B) in a space H^m or W_p^m the optimal restriction is $B_M < \infty$, where $M = M(m,p)$ is given by an involved explicit formula, we for simplicity demand that $B_M < \infty$ for all M.

Now we pass to the Burgers equation. We decompose a solution u to the problem (1.1.13) as

$$u(t,x) = w(t,x) + v(t,x), \tag{1.3.6}$$

where v satisfies (1.3.1). Then w solves the Cauchy problem:

$$\begin{cases} w_t - \nu w_{xx} = -\frac{1}{2}((w+v)^2)_x \,, \\ \qquad w(0) = 0. \end{cases} \tag{1.3.7}$$

The advantage of eq. (1.3.7) compared to (B) is that all its coefficients are regular.

For $m \in \mathbf{N}_0 = \mathbf{N} \cup \{0\}$ and $1 \le p \le \infty$ we denote by W_p^m the Sobolev space of zero mean functions $h(x)$, $x \in S^1$, having finite norm

$$|h|_{m,p} = |\partial^m h|_p. \tag{1.3.8}$$

We start with a classical result; for its proof see [**32, 125**]. The inequalities on $\mathbb{R}$ imply the ones on S^1 by a classical localisation argument.

LEMMA 1.3.3. *(the Gagliardo–Nirenberg inequality). Let $m \in \mathbf{N}$ and $\beta \in \mathbf{N}_0$, $\beta \le m-1$. Let also $p, q, r \in [1, \infty]$. Then there exists $C = C(\beta, p, q, r, m) > 0$ such that*

$$|h|_{\beta,r} \le C |h|_{m,p}^{\theta} |h|_q^{1-\theta}, \tag{1.3.9}$$

where the exponent θ belongs to the interval $[\beta/m, 1)$ and satisfies

$$\beta - \frac{1}{r} = \theta\Big(m - \frac{1}{p}\Big) - (1-\theta)\,\frac{1}{q}\,.$$

Moreover, if $p = 1$ or $p = \infty$, then in addition to the relation above it is required that $\theta = \beta/m$.

In particular, choosing $p = 2$ we get that provided $r \ge 2$ and $0 \le k \le m-1$,

$$|h|_{k,r} \le C \|h\|_m^{\theta} \|h\|^{1-\theta}, \qquad \theta = \frac{2rk + r - 2}{2rm}. \tag{1.3.10}$$

If $k \ge 1$, then applying (1.3.9) to h_x we obtain that provided $2m \ge k+1+(2m-2)/r$ and $1 \le k \le m-1$,

$$|h|_{k,r} \le C \|h\|_m^{\theta} |h|_{1,1}^{1-\theta}, \qquad \theta = \frac{2}{r}\,\frac{rk-1}{2m-1}. \tag{1.3.11}$$

Choosing $r = p = q = \infty$ we get that

$$|h|_{\beta,\infty} \le C |h|_{m,\infty}^{\theta} |h|_{\infty}^{1-\theta}, \quad \theta = \beta/m. \tag{1.3.12}$$

This relation is known as the Landau–Kolmogorov inequality (Kolmogorov's contribution to the result was the calculation of the optimal constants $C(\beta, m)$).

LEMMA 1.3.4. *For $m \in \mathbb{N}$ and $h \in H^{m+1}$, there exists $C(m) > 0$ such that*

(1.3.13)
$$N_m(h) := |\langle \partial_x^{2m} h, \partial_x h^2 \rangle| = |\langle \partial_x^{m+1} h, \partial_x^m h^2 \rangle| \le \|h\|_{m+1} N_m^0(h), \quad N_m^0(h) = \|\partial_x^m h^2\|.$$

Here, on one hand,

$$N_m^0(h) \le C(m) \|h\|_{m+1}^{\frac{2m-1}{2m+1}} |h|_{1,1}^{\frac{2m+3}{2m+1}}, \tag{1.3.14}$$

and on the other hand,

$$N_m^0(h) \le C(m) \|h\|_{m+1}^{\frac{2m+1}{2m+2}} \|h\|^{\frac{2m+3}{2m+2}}. \tag{1.3.15}$$

PROOF. Using Hölder's inequality we get:

(1.3.16)
$$N_m^0(h)^2 \le C(m) \sum_{k=0}^{m} \int_{S^1} (h^{(k)})^2 (h^{(m-k)})^2 dx \le C(m) \sum_{k=0}^{m} |h|_{k,2m/k}^2 |h|_{m-k,2m/(m-k)}^2.$$

Due to (1.3.11),

$$|h|_{k,2m/k} |h|_{m-k,2m/(m-k)} \le C \|h\|_{m+1}^{\frac{2m-1}{2m+1}} |h|_{1,1}^{\frac{2m+3}{2m+1}} \qquad \forall\, k,$$

so (1.3.14) follows. Next, due to (1.3.10),

$$|h|_{k,2m/k} |h|_{m-k,2m/(m-k)} \le C \|h\|_{m+1}^{\frac{2m+1}{2m+2}} \|h\|^{\frac{2m+3}{2m+2}} \qquad \forall\, k,$$

and (1.3.15) follows as well. □

PROBLEM 1.3.5. Consider $N_m^{0f}(h) = \|\partial_x^m f(h)\|$, where the function f satisfies (0.0.13). Verify that it satisfies the following versions of estimates (1.3.14), (1.3.15):

$$N_m^{0f}(h) \le C(m) \|h\|_{m+1}^{(2m-1)/(2m+1)} |h|_{1,1}^{(2m+3)/(2m+1)} \Big(1 + \delta(m) |h|_{1,1}^{(m-2)}\Big), \tag{1.3.17}$$

$$N_m^{0f}(h) \le C(m) \|h\|_{m+1}^{\frac{2m+1}{2m+2}} \|h\|^{\frac{2m+3}{2m+2}} \big(1 + \delta(m) \|h\|_{m+1}^{\frac{m-2}{2m+2}} \|h\|^{\frac{(m-2)(2m+1)}{2m+2}}\big), \tag{1.3.18}$$

where $\delta(1) = \delta(2) = 0$ and $\delta(j) = 1$ for $j \ge 3$.

Hint: Note that the integrands in an analogue of the estimate (1.3.16) for N_m^f contain not only squares of quadratic terms, but also a finite sum of squares of monomials of degree up to m, and each of them may be estimated by a calculation similar to that in the proof of Lemma 1.3.4, using that $|f'(h)| \le C|h|$ and $|f^{(k)}(h)| \le C_k$ for each $k \ge 2$ and each h.

Well-posedness of equation (B).

Let us fix any $m \in \mathbf{N}$. Our goal in this section is to prove that eq. (B), (1.1.5) is well posed in the spaces H^m.

THEOREM 1.3.6. *For $0 < T < \infty$, $u_0 \in H^m, m \in \mathbf{N}$, and $\xi \in \dot{X}_T^{m_*}$ with some $m_* > m$, the problem* (1.1.13) *has a unique solution $u(t,x)$ in X_T^m. The norm of u in X_T^m is bounded in terms of $m, T, \nu, \|u_0\|_m$ and $\|\xi\|_{X_T^{m_*}}$.*

Applying the result with any positive T we get that if $\xi \in \dot{X}_\infty^{m_*}$, then the problem has a unique solution $u(t,x)$ in X_∞^m.

PROOF. The constants and the exponents in the proof below depend on the parameters listed in the theorem's assertion.
Step 1: uniqueness. Consider any two solutions $u_1, u_2 \in X_T^m$ to (1.1.13). Then $\widetilde{u} := u_1 - u_2$ solves the equation

$$\widetilde{u}_t - \nu\widetilde{u}_{xx} = -\frac{1}{2}((u_1 + u_2)\widetilde{u})_x, \quad \widetilde{u}(0) = 0.$$

Taking the L_2 scalar product of this equation with $\widetilde{u}$ and integrating by parts we get:

$$\frac{1}{2}\frac{d}{dt}\|\widetilde{u}(t)\|^2 + \nu\|\widetilde{u}(t)\|_1^2 = \frac{1}{2}\int (u_1(t) + u_2(t))\widetilde{u}(t)\widetilde{u}_x(t)dx.$$

Using the Cauchy-Schwarz inequality and the embedding of H^1 in L_∞ (see (1.2.14)), we get:

$$\begin{aligned}\frac{1}{2}\frac{d}{dt}\|\widetilde{u}(t)\|^2 + \nu\|\widetilde{u}(t)\|_1^2 &\leq \frac{1}{2}\|(u_1(t) + u_2(t))\widetilde{u}(t)\| \cdot \|\widetilde{u}_x(t)\| \\ &\leq C(u_1, u_2, \nu)\|\widetilde{u}(t)\|^2 + \nu\|\widetilde{u}(t)\|_1^2.\end{aligned}$$

Thus

$$\frac{d}{dt}\|\widetilde{u}(t)\|^2 \leq 2C(u_1, u_2, \nu)\,\|\widetilde{u}(t)\|^2.$$

Since $\widetilde{u}(0) = 0$, then $\widetilde{u} \equiv 0$ by Gronwall's lemma. So a solution for (1.1.13) is unique.
Step 2: a priori estimates. Suppose that $u \in X_T^m$ solves the problem (1.1.13), and let us write it as $u = v + w$, where v is a solution of (1.3.1). By our agreement concerning the constants and Lemma 1.3.1,

$$|v|_{X_T^m} \leq C'. \tag{1.3.19}$$

The function $w \in X_T^m$ solves (1.3.7), and the L_2 scalar product of the l.h.s. of (1.3.7) with w becomes:

$$\int (w_t - \nu w_{xx})w dx = \frac{1}{2}\frac{d}{dt}\|w(t)\|^2 + \nu\|w(t)\|_1^2,$$

whereas the scalar product of the r.h.s. with w is

$$-\frac{1}{2}\int ((w+v)^2)_x w dx = \frac{1}{2}\int (w^2 w_x + v^2 w_x + 2vww_x)dx = \frac{1}{2}\int (v^2 w_x + 2vww_x)dx.$$

Using firstly the Cauchy-Schwarz inequality, then Young's inequality and finally (1.2.14) for $k = 0$, we bound from above the term in the r.h.s. by

$$\frac{\nu}{4}\|w(t)\|_1^2 + C\|v\|_{X_T^1}^4 + \frac{\nu}{4}\|w(t)\|_1^2 + C\|v\|_{X_T^1}^2\|w(t)\|^2.$$

Thus we have

$$\frac{d}{dt}\|w(t)\|^2 + \nu\|w(t)\|_1^2 \leq C_1\|v\|_{X_T^1}^2\|w(t)\|^2 + C_2\|v\|_{X_T^1}^4.$$

Now Gronwall's lemma and (1.3.19) imply that

$$\|w(t)\|^2 \leq C_3, \quad 0 \leq t \leq T. \tag{1.3.20}$$

Taking the L_2 scalar product of (1.3.7) with $w^{(2m)}$ and integrating by parts in space, we get:

(1.3.21)

$$\frac{1}{2}\frac{d}{dt}\|w(t)\|_m^2 + \nu\|w(t)\|_{m+1}^2 \le \frac{1}{2}\left|\langle\frac{d^m}{dx^m}(w(t)+v(t))^2, \frac{d^{m+1}}{dx^{m+1}}w(t)\rangle\right|$$

$$\le \underbrace{\frac{1}{2}\left|\langle (w^2)_x(t), w^{(2m)}(t)\rangle\right|}_{=:I_1} + \underbrace{\left|\langle (wv)^{(m)}(t), w^{(m+1)}(t)\rangle\right|}_{=:I_2} + \underbrace{\frac{1}{2}\left|\langle (v^2)^{(m)}(t), w^{(m+1)}(t)\rangle\right|}_{=:I_3}.$$

Using firstly (1.3.15) and then Young's inequality we obtain:

$$I_1 \le \frac{\nu}{4}\|w(t)\|_{m+1}^2 + C_1'\|w(t)\|^{c_1'},$$

for a suitable positive exponent c_1'. Similarly, using (1.2.16), (0.0.15), Young's inequality and (1.3.19) we get:

$$\begin{aligned}I_2 &= 2\langle w(t)v(t), w_x(t)\rangle_m \\ &\le C_1\|w(t)v(t)\|_m\|w(t)\|_{m+1} \le C_2\|w(t)\|_m\|v(t)\|_m\|w(t)\|_{m+1} \\ &\le C_3\|v(t)\|_m\|w(t)\|^{\frac{1}{m+1}}\|w(t)\|_{m+1}^{\frac{2m+1}{m+1}} \le \frac{\nu}{4}\|w(t)\|_{m+1}^2 + C_4\|w(t)\|^{c_2'}.\end{aligned}$$

Similarly, $I_3 \le C\|w\|_{m+1}\|v\|_m^2 \le \frac{\nu}{4}\|w\|_{m+1}^2 + C$. Combining the three estimates with (1.3.20) we obtain:

$$\frac{1}{2}\frac{d}{dt}\|w(t)\|_m^2 + \frac{\nu}{4}\|w(t)\|_{m+1}^2 \le C',$$

and integrating in time we see that

$$\|w\|_{X_T^m}^2 + \int_0^T \|w(t)\|_{m+1}^2 dt \le C_m. \tag{1.3.22}$$

Step 3: Galerkin approximations. For $N \in \mathbf{N}$ let $H^{(N)}$ be the $2N$-dimensional space in (1.2.4). Now we define a sequence $\{w^N\}$ of Galerkin approximations for solutions of the problem (1.3.7). To do this, denote by Π_N the *Galerkin projection* which sends a function $u(x) = \sum u_s e_s(x)$ to

$$\Pi_N(u(x)) = \sum_{|s|\le N} u_s e_s(x) \in H^{(N)}. \tag{1.3.23}$$

Let us substitute in (B) $u := u^N = u^N(t) = \sum_{|s|\le N} u_s^N(t)e_s$, and apply to the obtained relation the projection Π_N. We get:

$$\partial_t u^N - \nu u_{xx}^N + \Pi_N\big(u^N u_x^N\big) = \partial_t \Pi_N \xi(t), \quad u^N(0) = \Pi_N u_0. \tag{1.3.24}$$

As before, we write a solution of this initial-value problem as $u^N = v^N + w^N$, where $v^N(t) = \Pi_N v(t)$ solves the problem (1.3.1) with ξ replaced by $\Pi_N\xi$ and u_0 replaced by $\Pi_N u_0$. Then $w^N(t) \in H^{(N)}$ is a solution of the following problem:

$$\partial_t w^N - \nu w_{xx}^N + \tfrac{1}{2}\Pi_N\big((w^N + v^N)^2\big)_x = 0, \qquad w^N(0) = 0. \tag{1.3.25}$$

This is an ODE in the space $H^{(N)}$, and its j-th component, $|j| \le N$, has the form

$$\frac{dw_j^N}{dt}(t) = -\nu(2\pi j)^2 w_j^N(t) - P_j\big(w^N(t) + v^N(t)\big), \quad w_j^N(0) = 0. \tag{1.3.26}$$

The nonlinearity P_j is a polynomial in the variables $(w_k^N)_{|k|\le N}$ and $(v_k^N(t))_{|k|\le N}$. So the equation in (1.3.25) defines in $H^{(N)}$ a smooth vector field, continuous in time. Accordingly, (1.3.25) has a unique maximal solution $w^N \in C^1([0,T^N[;H^{(N)})$ for some $0<T^N\le T$, where $\|w^N(t)\|\to\infty$ as $t\to T^N$ if $T^N<T$.

The system (1.3.25) is called the *Galerkin approximation of order N for the problem* (1.1.13), and the curves $\{w^N(t)\}$ – the *Galerkin approximations for a solution* of the problem. Taking the L_2–scalar product of eq. (1.3.25) with $w^N(t)$, noting that

$$\langle \Pi_N((w^N+v^N)^2)_x, w^N\rangle = \langle((w^N+v^N)^2)_x, w^N\rangle$$

and repeating the argument of Step 2, we see that w^N satisfies (1.3.20) with the same N–independent constant, since $\|v^N(t)\|_m \le \|v(t)\|_m$ for any m and every t.

Thus, $\limsup_{t\to T^N}\|w^N(t)\|<\infty$, and $T^N=T$ for each N. Repeating further the argument which follows (1.3.20), we see that each Galerkin approximation w^N also satisfies (1.3.22), uniformly in N.

Step 4: Convergence to a solution. Firstly let us analyse the convergence of v^N to v. Since $v\in X_T^m$, then $\{\|v^N(t)-v(t)\|_m,\ N\ge 1\}$, is a monotone sequence of continuous functions of $t\in[0,T]$, pointwise converging to zero. By Dini's Theorem 11.1.12 the convergence is uniform, so

$$v^N \to v \quad \text{in} \quad X_T^m \quad \text{as} \quad N\to\infty. \tag{1.3.27}$$

Since the Galerkin approximations w^N satisfy estimates (1.3.22) (where $m\ge 1$), and are solutions of the problem (1.3.25), then their time-derivatives are bounded in the space $L_2(0,T;H^{m-1})$, uniformly in N. Therefore the sequence $\{w^N\}$ remains bounded in the Hilbert space[3]

$$W^m = \{u\in L_2(0,T;H^{m+1}) : u_t\in L_2(0,T;H^{m-1})\}\subset X_T^m \tag{1.3.28}$$

(concerning the embedding on the right see Theorem 11.2.6). Since $W^m \Subset L_2(0,T;H^m)$ (see Theorem 11.2.4), there exists a subsequence $\{w^{N_j}\}$ converging to a function $w\in W^m$ weakly in W^m and strongly in $L_2(0,T;H^m)$.

As $\|((w+v)^2)_x\|_{m-1}\le C\|w+v\|_m^2$ and $v\in X_T^m$, then in view of (1.3.27) $u^N=w^N+v^N$ and $u=w+v$ satisfy

$$u^{N_j}\to u \quad \text{strongly in} \quad L_2(0,T;H^m) \quad \text{and weakly in} \quad W^m. \tag{1.3.29}$$

So

$$((u^{N_j})^2)_x \to (u^2)_x \quad \text{strongly in} \quad L_1(0,T;H^{m-1}). \tag{1.3.30}$$

Now consider the integrated in time equation in (1.3.26) and write it as

$$\Phi_j(w^N(\cdot);v^N(\cdot))(t)=0, \qquad t\in[0,T]. \tag{1.3.31}$$

Clearly if we show that the limiting curve $w(\cdot)\in W^m$ vanishes at zero and for each j satisfies (1.3.31) with $v^N(t)$ replaced by $v(t)$, then w is a solution of (1.3.7). To prove this, for any j consider the curves w^{N_l} with $N_l\ge j$. They vanish at zero, satisfy (1.3.31) since $\Pi_j\circ\Pi_{N_l}=\Pi_j$, converge to w weakly in W^m and satisfy (1.3.30). Using (1.3.27) we pass to the limit in (1.3.31). We deduce that $\Phi_j(w(\cdot);v(\cdot))(t)\equiv 0$ and $w(0)=0$. So indeed w is a solution of (1.3.7), and $u=w+v$ solves (1.1.13). It is bounded in X_T^m since both w and v are. ☐

[3]Here and in similar situations below, the space W^m is given the natural norm defined by the relation $|u|^2_{W^m}=|u|^2_{L_2(0,T;H^{m+1})}+|u_t|^2_{L_2(0,T;H^{m-1})}$. For properties of such spaces see **[118, 119]** and Appendix 11.2.

PROBLEM 1.3.7. Consider a solution of (1.1.13) where in the decomposition of the initial data $u_0(y)$ and of the force $\xi(t, y)$ into Fourier series in y, only the modes $e^{2\pi ikjy}, k \in \mathbb{Z}$, are nonzero, for some fixed $j \in \mathbf{N}$. Prove that the Fourier decomposition of the solution $u(t, y)$ satisfies the same property.

PROBLEM 1.3.8. Prove that the assertion of Theorem 1.3.6 remains true for solutions of equation (0.0.12), (0.0.13).
Hint: To prove the uniqueness use that

$$f(\phi^1) - f(\phi^2) = \int_0^1 \frac{\partial}{\partial t}\big(f(\phi^2 + t(\phi^1 - \phi^2))\big)dt = (\phi^1 - \phi^2)F(\phi^1, \phi^2), \tag{1.3.32}$$

where $F = \int_0^1 f'(t\phi^1 + (1-t)\phi^2)dt$. To establish the existence, repeat Steps 2-4 of the proof of Theorem 1.3.6. Now Steps 3 and 4 are identical to those for the Burgers equation, while Step 2 is a bit different: firstly we get estimate (1.3.20), next, using it and (1.3.18) get (1.3.22) with $m = 1$, and finally use the latter and (1.3.17) to obtain (1.3.22).

Now we specify the convergence of Galerkin approximations. As the limit u in (1.3.29) does not depend on the subsequence $\{u^{N_j}\}$ since this is a solution of (1.1.13) which is unique by Step 1, the convergence (1.3.29) holds not only when $N_j \to \infty$, but also when $N \to \infty$. Since any bounded subset of the space W^m is bounded in X_T^m and uniformly continuous in H^{m-1},[4] then by Theorem 11.2.2 it is precompact in X_T^{m-1}. So $W^m \Subset X_T^{m-1}$. Accordingly the convergence above also is strong in X_T^{m-1}, and we get:

PROPOSITION 1.3.9. *For any $T > 0$, $m \in \mathbf{N}$ and any $u_0 \in H^m$, $\xi \in \dot{X}_T^{m_*}$, where $m_* > m$, the Galerkin approximations u^N solving* (1.3.24) *converge to the solution $u(t)$ constructed in Theorem* 1.3.6, *weakly in X_T^m and strongly in X_T^{m-1} and in $L_2(0, T; H^m)$.*

This result is important for what follows since it provides a convenient finite-dimensional approximation for solutions of (B).

PROBLEM 1.3.10. Prove that if in (1.1.5) $\xi = 0$ and $u(0) = u_0 \in H^1$, then for any $\sigma > 0$ the solution $u(t, x)$ constructed in Theorem 1.3.6, satisfies the *balance relation:*

$$\tfrac{1}{2}\|u(\sigma)\|^2 - \tfrac{1}{2}\|u_0\|^2 + \nu \int_0^\sigma \|u(s)\|_1^2 \, ds = 0. \tag{1.3.33}$$

Using Theorem 1.3.6 for $m \geq 1$ we define the map

$$\mathcal{M} = \mathcal{M}^{T,\nu} : H^m \times \dot{X}_T^{m_*} \to X_T^m, \quad (u_0, \xi) \longmapsto u, \tag{1.3.34}$$

where u solves (1.1.13), and for $t \in [0, T]$ denote by $\mathcal{M}_t^{T,\nu}$ the mapping

$$\mathcal{M}_t = \mathcal{M}_t^{T,\nu} : H^m \times \dot{X}_T^{m_*} \to H^m, \quad (u_0, \xi) \longmapsto u(t) = \mathcal{M}^{T,\nu}_{|t=\text{const}}. \tag{1.3.35}$$

We will omit the upper index T, ν or both of them, except in the situations when the $T-$ or ν-dependence is relevant.

[4]Indeed, for any $\zeta \in B^R_{W^m}$, $R > 0$, and $t_2 > t_1$ we have

$$\|\zeta(t_2) - \zeta(t_1)\|_{m-1} = \|\int_{t_1}^{t_2} \dot{\zeta}\, ds\|_{m-1} \leq \int_{t_1}^{t_2} \|\dot{\zeta}\|_{m-1}\, ds \leq \Big(\int_{t_1}^{t_2} \|\dot{\zeta}\|^2_{m-1}\, ds\Big)^{1/2} \sqrt{t_2 - t_1} \leq R\sqrt{t_2 - t_1}.$$

Naturally, if $T_1 \geq T_2 \geq t$, then for any $u_0 \in H^m$ and $\xi \in \dot{X}^m_{T_1}$ we have

$$\mathcal{M}_t^{T_1}(u_0, \xi) = \mathcal{M}_t^{T_2}(u_0, \xi\mid_{[0,T_2]}).$$

This allows to define mappings

$$\mathcal{M}_t^{\infty} : H^m \times \dot{X}^{m_*}_{\infty} \to H^m, \qquad t \geq 0,$$

by the relation

$$\mathcal{M}_t^{\infty}(u_0, \xi) = \mathcal{M}_t^{T}(u_0, \xi\mid_{[0,T]}), \qquad T \geq t. \tag{1.3.36}$$

THEOREM 1.3.11. *For any $m \in \mathbf{N}$ and $t \in [0,T]$ the mappings $\mathcal{M}$ and $\mathcal{M}_t$ are locally Lipschitz (and therefore continuous).*

By this result and (11.2.3) the mappings $\mathcal{M}_t^{\infty}$, defined in (1.3.36), are continuous.

PROBLEM 1.3.12. Prove the theorem above by considering solutions $u_j(t,x) = \mathcal{M}^t(u_{0j}, \xi_j)$, $j = 1,2$, and decomposing $u_j = w_j + v_j$ as in (1.3.6).
Hint: Denote $v_3 = v_1 - v_2$ and $w_3 = w_1 - w_2$. Then an estimate for v_3 in terms of $(\|u_{01} - u_{02}\|_m + \|\xi_1 - \xi_2\|_{X_T^{m_*}})$ follows from Lemma 1.3.1, while w_3 satisfies (1.3.38) with $v = \frac{1}{2}(u_1 + u_2)$, $z = -\frac{1}{2}\frac{\partial}{\partial x}(u_1 + u_2)v_3$ and $w_3(0) = 0$. So an estimate for w_3 follows from Proposition 1.3.14.

Another proof of Theorem 1.3.11 is given below in Proposition 1.8.4.

PROBLEM 1.3.13. Prove an analogue of Theorem 1.3.11 for problem (0.0.12), (0.0.13) with $\eta = \xi_t(t,x)$.
Hint: Keeping the notation suggested in Problem 1.3.12 and following the hint to its proof, note that the norm of v_3 in X_T^m may still be estimated in terms of $(\|u_{01} - u_{02}\|_m + \|\xi_1 - \xi_2\|_{X_T^{m_*}})$, using Lemma 1.3.1. In view of (1.3.32) the function $w_3(t,x)$ satisfies (1.3.38) with $v = F(u_1, u_2)$, $z = -\partial_x(v_3 F(u_1,u_2))$ and $w_0 = 0$. By a Moser-like inequality (see [**145**, Chapter 13, Proposition 3.9]), $\|F(u_1,u_2)\|_m \leq C\big(\|u_1\|_m + \|u_2\|_m\big)$ (we use that $H^m \subset L_\infty$ since $m \geq 1$). Moreover, in view of an analogue of the equality (1.3.32) the mapping $H^m \times H^m \ni (u_1,u_2) \mapsto F(u_1,u_2) \in H^m$ is continuous. So since $u_1, u_2 \in X_T^m$ by the result of Problem 1.3.8, then $F(u_1(t), u_2(t)) \in X_T^m$ and we may apply Proposition 1.3.14 to estimate $\|w_3\|_{X_T^m}$.

Finally, let $0 < T_1 < T$. Then under the assumption of Theorem 1.3.6, for $T_1 \leq t \leq T$ a solution $u(t) = \mathcal{M}(u_0, \xi)$ solves (B) with the initial data $u(T_1)$. So the mappings $\mathcal{M}_t$ satisfy the *cocycle property:*

$$\mathcal{M}_{T_1+\tau}(u_0, \xi) = \mathcal{M}_\tau\Big(\mathcal{M}_{T_1}(u_0,\xi), \big(\xi\mid_{[T_1,T]} - \xi(T_1)\big) \circ i_{T_1}\Big), \tag{1.3.37}$$

where i_{T_1} is the shift operator, so $\big(\xi\mid_{[T_1,T]} - \xi(T_1)\big) \circ i_{T_1}(t) = \xi(t + T_1) - \xi(T_1)$ (and the curve vanishes at $t = 0$, as it should).

We end this section with a related result, needed later:

PROPOSITION 1.3.14. *Let $m \in \mathbf{N}$.*

i) If $v \in L_\infty(0,T;H^m)$, $z \in L_2(0,T;H^{m-1})$ and $w_0 \in H^m$, then the linearised problem

$$w_t - \nu w_{xx} + (vw)_x = z, \qquad w(0,x) = w_0(x), \tag{1.3.38}$$

has a unique solution $w \in W^m$ (see (1.3.28)). This solution satisfies

$$\|w\|_{W^m} \leq C\big(\|z\|_{L_2(0,T;H^{m-1})} + \|w_0\|_m\big), \quad C = C(m, \nu, T, \|v\|_{L_\infty(0,T;H^m)}). \tag{1.3.39}$$

ii) If $v \in W^m, w_0 \in H^{m+1}$ and $z \in L_2(0,T;H^m)$, then

$$\|w\|_{W^{m+1}} \le C\big(\|z\|_{L_2(0,T;H^m)} + \|w_0\|_{m+1}\big), \quad C = C(m,\nu,T,\|v\|_{W^m}). \tag{1.3.40}$$

PROOF. The constants below in the proof depend on the same parameters as C in (1.3.39).

Uniqueness follows by the same argument as in Theorem 1.3.6. To prove existence we start with an a priori estimate. Let w be a smooth solution of (1.3.38). Multyplying the equation by $w(t)$ in H^m and using the algebra property (1.2.16) we have:

$$\begin{aligned}&\frac{1}{2}\frac{\partial}{\partial t}\|w\|_m^2 + \nu\|w\|_{m+1}^2 = \langle vw, w_x\rangle_m + \langle z, w\rangle_m \\ &\le C\|vw\|_m\|w\|_{m+1} + \frac{\nu}{4}\|w\|_{m+1}^2 + C\|z\|_{m-1}^2 \le \frac{\nu}{2}\|w\|_{m+1}^2 + C_1\|w\|_m^2 + C\|z\|_{m-1}^2.\end{aligned}$$

Using the Gronwall lemma we firstly estimate $\|w(t)\|_m$ for $0 \le t \le T$, next integrating the inequality in t from 0 to T we estimate $\|w\|_{L_2(0,T;H^{m+1})}$, and finally expressing $\dot{w}$ from the equation we get that w satisfies (1.3.39). Now Galerkin's method implies that the problem has a unique solution, satisfying (1.3.39). This proves i).

Since $W^m \subset X_T^m$, then to establish ii) it remains to show that if v, w_0 and z are smoother as specified in the assertion, then the solution w also is smoother and satisfies (1.3.40). Doing that, by the linearity of the problem, we may assume that

$$\|z\|_{L_2(0,T;H^m)} + \|w_0\|_{m+1} \le 1. \tag{1.3.41}$$

Consider the term $(vw)_x =: I$ of the equation. By (1.2.16), $\|I\|_m \le C\|v\|_{m+1}\|w\|_{m+1}$. Let us improve this estimate. We have:

$$\|I^2\|_m \le \|v\|_{m+1}^2|w|_\infty^2 + \|w\|_{m+1}^2|v|_\infty^2 + C\sum_{\substack{\alpha_1+\alpha_2=m+1\\ \alpha_1,\alpha_2\ge 1}}\int\big(\partial_x^{\alpha_1}v\,\partial_x^{\alpha_2}w\big)^2\,dx.$$

By Hölder's inequality, the integral $\int\big(\partial_x^{\alpha_1}v\,\partial_x^{\alpha_2}w\big)^2\,dx$ in the r.h.s. is bounded by $C_1|\,\partial_x^{\alpha_1}v|_{p_1}^2|\partial_x^{\alpha_2}w|_{p_2}^2$, where $p_i = 2(m+1)/\alpha_i$, $i=1,2$. Using the Gagliardo–Nirenberg inequality (1.3.9) with $q=\infty$ we find that $|\partial_x^{\alpha_1}v|_{p_1} \le C|v|_\infty^{1-\theta_1}\|v\|_{m+1}^{\theta_1}$ and $|\partial_x^{\alpha_2}w|_{p_2} \le C|w|_\infty^{1-\theta_2}\|w\|_{m+1}^{\theta_2}$, where $\theta_1+\theta_2=1$ and $\theta_1,\theta_2>0$. From here and the above, using Young's inequality, we get:

$$\begin{aligned}\|I(t)\|_m^2 &\le C\big(1+|v(t)|_\infty^2+|w(t)|_\infty^2\big)\big(\|v(t)\|_{m+1}^2+\|w(t)\|_{m+1}^2\big)\\ &\le C_1\big(\|v(t)\|_{m+1}^2+\|w(t)\|_{m+1}^2\big),\end{aligned}$$

$C_1 = C_1(m,\nu,T,\|v\|_{W^m})$, where we used (1.3.39) and (1.3.41) to get the second inequality. Multiplying equation (1.3.38) by $w(t)$ in H^{m+1}, integrating by parts and using the last estimate we obtain the relations:

$$\begin{aligned}\frac{1}{2}\frac{\partial}{\partial t}\|w\|_{m+1}^2 + \nu\|w\|_{m+2}^2 &\le \|I\|_m\|w\|_{m+2} + \|z\|_m\|w\|_{m+2}\\ &\le \frac{\nu}{2}\|w\|_{m+2}^2 + C_2\big(\|w\|_{m+1}^2+\|v\|_{m+1}^2+\|z\|_m^2\big),\end{aligned}$$

$C_2 = C_2(m,\nu,T,\|v\|_{W^m})$. Now Gronwall's inequality implies (1.3.40). □

1.4. Strong solutions of the stochastic Cauchy problem.

Notion of a strong solution and its existence.

In Sections 1.4-1.6 m is a fixed natural number and we assume that

$$B_{m_*} < \infty \text{ for some } m_* > m. \tag{1.4.1}$$

So by Theorem 1.2.3 $\xi^\omega \in \dot{X}_\infty^{m_*}$ for each ω. We continue to study the mappings $\mathcal{M}$ and $\mathcal{M}_t$. Sometimes we will write them as

$$\mathcal{M}(u_0, \xi) = u(\cdot; u_0, \xi), \quad \mathcal{M}_t(u_0, \xi) = u(t; u_0, \xi), \tag{1.4.2}$$

– this is another notation for the solution of (1.1.13), constructed in Theorem 1.3.6.

DEFINITION 1.4.1. Let $0 < T \leq \infty$ and $u_0 \in H^m$ be a random variable, independent from the Wiener process $\xi^\omega(t)$. A random process $u^\omega(t) \in H^m$,[5] $0 \leq t \leq T$ defined on $(\Omega, \mathcal{F}, \mathbf{P})$, is called a *strong solution* of the Cauchy problem (1.1.13) if there exists a null–set Q such that for each $\omega \in \Omega \backslash Q$ the function $u^\omega(t, x)$ solves the problem (1.1.13).

A random process $u^\omega(t) \in H^m$, $t \geq 0$, is called a strong solution of equation (1.1.12) in H^m if it solves the problem (1.1.13) for some u_0 (this implies that the r.v. $u^\omega(0)$ is independent from the corresponding processes β_s^ω).

Note that we impose no restriction on the law $\mathcal{D}(u_0) \in \mathcal{P}(H^m)$. In particular, we do not assume that $\mathbb{E}\|u_0\|_m < \infty$.

Theorems 1.2.3, 1.3.6 and 1.3.11 jointly imply that for each ω the Cauchy problem (1.1.13) has a unique solution $u^\omega(\cdot) = \mathcal{M}_\cdot^T(u_0, \xi\,|_{[0,T]}) \in X_T^m$, which is a measurable function of ω, and therefore it is a random process. We recall (see in Section 1.2) that the random force ξ^ω is defined only for a.e. ω and its extension to all $\omega \in \Omega$ is not unique. Changing the extension we get another version of a strong solution of (1.1.13), coinciding with the one above for a.e. ω.

By Proposition 1.3.9 we get:

THEOREM 1.4.2. *Let $B_{m_*} < \infty$ for some $m_* > m \in \mathbf{N}$, $0 < T \leq \infty$ and $u_0 \in H^m$ be a r.v., independent from the process $\xi^\omega(t)$. Then the Cauchy problem (1.1.13) has a unique strong solution $u^\omega(t) \in H^m$, $0 \leq t \leq T$, and*

$$u^\omega = \mathcal{M}(u_0^\omega, \xi^\omega\,|_{[0,T]}) \qquad \forall \omega \in \Omega\,. \tag{1.4.3}$$

For each ω, $u^\omega(\cdot)$ is a strong in X_T^{m-1} and weak in X_T^m limit of the sequence of Galerkin approximations $\{u^{N,\omega}(\cdot)\}$ formed by solutions of the problem (1.3.24) *with $\xi = \xi^\omega$.*

The solution (1.4.3) is unique in the sense that any other strong solution coincides with it, a.s. We will often write it as

$$\mathcal{M}(u_0^\omega, \xi^\omega) = u(\cdot; u_0^\omega), \tag{1.4.4}$$

as the random force ξ usually is fixed.

Since u_0 and ξ are independent, then for any $0 \leq t \leq T$,

$$\mathcal{D}u(t; u_0^\omega) = \mathcal{M}_t \circ (\mathcal{D}(u_0) \times \mathcal{D}(\xi)).$$

---

[5] We recall that all random processes in this book are assumed to have continuous trajectories, for all ω.

So for a measurable non-negative function $f(u)$, using Fubini's theorem, we have

$$\mathbb{E}f\big(u(t;u_0^\omega)\big) = \int\int f\big(\mathcal{M}_t(v,\zeta)\big)\mathcal{D}(u_0)(dv)(\mathcal{D}(\xi)(d\zeta)$$
$$= \int \mathcal{D}(u_0)(dv)\int f\big(\mathcal{M}_t(v,\zeta)\big)\mathcal{D}(\xi)(d\zeta) = \int \mathbb{E}f(u(t;v))\mathcal{D}(u_0)(dv).$$

That is,

$$\mathbb{E}f\big(u(t;u_0^\omega)\big) = \int \mathbb{E}f(u(t;v))\mathcal{D}(u_0)(dv), \tag{1.4.5}$$

where both parts may equal $+\infty$.

To calculate $u^\omega(t;u_0^\omega)$ for some $t\ge 0$ we may use Theorem 1.4.2 with any $T\ge t$. Choosing there $T=t$ we get:

COROLLARY 1.4.3. *For any $t\ge 0$, $u^\omega(t;u_0^\omega)\in H^m$ as in Theorem 1.4.2 is a continuous vector–function of $u_0^\omega\in H^m$ and $\xi^\omega\mid_{[0,t]}\in \dot X_t^{m_*}$.*

This observation naturally leads to the important notions of a filtered probability space and random processes, adapted to the filtration. These objects are discussed in Appendix 11.6.

Sometimes we will consider solutions $u(t)$ of (1.1.12), defined for $s\le t\le\infty$, $s\ge 0$, with an initial data prescribed at time s:

$$u(s) = u_0^\omega \in H^m, \quad m\ge 1. \tag{1.4.6}$$

A strong solution for problem (1.1.12), (1.4.6) is defined similarly to that of problem (1.1.13). It exists and is unique for the same reasons as before, will be denoted $u_s(t;u_0^\omega)$, and may be written as

$$u_s(t;u_0^\omega) = \mathcal{M}\big(u_0^\omega, \xi^\omega(\cdot)-\xi^\omega(s)\big)(t-s)\in C([s,\infty],H^m) \tag{1.4.7}$$

(note that due to Problem 11.5.1 the random process $\tau\mapsto\xi^\omega(\tau+s)-\xi(s)=:\xi_s^\omega(\tau)$ is another Wiener process in H^m, distributed as ξ^ω). As in Corollary 1.4.3, $u_s(t;u_0^\omega)$, $t\ge s$, is a function of u_0^ω and $\xi^\omega\mid_{[s,t]}$. By the cocycle property (1.3.37), for $t\ge s$ the solution $u(t;u_0^\omega)$ may be written as

$$u(t;u_0^\omega,\xi^\omega) = u_s(t-s;u(s;u_0^\omega,\xi_s^\omega)). \tag{1.4.8}$$

Basic estimates for strong solutions.

Now we obtain some estimates for moments of Sobolev norms of strong solutions $u^\omega(t)$ of (1.1.13), constructed in Theorem 1.4.2. These results are preliminary and will be significantly strengthened below in Section 2.2.

THEOREM 1.4.4. *Let $B_{1_*}<\infty$ for some $1_*>1$ and $u_0\in H^1$ be a r.v., independent from the process $\xi^\omega(t)$. Consider the solution $u(t)=u(t;u_0)$.*

1) If $\mathbb{E}e^{\sigma\|u_0\|^2}<\infty$ for some $\sigma\in(0,b_{max}^{-2}\nu]$ (see (1.1.9)), then

$$\mathbb{E}e^{\sigma'\|u(t)\|^2}\le e^{-t}\,\mathbb{E}e^{\sigma'\|u_0\|^2}+C \quad \textit{for}\quad \sigma'=\sigma/3,\quad t\ge 0, \tag{1.4.9}$$

where $C=C(\sigma,\nu,B_0)$.

2) If in addition $B_m<\infty$ for some $m\in\mathbf{N}$ and $\mathbb{E}\|u_0\|_m^2<\infty$, then for all $t\ge 0$

$$\mathbb{E}\|u(t)\|_m^2\le C_m\big(1+\mathbb{E}\|u_0\|_m^2+\mathbb{E}e^{\sigma'\|u_0\|^2}\big), \tag{1.4.10}$$

and

$$\mathbb{E}\int_0^t \|u(s)\|_{m+1}^2\,ds \le \mathbb{E}\|u_0\|_m^2 + tC_m\big(1+\mathbb{E}e^{\sigma'\|u_0\|^2}\big). \tag{1.4.11}$$

The constant C_m *depends on* m, ν, σ' *and* B_m.

To prove the inequalities we will approximate $u(t)$ by solutions $u^N(t)$ of the Galerkin system (1.3.24), derive for them analogues of the required relations using Ito's formula (11.8.11), and next obtain (1.4.9) – (1.4.11) by passing to the limit.

PROOF. First we assume that u_0 is a non-random vector.

1) The Galerkin equation (1.3.24) is an SDE of the form (11.8.1) in the $2N$–dimensional space $H^{(N)}$. The Euclidean structure in $H^{(N)}$ is inherited from the space H, and we write the scalar product in $H^{(N)}$ as $\langle\cdot,\cdot\rangle$. Denoting the deterministic part of the vector field in (1.3.24) by $f_N(u^N) = \nu u^N_{xx} - \Pi_N(u^N u^N_x)$, we have

$$\langle f_N(u^N), u^N\rangle = -\nu\|u^N\|_1^2.$$

Thus the equation satisfies condition (11.8.8) with $\alpha = \nu$, and so by (11.8.9)

$$\mathbb{E}e^{\sigma\|u^N(t)\|^2} \le e^{t\sigma B_0}e^{\sigma\|u_0\|^2}, \tag{1.4.12}$$

since $\sum_{|j|\le N} b_j^2 \le B_0$. Now consider the function $F(u^N) = e^{\sigma'\|u^N\|^2}$, $\sigma' = \sigma/3$. Then

$$\nabla F(u^N) = 2\sigma' F(u^N)u^N,$$

$$\frac12\sum_{|j|\le N} b_j^2\frac{\partial^2}{\partial u_j^2}F(u^N) = \sigma' F(u^N)\Big[\sum_{|j|\le N} b_j^2 + 2\sigma'\sum_{|j|\le N} b_j^2u_j^2\Big]. \tag{1.4.13}$$

So

$$\begin{aligned}|\langle\nabla F(u^N), f(u^N)\rangle| + \|\nabla F(u^N)\| &= \nu|\langle\nabla F(u^N),(u^N)_{xx}\rangle| + \|\nabla F(u^N)\|\\ &\le C_N F(u^N)(1+\|u^N\|^2).\end{aligned}$$

From here and (1.4.12) we see that (11.8.12) holds, so Ito's formula applies to $F(u^N(t))$. Using it in combination with (1.4.13) we get

$$\frac{\partial}{\partial t}\mathbb{E}e^{\sigma'\|u^N(t)\|^2} \le \mathbb{E}\left[\sigma' e^{\sigma'\|u^N(t)\|^2}\Big(\|u^N\|^2(-2\nu + 2\sigma'(b_{max})^2) + B_0\Big)\right].$$

Since $\sigma'(b_{max})^2 \le \frac13\nu$, then the second factor under the expectation sign is smaller than $-\frac43\nu\|u^N\|^2 + B_0$. Considering separately the cases when $\nu\|u^N\|^2$ is larger or smaller than $\max((\sigma')^{-1}, 3B_0)$, we get that the r.h.s. is bounded by $-\mathbb{E}e^{\sigma'\|u^N(t)\|^2} + C(\sigma',\nu,B_0)$. By the differential inequality above, this bound and the Gronwall lemma we find that

$$\mathbb{E}e^{\sigma'\|u^N(t)\|^2} \le e^{-t}e^{\sigma'\|u_0\|^2} + C(\sigma',\nu,B_0), \tag{1.4.14}$$

uniformly in N.

By Theorem 1.4.2, $\|u^{N,\omega}(t)\| \to \|u^\omega(t)\|$ as $N\to\infty$, a.s. So Fatou's lemma and the last inequality imply (1.4.9).

2) Assume firstly that $B_{m_*} < \infty$ for some $m_* > m$. Let us consider the function $F(u^N) = \|u^N\|_m^2$. Since on $H^{(N)}$ the norm $\|\cdot\|_m$ is equivalent to $\|\cdot\|$, then again

in view of (1.4.12) assumption (11.8.12) holds for this function for any $T>0$ (with $C_T = C_T(N)$) and Ito's formula applies. Since

$$\frac{1}{2}\sum_{|j|\le N} b_j^2 \frac{\partial^2}{\partial u_j^2}F(u^N) = \sum_{|j|\le N}(2\pi j)^{2m} b_j^2 =: B_m^N \le B_m,$$

and

$$\begin{aligned}\langle \nabla F(u^N), f_N(u^N)\rangle &= -2\nu\langle(-\Delta)^m u^N, (-\Delta)u^N\rangle - \langle(-\Delta)^m u^N, \partial_x (u^N)^2\rangle \\ &= -2\nu\|u^N\|_{m+1}^2 - \langle u^N, \partial_x (u^N)^2\rangle_m,\end{aligned}$$

then (11.8.11) gives us that

$$\frac{\partial}{\partial t}\mathbb{E}\|u^N(t)\|_m^2 = -2\nu\mathbb{E}\|u^N\|_{m+1}^2 - \mathbb{E}\langle u^N, \partial_x(u^N)^2\rangle_m + B_m^N. \tag{1.4.15}$$

By (1.3.13), (1.3.15) and Young's inequality,

(1.4.16)

$$|\langle(-\Delta)^m u^N, \partial_x (u^N)^2\rangle| \le C'_m \|u^N\|^{\frac{2m+3}{2m+2}} \|u^N\|_{m+1}^{\frac{4m+3}{2m+2}} \le \nu\|u^N\|_{m+1}^2 + C_m\|u^N\|^M,$$

for some $M(m)>0$. So

$$\frac{\partial}{\partial t}\mathbb{E}\|u^N(t)\|_m^2 \le -\nu\mathbb{E}\|u^N\|_{m+1}^2 + B_m + C_m\mathbb{E}\|u^N\|^M. \tag{1.4.17}$$

Thus

$$\mathbb{E}\|u^N(t)\|_m^2 \le e^{-\nu t}\|u_0\|_m^2 + \int_0^t e^{-\nu(t-s)}\big[B_m + C_m\mathbb{E}\|u^N(s)\|^M\big]ds,$$

and therefore

(1.4.18)

$$\mathbb{E}\|u^N(t)\|_m^2 \le e^{-\nu t}\|u_0\|_m^2 + \min(\nu^{-1}, t)\big(B_m + C_m \sup_{0\le s\le t}\mathbb{E}|u^N(s)|^M\big), \qquad t\ge 0.$$

Since $x^M \le C(M,\sigma')e^{\sigma' x^2}$ for $x\ge 0$, then from here and (1.4.9) we get that

$$\mathbb{E}\|u^N(t)\|_m^2 \le C_m\big(1 + \|u_0\|_m^2 + e^{\sigma'\|u_0\|^2}\big). \tag{1.4.19}$$

Finally, integrating (1.4.17) and using (1.4.9) we find that also

$$\mathbb{E}\int_0^t \|u^N(s)\|_{m+1}^2 ds \le \|u_0\|_m^2 + tC_m\big(1 + e^{\sigma'\|u_0\|^2}\big).$$

As $u^N \to u$ in X_t^{m-1} a.s. by Theorem 1.4.2, then (1.4.19) jointly with Theorem 11.1.4 imply (1.4.10). Similarly the estimate above and Theorem 11.1.4 applied to the probability $t^{-1}\mathbf{P}(d\omega)ds$, imply (1.4.11).

To remove the assumption $B_{m_*} < \infty$ we denote $\xi^M(t) = \Pi_M(\xi(t))$ (see (1.3.23)) and consider the corresponding solutions $u(t; u_0, \xi^M)$. By Theorem 1.3.11 with $m=1$, for any $T>0$ they converge to $u(t)$ in X_T^1 as $M\to\infty$ a.s., and satisfy (1.4.9)-(1.4.11) by the above. Applying Theorem 11.1.4 we get the validity of these estimates for the solution u.

3) We have established the estimates (1.4.9)-(1.4.11) for non-random initial vectors u_0. To get the required relations with random u_0 it remains to integrate both parts of the obtained inequalities with respect to the measure $\mathcal{D}(u_0)$, see (1.4.5). □

PROBLEM 1.4.5. Prove that the assertions of Theorem 1.4.4 remain true for solutions of equation (0.0.12), (0.0.13) if the initial data $u_0 \in H^1$ is a non-random vector.
Hint: Proofs of (1.4.9) and (1.4.10), (1.4.11) with $m = 1$ go as previously. Then for any integer $k \geq 2$ we apply Ito's formula to $G_k(u^N)=\|u^N\|_1^{2k}$ and next use Gronwall's lemma to show that $\mathbb{E}G_k(u^N(t))$ is bounded, uniformly in t and N. Thus we get upper bounds for all moments of $\|u^N(t)\|_1$, and therefore also of $|u^N(t)|_{1,1}$. Consequently, using (1.3.17) instead of (1.3.15) we derive estimate (1.4.16) for $m \geq 2$, and thus prove (1.4.10), (1.4.11).

The relations obtained when proving the second assertion of the theorem for $m = 0, 1$ are important for our book. We discuss them below, in Chapter 6.

The balance of energy relation.

Choosing in the calculation at Step 2) of the proof $F(u) = \|u\|^2$ and using that $\langle u^N, \partial_x(u^N)^2\rangle = 0$, we get for the Galerkin approximations $u^N(t)$ the relation

$$\frac{\partial}{\partial t}\mathbb{E}\|u^N(t)\|^2 = -2\nu\mathbb{E}\|u^N\|_1^2 + B_0^N. \tag{1.4.20}$$

PROPOSITION 1.4.6. *1) For any $m \in \mathbf{N}$ denote $M(m) = 4m + 4$. Assume that $B_M < \infty$ and a r.v. $u_0 \in H^M$ satisfies*

$$\mathbb{E}e^{\sigma\|u_0\|^2} < \infty, \qquad \mathbb{E}\|u_0\|_M^2 < \infty, \tag{1.4.21}$$

for some $\sigma > 0$. Let $u(t, x)$ be a solution of (1.1.13). Then the function $t \mapsto \mathbb{E}\|u(t)\|_m^2$ is C^1-smooth and satisfies

$$\frac{d}{dt}\mathbb{E}\|u(t)\|_m^2 = -2\nu\mathbb{E}\|u\|_{m+1}^2 - \mathbb{E}\langle u, \partial_x u^2\rangle_m + B_m. \tag{1.4.22}$$

2) If (1.4.21) holds with $M = 2$, then relation (1.4.22) is valid with $m = 0$ and takes the form of the energy balance:

$$\frac{d}{dt}\mathbb{E}\tfrac{1}{2}\|u(t)\|^2 = -\nu\mathbb{E}\|u\|_1^2 + \tfrac{1}{2}B_0. \tag{1.4.23}$$

PROOF. 1) By the interpolation inequality (0.0.15),

$$\mathbb{E}\|u\|_{m+1}^4 \leq \mathbb{E}\big(\|u\|_0^3\|u\|_M\big) \leq \tfrac{1}{2}\mathbb{E}\|u\|_0^6 + \tfrac{1}{2}\mathbb{E}\|u\|_M^2.$$

Since $|\langle u^N, \partial_x(u^N)^2\rangle_m| \leq C\|u^N\|_{m+1}^3$, then using (1.4.14) and (1.4.19) we immediately see that the r.v.'s

$$f_{N,m}^\omega(\tau) := 2\nu\|u^N\|_{m+1}^2 + \langle u^N, \partial_x(u^N)^2\rangle_m, \quad \|u^N(t)\|_m^2 - \|u_0^N\|_m^2$$

are uniformly in N integrable as functions of $(\tau, \omega) \in [0, t] \times \Omega$, $t > 0$. That is, they satisfy (11.1.4) with $G(r) = r^{4/3}$ and same constant C. Integrating (1.4.15) from 0 to t and applying to the obtained equality Theorem 11.1.8 with (X, μ) equal to $([0, t] \times \Omega, d\tau \times \mathbf{P})$, we pass to the limit as $N \to \infty$ and get that

$$\mathbb{E}\|u(t)\|_m^2 - \mathbb{E}\|u(0)\|_m^2 = -\int_0^t \mathbb{E}\big(2\nu\|u\|_{m+1}^2 + \langle u, \partial_x u^2\rangle_m\big)d\tau + tB_m. \tag{1.4.24}$$

By Theorem 1.3.6 the integrand $f_m^\omega(s) = 2\nu\|u(s)\|_{m+1}^2 + \langle u(s), \partial_x u(s)^2\rangle_m$ is continuous in s for each ω. Since it is uniformly in s integrable as a function of ω, then using again Theorem 11.1.8 we find that $\mathbb{E}f_m(s)$ is a continuous function of s. Now (1.4.22) follows from (1.4.24).

2) Since

$$\|u\|_1^3 \le \|u\|^{3/2}\|u\|_2^{3/2} \le C(\|u\|_2^2 + \|u\|^4)$$

and $f^{\omega}_{N,0}(t)=2\nu\|u^N\|_1^2$, then under the assumptions of the second assertion the collection of the r.v.'s $\{f^{\omega}_{N,0}(t), N \ge 1\}$, is uniformly integrable. Using (1.4.20) and repeating the argument above we get (1.4.23). □

By analogy with the hydrodynamics, see [**103**, Section 2.1.9], the quantity $\frac{1}{2}\mathbb{E}[\|u(t)\|^2]$ is called the *(averaged) energy* of a solution $u(t)$, and $\nu\mathbb{E}[\|u(t)\|_1^2]$ – the *(averaged) energy dissipation rate*. Accordingly, relation (1.4.23) is called the *balance of energy* for the stochastic Burgers equation.

The vorticity stretching.

A natural analogue of the vorticity for a 1d velocity field u is its derivative $\zeta = u_x(t,x)$. The balance of its "energy" is given by relation (1.4.22) with $m = 1$. There

$$\langle u, \partial_x u^2\rangle_1 = 2\langle u_x, u_x^2\rangle + 2\langle u_x, uu_{xx}\rangle.$$

Since $uu_xu_{xx} = \frac{1}{2}u\partial_x u_x^2$, then integrating by parts we get $\langle u_x, uu_{xx}\rangle = -\frac{1}{2}\int u_x^3$. So $\langle u, \partial_x u^2\rangle_1 = \int u_x^3$, and we get that

$$\text{(1.4.25)} \qquad \frac{d}{dt}\mathbb{E}\Big(\frac{1}{2}\int \zeta^2(t,x)dx\Big) = -\nu\mathbb{E}\int \zeta_x^2(t,x)dx - \frac{1}{2}\mathbb{E}\int \zeta^3(t,x)dx + \frac{1}{2}B_1.$$

The term $-\frac{1}{2}\mathbb{E}\int \zeta^3 dx$ is called the (averaged) 1d vorticity stretching. It admits an important interpretation in terms of 1d turbulence, discussed below in Chapter 6.

1.5. Weak solutions and space–homogeneous solutions.

In this section, as everywhere in our book, "weak solutions" are weak in the sense of stochastic analysis (rather than in the sense of partial differential equations). Now we explain what it means. We recall (1.4.1).

Weak solutions of (B).

Consider the problem (1.1.13), where u_0 is a r.v. in H^m, independent from the random force ξ.

DEFINITION 1.5.1. Let $(\tilde\Omega, \tilde{\mathcal{F}}, \tilde{\mathbf{P}})$ be some probability space, let $0 < T \le \infty$ and let $\tilde u^{\tilde\omega}(t,x), 0 \le t \le T$, be a random field on $\tilde\Omega$ which defines a random process $\tilde u^{\omega}(t) \in H^m$. Then $\tilde u$ is called a weak solution of (1.1.13) in H^m if on $\tilde\Omega$ there exist a set of standard independent Wiener processes $\{\tilde\beta_s^{\tilde\omega}(t), s \in \mathbb{Z}^*\}$, and a r.v. $\tilde u_0^{\tilde\omega} \in H^m$, independent from these Wiener processes and distributed as u_0^{ω}, such that $\tilde{\mathbf{P}}$–a.s. $\tilde u^{\tilde\omega}(t,x)$ solves the problem (1.1.13) in H^m, where the processes β_s^{ω} are replaced by $\tilde\beta_s^{\tilde\omega}$ and u_0 is replaced by $\tilde u_0$.

A random field $\tilde u^{\tilde\omega}(t,x)$ as above is a weak solution of equation (1.1.12) if it solves (1.1.13) for some u_0 (in particular, it means that $\tilde u^{\tilde\omega}(0)$ is independent from the corresponding Wiener processes $\{\tilde\beta_s^{\tilde\omega}(t), s \in \mathbb{Z}^*\}$).

Clearly every strong solution is a weak solution, so the latter exists by Theorem 1.4.2. We recall that for $0 < T \le \infty$, $\pi_T^{m_*}$ is the law of the process $\xi\mid_{[0,T]} \in \dot X_T^{m_*}$.

THEOREM 1.5.2. *If* $0 < T \le \infty$ *and* u_0^ω *is a r.v. in* H^m, *then:*

1) A weak solution of the Cauchy problem (1.1.13) *exists and is unique in the sense that if* $\tilde{u}^1$ *and* $\tilde{u}^2$ *are two weak solutions of* (1.1.13) *(possibly defined on different probability spaces), then* $\mathcal{D}(\tilde{u}^1) = \mathcal{D}(\tilde{u}^2)$.

2) If $\tilde{u}(t,x)$, $0 \le t \le T$, *is a weak solution of* (1.1.13), *then* $\mathcal{D}\tilde{u}(t) = \mathcal{M}_t \circ (\mathcal{D}(u_0) \times \pi_T^{m_*})$ *and* $\mathcal{D}\tilde{u}(\cdot) = \mathcal{M} \circ (\mathcal{D}(u_0) \times \pi_T^{m_*})$, *which is the law of a strong solution for* (1.1.13) *(here and in similar situations below* $\pi_\infty^{m_*} = \pi^{m_*}$*).*

PROOF. Consider a weak solution $\tilde{u}$ (we already know that it exists). By Theorem 1.4.2, $\widetilde{u}^\omega(t) = \mathcal{M}_t(\widetilde{u}_0^\omega, \widetilde{\xi}^\omega)$. So

$$\mathcal{D}(\widetilde{u}(t)) = \mathcal{M}_t \circ \mathcal{D}(\widetilde{u}_0^\omega, \widetilde{\xi}^\omega) = \mathcal{M}_t \circ (\mathcal{D}u_0^\omega \times \pi_T^{m_*}).$$

Similarly $\mathcal{D}\tilde{u}(\cdot) = \mathcal{M} \circ (\mathcal{D}u_0^\omega \times \pi_T^{m_*})$. Therefore the law of a weak solution is defined in a unique way, which proves item 1), and both assertions of item 2) also hold. □

EXAMPLE 1.5.3. Consider a strong solution $u(t; u_0^\omega)$ of (1.1.13). For any $t_1 > 0$ consider $v^\omega(t) = u(t_1 + t; u_0^\omega)$. Then v is a solution to the Cauchy problem

$$v_t + v v_x - \nu v_{xx} = \widetilde{\eta}^\omega, \qquad v(0,x) = v_0,$$

where $\widetilde{\eta}^\omega(t,x) = \eta^\omega(t_1 + t, x)$ and $v_0 = u(t_1; u_0^\omega)$. We have

$$\widetilde{\eta}^\omega(t,x) = \partial_t(\xi^\omega(t_1 + t, x)) = \partial_t(\xi^\omega(t_1 + t, x) - \xi^\omega(t_1, x)) = \partial_t(\widetilde{\xi}^\omega(t,x)).$$

where

$$\widetilde{\xi}^\omega(t,x) := \sum_{s \in \mathbb{Z}^*} b_s(\beta_s^\omega(t + t_1) - \beta_s^\omega(t_1)) e_s(x).$$

The processes $\{\beta_s^\omega(t+t_1) - \beta_s^\omega(t_1)\}$ are standard independent Wiener processes (see Appendix 11.5). By Corollary 1.4.3 v_0 is a function of u_0 and the process $\xi_{[0,t_1]}$, where the latter is independent from the process $\tilde{\xi}$ (see Appendix 11.5). The r.v. u_0 also is, so v_0 is independent from $\tilde{\xi}$ as well. Accordingly, $v(t,x)$ is a weak solution of (1.1.13) with the random initial data $v_0 = u(t_1)$.

The notion of weak solutions is fundamental for the theory of stochastic differential equations and stochastic PDEs, since often it simplifies significantly derivation of various properties of an equation and their discussion; the reader will see examples below in Section 1.6. Moreover, some important notions and properties of solutions and equations cannot be formulated without evoking weak solutions. For example, such is the notion of space-homogeneous solutions, introduced below in this section. As it follows from the definition, the existence of a strong solution implies that of a weak one. But as the very name of the latter suggests, for equations more complicated than the stochastic Burgers one it may be easier to establish the existence of the latter than that of the former. After that some tools from the arsenal of stochastic equations often allow to obtain the strong solubility of an equation as a consequence of the weak one, see [**80**, Section 5.3.D].

Weak solutions of a stochastic equation may be equivalently understood as its *martingale solutions.* The corresponding techniques provide a powerful tool to study the equations, especially under various limiting transformations. We do not touch this approach in our book and instead refer the reader to [**142**] and [**80**, Section 5.4].

Space-homogeneous random fields and solutions.

Let Z be a Banach space of functions $u(x), x \in S^1$, invariant with respect to space-translations $u(\cdot) \mapsto u(\cdot + y)$, $y \in S^1$. Let $\mathcal{B} = \mathcal{B}_Z$ be its Borel sigma-algebra.

DEFINITION 1.5.4. A measure on $(Z, \mathcal{B})$ is called space-homogeneous if it is invariant with respect to all space-translations. A random field $u^\omega(x)$ which defines a random variable $(\Omega, \mathcal{F}, \mathbf{P}) \to (Z, \mathcal{B})$ is called space-homogeneous if its law is a space-homogeneous measure. A weak solution $u^\omega(t,x)$ of the stochastic Burgers equation in a space H^m is called space-homogeneous if for each $t \geq 0$ the random field $u^\omega(t, \cdot)$ is space-homogeneous in H^m.

PROBLEM 1.5.5. Prove that if $b_s = b_{-s}$ for each s, then for every t the random field $\xi^\omega(t, \cdot)$ in (1.1.6) is space-homogeneous.
Hint: Note that $\xi^\omega(t, x+y)$ may be written as $\sum_{s \in \mathbf{N}} \big(b_s(\beta'_s(t)e_s + \beta'_{-s}(t)e_{-s})\big)$, where

$$\begin{pmatrix} \beta'_s(t) \\ \beta'_{-s}(t) \end{pmatrix} = U_{s,y} \begin{pmatrix} \beta_s(t) \\ \beta_{-s}(t) \end{pmatrix},$$

and $U_{s,y}$ is some orthogonal time-independent 2×2–matrix. Derive from here that $\{\beta'_s, s \in \mathbb{Z}^*\}$ is a collection of standard independent Wiener processes.

THEOREM 1.5.6. *Let $b_s \equiv b_{-s}$, let $u_0^\omega(x)$ be a space-homogeneous random field, $u^\omega \in H^m$ a.s., independent from $\xi^\omega(t,x)$, and $u(t,x)$ be a weak solution of the Cauchy problem* (1.1.13). *Then $u(t,x)$ is space-homogeneous.*

PROOF. The function $u(t,x)$ satisfies (1.1.13), where the r.v.'s ξ and u_0 are replaced by other ones with the same distributions. To simplify notation we still denote them ξ and u_0. Then $u(t, x+y)$ solves (1.1.13), where ξ and u_0 are replaced by $\xi(t, x+y)$ and $u_0(x+y)$. By the assertion of Problem 1.5.5 the random field ξ is space-homogeneous. So $\xi(t, x+y)$ and $u_0(x+y)$ are distributed as ξ and u_0. Hence $u(t, x+y)$ is another weak solution of the Cauchy problem in H^m, and by Proposition 1.5.2 its distribution equals to that of $u(t,x)$. So u is space-homogeneous. □

Note that if $u^\omega(x)$ is a space-homogeneous random field such that $u^\omega \in H^1$ a.s., then $0 = \mathbb{E} \int u(x)\, dx = \int \mathbb{E} u(x)\, dx$. So $\mathbb{E}u(x) = 0$ for all x. Therefore the solution u, constructed in the theorem above, satisfies

$$\mathbb{E}u(t,x) = 0 \qquad \forall t, \ \forall x.$$

1.6. The two Markov semigroups and the Markov property.

We keep the agreement (1.4.1), and recall that a measurable map $F : (X, \mathcal{B}_X) \to (Y, \mathcal{B}_Y)$ induces the mapping of the spaces of measures

$$F\circ : \mathcal{P}(X, \mathcal{B}_X) \to \mathcal{P}(Y, \mathcal{B}_Y), \qquad \mu \longmapsto F \circ \mu, \quad (F \circ \mu)(Q) = \mu(F^{-1}Q) \ \forall Q \in \mathcal{B}_Y,$$

which is linear in the sense (11.3.4). If X and Y are complete metric spaces (e.g. closed subsets of some Banach spaces), equipped with the Borel sigma-algebras $\mathcal{B}_X$ and $\mathcal{B}_Y$, we call the Borel-measurable mapping from X to Y measurable, and abbreviate $\mathcal{P}(X, \mathcal{B}_X)$ and $\mathcal{P}(Y, \mathcal{B}_Y)$ as $\mathcal{P}(X)$ and $\mathcal{P}(Y)$.

Consider the Polish space $\dot{X}_\infty^{m_*}$ (see (11.2.1)) with its Borel sigma-algebra, and the space $C_b(\dot{X}_\infty^{m_*})$ of continuous bounded functions on $\dot{X}_\infty^{m_*}$ endowed with the uniform norm, $|f|_{C_b(\dot{X}_\infty^{m_*})} = \sup |f|$. By Theorem 1.2.6 the Wiener process

ξ^ω defines a random variable $\xi : \Omega \to \dot{X}_\infty^{m_*}$, whose law, denoted $\pi_\infty^{m_*} =: \pi^{m_*}$, is a Gaussian measure in $\dot{X}_\infty^{m_*}$, depending only on the coefficients $\{b_s\}$, i.e. $\pi^{m_*} = \pi_{\{b_s\}}^{m_*}$. For every $f \in C_b(\dot{X}_\infty^{m_*})$ we have

$$\langle f, \pi^{m_*} \rangle = \int_\Omega f(\xi^\omega) \mathbb{P}(d\omega) = \mathbb{E} f(\xi).$$

The semigroup for measures.

Consider the Cauchy problem (1.1.13), a probability measure $\mu_0 \in \mathcal{P}(H^m)$ and a random variable $u_0^\omega \in H^m$, independent from ξ^ω, such that $\mathcal{D}u_0^\omega = \mu_0$ (it exists by Lemma 11.3.4). We have:

$$\mathcal{P}(H^m \times \dot{X}_\infty^{m_*}) \ni \mathcal{D}(u_0^\omega, \xi^\omega) = \mathcal{D}u_0^\omega \times \mathcal{D}\xi^\omega = \mu_0 \times \pi^{m_*}.$$

We will study how the law of a solution $u^\omega(t,x)$ to the Cauchy problem depends on μ_0. Recalling the notation (1.4.2), (1.4.4) and agreement (1.4.1) we define the operators S_t^*:

$$S_t^* : \mathcal{P}(H^m) \to \mathcal{P}(H^m), \quad \mu_0 \mapsto \mathcal{M}_t^\infty \circ (\mu_0 \times \pi^{m_*}) := \mathcal{D}(u^\omega(t; u_0^\omega)), \ \ t \geq 0. \tag{1.6.1}$$

For any $T \geq t$ we also have

$$S_t^*(\mu_0) = \mathcal{M}_t^T(\mu_0 \times \pi_T^{m_*}),$$

see (1.3.35), so the mappings S_t^* may be defined in terms of $\mathcal{M}_t^T$ with finite T. The maps $\mathcal{M}_t^T$ are continuous by Theorem 1.3.11 (and by (1.3.36), (11.2.3) if $T = \infty$), so the operators S_t^* are well-defined. They are linear on the cone of probability measures $\mathcal{P}(H^m)$ in the sense of the relation (11.3.4).

THEOREM 1.6.1. *We have $S_0^* = Id$ and for all $t_1, t_2 \geq 0$:*

$$S_{t_1}^* \circ S_{t_2}^* = S_{t_1+t_2}^*. \tag{1.6.2}$$

So $\{S_t^*\}_{t \geq 0}$ is a semigroup of linear mappings in the cone $\mathcal{P}(H^m)$ (in the sense of (11.3.4)). It is called the *Markov semigroup for measures for equation* (B).

PROOF. The first statement is obvious. To prove the second one, we take any random initial data u_0, denote $\mathcal{D}(u_0) = \mu_0$ and for $t_1, t_2 \geq 0$ put $\nu_1 = \mathcal{D}(u^\omega(t_1; u_0^\omega))$, $\nu_2 = \mathcal{D}(u^\omega(t_1+t_2; u_0^\omega))$. Consider $v(t) = u(t_1+t; u_0^\omega)$. By Example 1.5.3, $v(t,x)$ is a weak solution of (1.1.13) with the random initial data $v_0 = u(t_1)$. Then $\mathcal{D}(v_0) = \nu_1$ and by Theorem 1.5.2

$$\nu_2 = \mathcal{D}(v(t_2)) = S_{t_2}^*(\nu_1) = S_{t_2}^*\Big(S_{t_1}^*(\mu_0)\Big).$$

Since $\nu_2 = S_{t_1+t_2}^*(\mu_0)$, we obtain (1.6.2). □

Consider the case when $u_0 = v \in H^m$ is a non-random vector. Then u_0 is independent from the force ξ, and $\mathcal{D}(u_0)$ is the Dirac measure $\delta_v \in \mathcal{P}(H^m)$. So $S_t^*(\delta_v) = \mathcal{D}(u(t;v))$. The mapping

$$\Sigma : \mathbb{R}^+ \times H^m \to \mathcal{P}(H^m), \quad (t,v) \mapsto \Sigma_t(v) = S_t^*(\delta_v),$$

is called the *transition probability* for eq. (B). It is weakly measurable in the sense that for any Borel set $Q \subset H^m$ the function $(t,v) \mapsto \Sigma_t(v)(Q)$ is measurable with respect to the Borel sigma-algebra $\mathcal{B}_{\mathbb{R}^+ \times H^m} = \mathcal{B}_{\mathbb{R}^+} \times \mathcal{B}_{H^m}$. Indeed, since

$$\Sigma_t(v)(Q) = \pi^{m_*}\{\xi : \mathcal{M}_t(v,\xi) \in Q\},$$

and the mapping $(t, v, \xi) \mapsto \mathcal{M}_t(v, \xi)$ is continuous, then the set

$$\{(t, v, \xi) : \mathcal{M}_t(v, \xi) \in Q\} \subset \mathbb{R}^+ \times H^m \times \dot{X}_\infty^{m_*}$$

is measurable, so the measurability of the function $(t, v) \mapsto \Sigma_t(v)(Q)$ follows from Lemma 11.3.6. Accordingly, for $t \geq 0$ the functions $H^m \ni v \mapsto \Sigma_t(v)(Q)$ also are measurable, see in Section 11.3 the paragraph above Fubini's Theorem 11.3.5. The transition probability allows to represent the operators S_t^* in the form

$$S_t^*(\mu) = \int \Sigma_t(v)(\cdot)\, \mu(dv), \tag{1.6.3}$$

where $\left(\int \Sigma_t(v)(\cdot)\, \mu(dv)\right)(Q) = \int \Sigma_t(v)(Q)\, \mu(dv)$ for $Q \in \mathcal{B}(H^m)$ – this relation defines a Borel measure on H^m by Lemma 11.3.7. We postpone a proof of (1.6.3) till after Theorem 1.6.2.

The semigroup for functions.

For $t \geq 0$, consider the mapping S_t which sends a function $f \in C_b(H^m)$ to a bounded function $S_t f$ on H^m, defined by the relation

$$\big(S_t f\big)(v) = \mathbb{E}[f(u(t; v))], \quad f \in C_b(H^m),\ v \in H^m. \tag{1.6.4}$$

Then

$$S_t f\,(v) = \langle f, \mathcal{D}(u(t; v))\rangle = \langle f, \mathcal{M}_t \circ (\delta_v \times \pi^{m_*})\rangle = \langle f, \Sigma_t(v)\rangle. \tag{1.6.5}$$

THEOREM 1.6.2. *For all $t \geq 0$ we have:*

1. *$S_t(C_b(H^m)) \subset C_b(H^m)$, and for any $f \in C_b(H^m)$ the function $(t, v) \mapsto S_t f(v)$ is continuous on $\mathbb{R}_+ \times H^m$.*
2. *The mapping S_t is linear and positive (i.e. it preserves the cone of non-negative functions).*
3. *The mapping $S_t : C_b(H^m) \to C_b(H^m)$ is a linear operator of unit norm.*
4. *For $f \in C_b(H^m)$ and $\mu \in \mathcal{P}(H^m)$ the following duality relation holds:*

$$\langle S_t f, \mu\rangle = \langle f, S_t^*\, \mu\rangle. \tag{1.6.6}$$

PROOF. 1) If $f \in C_b(H^m)$, then by (1.6.4) :

$$S_t f(v) = \mathbb{E}[f(u(t; v))] = \mathbb{E}[f \circ \mathcal{M}_t(v, \xi^\omega)].$$

If $v_n \to v \in H^m$, then for every ω, $f \circ \mathcal{M}_t(v_n, \xi^\omega) \to f \circ \mathcal{M}_t(v, \xi^\omega)$ in H^m (since $f \circ \mathcal{M}_t(\cdot, \xi^\omega)$ is continuous on H^m, $m \geq 1$). Therefore by Lebesgue's dominated convergence theorem we get $S_t f(v_n) \to S_t f(v)$. So $S_t f \in C_b(H^m)$. The proof of the second assertion is identical.

2) Obvious.

3) Clearly we have $S_t \mathbf{1} = \mathbf{1}$, where $\mathbf{1}(u) \equiv 1$. Take $f \in C_b(H^m)$ and denote $d = \|f\|_{C_b}$. Then

$$-d\mathbf{1} \leq f \leq d\mathbf{1}.$$

Since S_t is positive, we get that $-d\mathbf{1} \leq S_t f \leq d\mathbf{1}$, and therefore $\|S_t f\|_{C_b} \leq \|f\|_{C_b}$. Thus $|S_t|_{C_b, C_b} \leq 1$. Finally, since $S_t \mathbf{1} = \mathbf{1}$, then $|S_t|_{C_b, C_b} = 1$.

4) We have

$$\langle f, S_t^*(\mu)\rangle = \langle f, \mathcal{M}_t \circ (\mu \times \pi^{m_*})\rangle = \langle f \circ \mathcal{M}_t, \mu \times \pi^{m_*}\rangle.$$

By Fubini's theorem,

$$\langle f \circ \mathcal{M}_t, \mu \times \pi^{m_*} \rangle = \int_{H^m} \underbrace{\int_{\dot{X}_\infty^m} (f \circ \mathcal{M}_t)(v, \xi) \pi^{m_*}(d\xi)}_{S_t f(v)} \mu(dv) = \langle S_t f, \mu \rangle.$$

Thence the result. □

Now we can easily prove (1.6.3) Indeed, since we know that the integral in the r.h.s. defines a Borel measure on H^m, then it suffices to check that for any $f \in C_b(H^m)$ we have $\langle S_t^* \mu, f \rangle = \int \langle \Sigma_t(v), f \rangle \, \mu(dv)$. But by (1.6.6)

$$\langle S_t^* \mu, f \rangle = \langle S_t f, \mu \rangle = \int \langle f, \Sigma_t(v) \rangle \, \mu(dv),$$

which proves the result.

Let us show that the family of operators $\{S_t, t \geq 0\}$ is a semigroup.

THEOREM 1.6.3. *We have $S_0 = Id$ and for all $t_1, t_2 \geq 0$,*

$$S_{t_1} \circ S_{t_2} = S_{t_1 + t_2}. \tag{1.6.7}$$

PROOF. The first assertion is obvious. To prove the second, consider $t_1, t_2 > 0$, $f \in C_b(H^m)$ and $v \in H^m$. By Theorem 1.6.2, $(S_{t_1+t_2} f)v = \langle f, S_{t_1+t_2}^*(\delta_v) \rangle$ and by Theorem 1.6.1, $\langle f, S_{t_1+t_2}^*(\delta_v) \rangle = \langle f, S_{t_1}^* \circ S_{t_2}^*(\delta_v) \rangle$. Using once again Theorem 1.6.2, we get that

$$\langle f, S_{t_1}^* \circ S_{t_2}^*(\delta_v) \rangle = \langle S_{t_1} f, S_{t_2}^*(\delta_v) \rangle = \langle S_{t_2} \circ S_{t_1} f, \delta_v \rangle.$$

Therefore $(S_{t_1+t_2} f)(v) = (S_{t_2} \circ S_{t_1} f)(v)$ for any v, and the second assertion follows. □

The semigroup $\{S_t\}_{t \geq 0}$ is called the *Markov semigroup for functions for equation* (B). Note that in view of (1.6.5) and (1.6.3) both Markov semigroups $\{S_t\}$ and $\{S_t^*\}$ may be defined in terms of the transition probability.

Kolmogorov-Chapman relation and Markov property.

THEOREM 1.6.4. *(the Kolmogorov-Chapman equality.) For all $t_1, t_2 \geq 0$, $v \in H^m$ and any Borel set $\Gamma \subset H^m$,*

$$\Sigma_{t_1+t_2}(v)(\Gamma) = \int \Sigma_{t_2}(u)(\Gamma) \left(\Sigma_{t_1}(v)(du) \right). \tag{1.6.8}$$

If $f \in C_b(H^m)$, then

$$S_{t_1+t_2} f(v) = \int_{H^m} (S_{t_2} f)(u) \left(\Sigma_{t_1}(v)(du) \right). \tag{1.6.9}$$

PROOF. The l.h.s. of (1.6.8) equals

$$S_{t_1+t_2}^*(\delta_v)(\Gamma) = \left(S_{t_2}^* \Sigma_{t_1}(v) \right)(\Gamma) = \int \Sigma_{t_2}(u)(\Gamma) \, \Sigma_{t_1}(v)(du),$$

where the second equality is (1.6.3). So (1.6.8) holds. Both sides of (1.6.8) define Borel measures on H^m (see Lemma 11.3.7). Integrating against them the function f we get (1.6.9). □

Theorem 1.6.4 can be generalised. The following result will be needed later:

PROPOSITION 1.6.5. *For any $t_1, t_2 \geq 0$ and $v \in H^m$ denote by $u^{[0,t_1]}(\cdot, v)$ the restriction of $u(t, v)$ to the interval $[0, t_1]$. Then for each $f \in C_b(H^m)$ and every bounded measurable map Φ on $X^m_{t_1}$, we have*

$$\mathbb{E}[\Phi(u^{[0,t_1]}(\cdot; v)) f(u(t_1 + t_2; v))] = \mathbb{E}\left[\Phi(u^{[0,t_1]}(\cdot; v))\big(S_{t_2} f(u(t_1; v))\big)\right]. \tag{1.6.10}$$

Note that if $\Phi \equiv 1$, then (1.6.10) coincides with (1.6.9).

PROOF. We proceed similarly to the proof of Theorem 1.6.1. By Theorem 1.5.2 the r.h.s. of (1.6.10) will not change if we replace there $u(t; v)$ by another weak solution of problem (1.1.13) with $u_0 = v$. To construct a suitable weak solution, consider two collections of standard independent Wiener processes $\{\beta^{\omega_i}_s,\ s \in \mathbb{Z}^*\}$, $i = 1, 2$, defined on probability spaces $(\Omega_i, \mathcal{F}_i, \mathbb{P}_i)$, $i = 1, 2$, and set $(\Omega, \mathcal{F}, \mathbb{P}) = (\Omega_1 \times \Omega_2, \mathcal{F}_1 \times \mathcal{F}_2, \mathbb{P}_1 \times \mathbb{P}_2)$. For $t_1, t_2 \geq 0$ as in the assertion and $\omega = (\omega_1, \omega_2) \in \Omega$, we put:

$$\beta^\omega_s(t) = \begin{cases} \beta^{\omega_1}_s(t), & 0 \leq t \leq t_1 \\ \beta^{\omega_1}_s(t_1) + \beta^{\omega_2}_s(t - t_1), & t_1 \leq t \leq t_2. \end{cases} \tag{1.6.11}$$

Then (β_s) are standard independent Wiener processes on $\Omega_1 \times \Omega_2$. Let us use them to build a force ξ in (1.1.13). Then for $0 \leq t \leq t_1$ the solution $u^\omega(t; v) = u^{\omega_1}(t; v)$ does not depend on ω_2, and for $t_1 \leq \tau \leq t_2$ the equation for $u^\omega(\tau; v)$ depends only on ω_2. Thus, $u^\omega(t_1 + t; v) = u^{\omega_2}(t; u^{\omega_1}(t_1; v))$ and the integral on the l.h.s. of (1.6.10) can be rewritten as

$$\begin{aligned} &\int_\Omega \Phi(u^{[0,t_1]}(\cdot, v)) f(u^\omega(t_1 + t_2; v)) \mathbb{P}(d\omega) \\ &= \int_{\Omega_1}\int_{\Omega_2} \Phi(u^{[0,t_1],\omega_1}(\cdot; v)) f(u^{\omega_1,\omega_2}(t_1 + t_2; v)) \mathbb{P}_1(d\omega_1)\mathbb{P}_2(d\omega_2) \\ &= \int_{\Omega_1}\left(\int_{\Omega_2} f(u^{\omega_2}(t_2; u^{\omega_1}(t_1; v)))\mathbb{P}_2(d\omega_2)\right) \Phi(u^{[0,t_1],\omega_1}(\cdot; v))\mathbb{P}_1(d\omega_1) \\ &= \int_\Omega (S_{t_2} f)(u(t_1; v))\, \Phi(u^{[0,t_1]}(\cdot; v))\mathbb{P}(d\omega), \end{aligned} \tag{1.6.12}$$

where we used Fubini's theorem to pass from the second line to the third. This proves (1.6.10). □

Now consider relation (1.4.8). It implies for the family of solutions $u(t; u^\omega_0)$ of equation (1.1.12) the *Markov property*, which heuristically may be stated as follows: *for $\tau \geq s \geq 0$, the solution $u(\tau; u^\omega_0)$ depends on the past $0 \leq t \leq s$ only through $u(s; u^\omega_0)$*. More precisely, in Appendix 11.6, following closely the proof of Proposition 1.6.5, we apply a functional f to both parts of (1.4.8), multiply them by a function $\Phi(\xi^\omega_{[0,s]})$ of the restriction of ξ^ω to the interval $[0, s]$, take the expectation and arrive at the equality

$$\mathbb{E}(f(u(\tau; u^\omega_0))\, \Phi(\xi^\omega_{[0,s]})) = \mathbb{E}\big((S_{\tau-s} f)(u(s; u^\omega_0))\, \Phi(\xi^\omega_{[0,s]})\big).$$

As we explain in the Appendix, this relation shows that the family of solutions $u^\omega(t; v)$, $v \in H^m$, defines in H^m a family of Markov processes (or simply a Markov process) . The semigroups for measures and for functions constructed above are the two semigroups canonically associated with this Markov process, and the transition probability Σ is that of the process.

We placed the rigorous verification of the fact that solutions $u(t;u_0)$ make a family of Markov processes to an appendix since the corresponding exact definition involves additional notions and structures not used in the main text, while we have already constructed the objects from the theory of Markov processes needed for the purposes of this book (namely, the two Markov semigroups and the Kolmogorov–Chapman relation).

The properties given above allow us to apply to eq. (B) the rich theory of Markov processes, and this is the reason why it is easier to study stochastic PDEs with white noise forces rather than with the majority of other types of random forces.

1.7. Weak convergence of measures.

Here we recall some definitions and results concerning the weak convergence of measures [**16, 22, 51, 139, 147**]. Let $\mathcal{O}$ be a Polish (i.e. a complete separable metric) space. For example, $\mathcal{O}$ can be a closed subset of a separable Banach space X. As usual, we endow $\mathcal{O}$ with the Borel sigma-algebra, and denote by $\mathcal{P}(\mathcal{O})$ the space of probability Borel measures on $\mathcal{O}$.

DEFINITION 1.7.1. A sequence $\{\mu_n\} \subset \mathcal{P}(\mathcal{O})$ is said to be weakly converging to $\mu \in \mathcal{P}(\mathcal{O})$ (denoted by $\mu_n \rightharpoonup \mu$) if and only if

$$\forall f \in C_b(\mathcal{O}),\ \langle \mu_n, f\rangle \to \langle \mu, f\rangle.$$

Due to Theorem 11.3.4, measures in $\mathcal{P}(\mathcal{O})$ may be realized as distributions of random variables in $\mathcal{O}$. The following fundamental result, known as *Skorokhod's representation theorem*, states that a weakly converging sequence of measures may be realized as the distributions of a sequence of a.s. converging random variables:

THEOREM 1.7.2. *Let a sequence of Borel measures $\{\mu_n, n \geq 1\}$ on a Polish space $\mathcal{O}$ weakly converge to a measure $\mu_0 \in \mathcal{P}(\mathcal{O})$. Then there exists a collection of random variables $\{\xi_n, n \geq 0\}$, defined on the same probability space and valued in $\mathcal{O}$, such that $\mathcal{D}\xi_n = \mu_n$ for each $n \geq 0$, and $\xi_n \xrightarrow[n\to\infty]{} \xi_0$, a.s.*

For a proof see [**16**, Section 6] and [**22**, Section 2.6].

PROBLEM 1.7.3. Prove the converse result: if the random variables $\{\xi_n, n \geq 0\}$ are as in the theorem's assertion, then the laws $\{\mathcal{D}\xi_n, n \geq 1\}$ weakly converge to $\mathcal{D}\xi_0$.
Hint: use Lebesgue's dominated convergence theorem.

The convergence $\mu_n \rightharpoonup \mu$ does not imply that $\mu_n(Q)$ converges to $\mu(Q)$ for any Borel set Q. However, $\mu_n \rightharpoonup \mu$ in $\mathcal{P}(\mathcal{O})$ if and only if

$$\mu(Q) \geq \lim_{n\to+\infty} \sup \mu_n(Q) \text{ for each closed set } Q \subset \mathcal{O}. \tag{1.7.1}$$

This assertion is know as the *portmanteau theorem*, see [**16, 51, 130, 139**].

We denote

$$L(\mathcal{O}) := \{f \in C_b(\mathcal{O}) : \mathrm{Lip}(f) < \infty\}, \tag{1.7.2}$$

where $\mathrm{Lip}(f)$ denotes the Lipschitz constant of f, and endow the space $L(\mathcal{O})$ with the norm

$$\|f\|_{L(\mathcal{O})} := \|f\|_{C_b(\mathcal{O})} + \mathrm{Lip}(f)\,. \tag{1.7.3}$$

PROBLEM 1.7.4. Prove that the space $L(\mathcal{O})$ with the norm (1.7.3) is Banach, and show that the Banach space $L([0,1])$, where the interval $[0,1]$ is given the usual distance, is not separable.

DEFINITION 1.7.5. For $\mu, \nu \in \mathcal{P}(\mathcal{O})$ the Lipschitz-dual distance between μ and ν is defined by:[6]

$$\|\mu - \nu\|^*_{L(\mathcal{O})} := \sup_{\bar{B}}(\langle f, \mu\rangle - \langle f, \nu\rangle) = \sup_{\bar{B}} |\langle f, \mu\rangle - \langle f, \nu\rangle| \leq 2, \tag{1.7.4}$$

where $\bar{B} = \bar{B}^1_{L(\mathcal{O})}$ is the closed ball $\{\|f\|_{L(\mathcal{O})} \leq 1\}$. If the metric space is clear from the context, we abbreviate $\|f\|_{L(\mathcal{O})}$ to $\|f\|_L$ and $\|\cdot\|^*_{L(\mathcal{O})}$ to $\|\cdot\|^*_L$.

PROBLEM 1.7.6. Let $F : X \to Y$ be a C–Lipschitz mapping of Polish spaces. Show that the corresponding mapping $F\circ : \mathcal{P}(X) \to \mathcal{P}(Y)$ is C'-Lipschitz, where $C' = \max(C, 1)$:

$$\|F \circ \mu - F \circ \nu\|^*_{L(Y)} \leq C' \|\mu - \nu\|^*_{L(X)}. \tag{1.7.5}$$

In particular, if $i : X \to Y$ is an embedding of Banach spaces and $\|i(x)\|_Y \leq \|x\|_X$, then

$$\|i \circ \mu - i \circ \nu\|^*_{L(Y)} \leq \|\mu - \nu\|^*_{L(X)}.$$

The following celebrated theorem, essentially due to Kantorovich, characterises the weak convergence of measures in terms of the Lipschitz-dual distance:

THEOREM 1.7.7. *Let $\mathcal{O}$ be a Polish space. Then:*

(1) $(\mathcal{P}(\mathcal{O}), \|\cdot\|^*_{L(\mathcal{O})})$ *is a complete metric space.*
(2) *If $\{\mu_n\} \subset \mathcal{P}(\mathcal{O})$ and $\mu \in \mathcal{P}(\mathcal{O})$, then $\mu_n \rightharpoonup \mu$ if and only if $\|\mu_n - \mu\|^*_{L(\mathcal{O})} \to 0$.*

For a proof see [**51**, Section 11.3] and [**103**, Section 1.2.3].

The Lipschitz-dual distance metrizes the weak convergence of measures. It is not the only such a distance, but it is especially convenient, also for the purposes of this book, since it is well-suited to study the distances between the laws of random variables. Namely, if μ_1 and μ_2 are measures on a Polish space $\mathcal{O}$ which are the laws of random variables ξ_1 and ξ_2 then, firstly, if $\operatorname{dist}(\xi_1, \xi_2) \leq \varepsilon$ a.s., then for any $f \in \bar{B}^1_{L(\mathcal{O})}$ we have

$$\big|\langle f, \mu_1\rangle - \langle f, \mu_2\rangle\big| \leq \mathbb{E}\big|(f(\xi_1) - f(\xi_2))\big| \leq \varepsilon$$

as $\operatorname{Lip} f \leq 1$; so $\|\mu_1 - \mu_2\|^*_L \leq \varepsilon$. Secondly, if $\mathbf{P}(\xi_1 \neq \xi_2) \leq \varepsilon$, then for any $f \in \bar{B}^1_{L(\mathcal{O})}$

$$\big|\langle f, \mu_1\rangle - \langle f, \mu_2\rangle\big| \leq \mathbb{E}\big|\chi_{\xi_1 \neq \xi_2}(f(\xi_1) - f(\xi_2))\big| \leq 2\varepsilon$$

as $|f| \leq 1$, so $\|\mu_1 - \mu_2\|^*_L \leq 2\varepsilon$.

PROBLEM 1.7.8. Prove that the metric space $(\mathcal{P}(\mathcal{O}), \|\cdot\|^*_{L(\mathcal{O})})$, where $\mathcal{O}$ is a Polish space, is separable, so by item 1 of Theorem 1.7.7 it is Polish.
Hint: Show that finite linear combinations with rational coefficients of δ-functions in the points of countable $(1/n)$-nets in $\mathcal{O}$ with $n = 1, 2, \ldots$ are dense in $(\mathcal{P}(\mathcal{O}), \|\cdot\|^*_{L(\mathcal{O})})$.

DEFINITION 1.7.9. Let M be a collection of measures in $\mathcal{P}(\mathcal{O})$.

---

[6] This is the distance in the dual Banach space $L(\mathcal{O})^*$, restricted to the space of measures $\mathcal{P}(\mathcal{O})$, naturally embedded in $L(\mathcal{O})^*$.

(1) The collection M is said to be tight if for every $\varepsilon > 0$ there exists a compact set $K_\varepsilon \subset \mathcal{O}$ such that

$$\mu(K_\varepsilon) \geq 1 - \varepsilon, \quad \forall \mu \in M.$$

(2) The collection M is said to be weakly relatively compact if any sequence $\{\mu_n\} \subset M$ has a subsequence, weakly converging in $\mathcal{P}(\mathcal{O})$.

The following fundamental theorem due to Prokhorov characterises weakly relatively compact sets [**16**], [**51**], [**139**].

THEOREM 1.7.10. *A collection of measures $M \subset \mathcal{P}(\mathcal{O})$, where $\mathcal{O}$ is a Polish space, is tight if and only if it is weakly relatively compact.*

Taking for M a one-point set $M = \{\mu\}$ we get existence for any $\varepsilon > 0$ of a compact set $K \Subset \mathcal{O}$ such that $\mu(K) \geq 1 - \varepsilon$. This is a special case of *Ulam's theorem* [**22, 51**], saying that for any Borel subset $Q \subset \mathcal{O}$ and any $\varepsilon > 0$ there exists a compact set $K \subset Q$, satisfying $\mu(Q \setminus K) \leq \varepsilon$.

Taking for $\mathcal{O}$ a metric compact K we can choose $K_\varepsilon = K$ for each ε. In this way we derive from Theorem 1.7.10 that the set $\mathcal{P}(K)$ is weakly compact.

Now we return to equation (B) (recalling (1.4.1)):

THEOREM 1.7.11. *The mapping $S_t^* : \mathcal{P}(H^m) \to \mathcal{P}(H^m)$ is weakly continuous for every $t \geq 0$, i.e. if $\mu_n \rightharpoonup \mu$, then $S_t^* \mu_n \rightharpoonup S_t^* \mu$.*

PROOF. For $\mu_n \rightharpoonup \mu$ and any $f \in C_b(H^m)$, by (1.6.6) we have: $\langle S_t^* \mu_n, f\rangle = \langle \mu_n, S_t f\rangle$ and $\langle S_t^* \mu, f\rangle = \langle \mu, S_t f\rangle$. Since $S_t f \in C_b(H^m)$ by Theorem 1.6.2, we get $\langle \mu_n, S_t f\rangle \to \langle \mu, S_t f\rangle$. Thus, $S_t^* \mu_n \rightharpoonup S_t^* \mu$. □

1.8. Appendix: More on the deterministic Burgers equation.

Analyticity of the flow of the deterministic equation (B).

Here we study the regularity of the mappings $\mathcal{M}$ and $\mathcal{M}_t$, using basic facts about analytic maps between Banach spaces. All of them may be found e.g. in [**131**, Appendix B] and in [**36**].

Consider the heat equation

$$w_t - \nu w_{xx} = z, \quad w_{|t=0} = 0, \ z \in X_T^{m-1}, \ m \geq 1.$$

By the result of Proposition 1.3.14 it has a unique solution $w \in X_T^m$ and

$$\|w\|_{X_T^m} \leq C \|z\|_{X_T^{m-1}}, \quad C = C(m, \nu, T). \tag{1.8.1}$$

Denote by $\mathcal{L}$ the operator

$$\mathcal{L} : Y_T^m := \dot{X}_T^m \cap C^1(0, T; H^{m-2}) \to X_T^{m-2}, \quad w \longmapsto w_t - \nu w_{xx}.$$

By (1.8.1), the mapping $\mathcal{L}^{-1}\,|_{X_T^{m-1}}$ is a well-defined continuous immersion

$$\mathcal{L}^{-1} : X_T^{m-1} \to Y_T^m.$$

We define the Banach space $Z_T^m = \mathcal{L}^{-1}(X_T^{m-1})$ and endow it with the norm $\|w\|_{Z_T^m} = \|\mathcal{L} w\|_{X_T^{m-1}}$. This space is continuously embedded in Y_T^m and

$$\mathcal{L} \text{ is an isometry between } Z_T^m \text{ and } X_T^{m-1}.$$

Now consider equation (1.3.7) with $v \in X_T^m$. By Theorem 1.3.6 its solution w belongs to X_T^m. Thence $\partial_x((w+v)^2) \in X_T^{m-1}$ and

$$w = \mathcal{L}^{-1}\left(-\frac{1}{2}\partial_x((w+v)^2)\right) \in Z_T^m.$$

If we define

$$\Phi(w,v) = w + \frac{1}{2}\mathcal{L}^{-1} \circ \partial_x((w+v)^2),$$

then (1.3.7) can be rewritten as:

$$\Phi(w,v) = 0, \tag{1.8.2}$$

Note that by Theorem 1.3.6 for each $v \in X_T^m$ the equation $\Phi(w,v) = 0$ has a unique solution $w \in Z_T^m$.

LEMMA 1.8.1. *The mapping $\Phi : Z_T^m \times X_T^m \to Z_T^m$ is analytic for all $m \geq 1$.*

PROOF. The mapping $X_T^m \to X_T^m, w \longmapsto w^2$ is analytic since it is quadratic and locally bounded, see [**36**]. Since $Z_T^m \subset \dot{X}_T^m$, then the mapping

$$Z_T^m \times X_T^m \to Z_T^m, \quad (w,v) \longmapsto (v+w)^2 \longmapsto \partial_x((w+v)^2) \longmapsto \mathcal{L}^{-1} \circ \partial_x((w+v)^2)$$

also is analytic. Thence Φ is analytic. □

At $(w,v) \in Z_T^m \times X_T^m$, the differential of Φ in w is the linear mapping $d_w\Phi(w,v) : Z_T^m \to Z_T^m$, given by:

$$d_w\Phi(w,v) : h \mapsto h + \mathcal{L}^{-1} \circ \partial_x\big((w+v)h\big)$$

(see [**36**] for basics about (Fréchet) differentials of maps between Banach spaces). It is continuous since Φ is analytic.

LEMMA 1.8.2. *For $(w,v) \in Z_T^m \times X_T^m$, $m \geq 1$, the linear mapping $d_w\Phi(w,v)$ is an isomorphism of Z_T^m.*

PROOF. We know that $d_w\Phi$ is continuous. Next, for any $w, h, g \in Z_T^m$ and $v \in X_T^m$, we have $d_w\Phi(w,v)h = g$ if and only if

$$h_t - \nu h_{xx} + \partial_x\Big((w+v)h\Big) = \mathcal{L}g.$$

Since $w, v \in X_T^m$ and $\mathcal{L}g \in X_T^{m-1}$, then by Proposition 1.3.14 the equation above has a unique solution $h \in \dot{X}_T^m$ and $\|h\|_{X_T^m} \leq C\|\mathcal{L}g\|_{X_T^{m-1}} = C\|g\|_{Z_T^m}$. So $d_w\Phi$ is an isomorphism. □

Finally, in view of Lemmas 1.8.1 and 1.8.2 the analytic version of the implicit function theorem [**131**, Appendix B] applies in the vicinity of any point (w,v) in $Z_T^m \times X_T^m$, satisfying (1.8.2). This application implies that the unique solution $w \in Z_T^m$ of equation (1.8.2) (where the latter is regarded as an equation on w with a parameter v) is analytic in $v \in X_T^m$:

$$\Phi(w,v) = 0 \ \text{ iff } \ w = W(v), \quad \text{where } W : X_T^m \to Z_T^m \text{ is analytic.} \tag{1.8.3}$$

Since the solution u of the Burgers equation decomposes as $u = v + w$, where by Lemma 1.3.1 v is a continuous linear function of (u_0, ξ) and w is an analytic function of v, we have:

THEOREM 1.8.3. *For any $m \in \mathbf{N}$ the mapping $\mathcal{M} : H^m \times \dot{X}_T^{m*} \to X_T^m$, defined in (1.3.34), is analytic (and therefore continuous).*

We endow the space $H^m \times \dot{X}_T^{m_*} =: \mathbf{X}$ with the Banach norm $\|(f,h)\|_{\mathbf{X}} = \|f\|_{H^m} + \|h\|_{\dot{X}_T^{m_*}}$. Analytic maps between Banach spaces are not necessarily locally bounded and locally Lipschitz, but the mapping $\mathcal{M}$ is:

PROPOSITION 1.8.4. *The mapping $\mathcal{M}$ in Theorem* 1.8.3 *is locally Lipschitz.*

PROOF. Denote by X_{Tc}^m and Z_{Tc}^m the naturally defined complexifications of the corresponding functional spaces. By the complex–analytic version of the implicit function theorem [**131**, Appendix B], we prove that for any $v' \in X_T^m$ the analytic mapping W in (1.8.3) extends to a complex-analytic map $W : B^{\varepsilon}_{X_{Tc}^m}(v') \to Z_{Tc}^m$, for some $\varepsilon > 0$, and there its norm is bounded by some constant K. Moreover, we can choose $\varepsilon = \varepsilon(\|v'\|_{X_{Tc}^m})$ and $K = K(\|v'\|_{X_{Tc}^m})$, where ε and K are positive continuous functions.[7] Hence, by the Cauchy inequality (see below relation (1.8.4)) the norm $\|dW(v')\|_{X_T^m, Z_T^m}$ is bounded by $(K/\varepsilon)(\|v'\|_{X_{Tc}^m})$. So the mapping W is locally Lipschitz. Since $\mathcal{M}(u_0,\xi) = u = v + W(v)$, where $(u_0,\xi) \mapsto v$ is a continuous linear operator, then $\mathcal{M}$ is locally Lipschitz as well. □

PROBLEM 1.8.5. Let X^c, Y^c be complex Banach spaces and $F : B^R_{X^c}(v) \to Y^c$ be a complex-analytic mapping, whose norm is bounded by 1. Then

$$\|dF(v)\|_{X^c,Y^c} \le R^{-1}. \tag{1.8.4}$$

Hint: Take any unit vector $x \in X^c$, any functional of unit norm $\xi \in X^{c*}$ and apply the usual Cauchy inequality to the holomorphic function

$$D \ni z \mapsto \xi\big(F(zx)\big), \quad D = \{z \in \mathbb{C} : |z| < 1\}.$$

Burgers equation with a regular force.

For the Burgers equation (B), where the force $\eta(t,x)$ is a regular function of (t,x),

$$\eta \in U^r = L_2(0,T;H^r), \quad r \ge 0,$$

an analogue of the results on well-posedness of (B) holds with a similar but easier proof:

THEOREM 1.8.6. *If $u_0 \in H^{r+1}$ and $\eta \in U^r$, where r is a non-negative integer, then the problem* (B), (1.1.11) *has a unique solution $u \in W^{r+1}$ (see* (1.3.28)*). The mapping*

$$\Psi : H^{r+1} \times U^r \to W^{r+1}, \quad (u_0,\eta) \mapsto u, \tag{1.8.5}$$

is locally bounded (i.e. is bounded on bounded sets) and locally Lipschitz.

PROBLEM 1.8.7. Prove the theorem, following the proof of Theorem 1.3.6 (firstly derive a priori estimates, then use Galerkin's method).

PROBLEM 1.8.8. Applying the inverse function theorem [**131**, Appendix B] to the mapping

$$W^{r+1} \to H^{r+1} \times U^r, \quad u(t,x) \mapsto (u(0), u_t + uu_x - \nu u_{xx})$$

prove that the mapping (1.8.5) is analytic. (Now the proof of the analyticity is easier than in Theorem 1.8.3 since by Proposition 1.3.14 the linearisations of the map above define isomorphisms between the space of solutions W^{r+1} and the space

[7] This is a general property of the implicit function theorem which readily follows from any of its proofs.

$H^{r+1}\times U^r$, while for the equation with the irregular forces (1.1.5), where solutions are considered in the space X_T^m, there is no natural analogue of this property.)

Consider the mapping

$$\bar S_1 : (u_0,\eta)\mapsto \Psi(u_0,\eta)\,|_{t=1}= u(1,x),$$

where u is a solution of the problem (B), (1.1.11). Due to Theorem 1.8.6 $\bar S_1$ defines a locally Lipschitz mapping $\bar S_1 : H^{r+1}\times U^r\to H^{r+1}$, $r\ge 1$. This assertion can be strengthened:

PROPOSITION 1.8.9. *Let $r\in\mathbf{N}$. Then the mapping $\bar S_1 : H^r\times U^r\to H^{r+1}$ is locally bounded and locally Lipschitz.*

This result is a manifestation of the smoothing property of the parabolic equation (B). Below in Section 2.1 we will give an example of much stronger smoothing exhibited by the equation.

SKETCH OF THE PROOF. Local boundedness follows from the local Lipschitz continuity since $\bar S_1(0,0)=0$. To prove the latter, for $u_{1,0},u_{2,0}\in B_{H^r}^{R_1}$ and $\eta_1,\eta_2\in B_{U^r}^{R_2}$ consider the increment

$$\begin{aligned}&\bar S_1(u_{1,0},\eta_1)-\bar S_1(u_{2,0},\eta_2)\\&\quad=\big(\bar S_1(u_{1,0},\eta_1)-\bar S_1(u_{2,0},\eta_1)\big)+\big(\bar S_1(u_{2,0},\eta_1)-\bar S_1(u_{2,0},\eta_2)\big)\\&\quad=:\xi_1+\xi_2.\end{aligned}$$

The constants below depend on m,ν,T,R_1,R_2. Denote $\Psi(u_{2,0},\eta_1)=:u_1$ and $\Psi(u_{2,0},\eta_2)=:u_2$. Then $u_1,u_2\in W^r\subset X_T^r$ and $\xi_2=v(1,x)$, where $v(t,x)$ satisfies

$$\partial_t v-\nu v_{xx}+\frac{1}{2}\frac{\partial}{\partial x}(v(u_1+u_2))=h,\tag{1.8.6}$$

with $h=\eta_1-\eta_2$, and $v(0)=0$. Applying Proposition 1.3.14. ii) with $m=r$ we find that

$$\|\xi_2\|_{r+1}=\|v(1)\|_{r+1}\le C\|v\|_{W^{r+1}}\le C_1\|\eta_2-\eta_1\|_{U^r}.$$

Similarly, denoting $\Psi(u_{1,0},\eta_1)=:u_1$ and $\Psi(u_{2,0},\eta_1)=:u_2$ we have that $\xi_1=\tilde v(1,x)$, where $\tilde v$ solves (1.8.6) with $h=0$ and $\tilde v(0,x)=u_{2,0}-u_{1,0}=:u^\Delta$, so $\|\tilde v\|_{W^r}\le C\|u^\Delta\|_r$ by Proposition 1.3.14.i). Next, setting $t\tilde v=w$ we see that w satisfies (1.8.6) with $h=\tilde v$ and $w_0=0$. So in view of the assertion ii) of the same proposition,

$$\|w\|_{W^{r+1}}\le C\|\tilde v\|_{W^r}\le C_1\|u^\Delta\|_r.$$

Accordingly,

$$\|\xi_1\|_{r+1}=\|\tilde v(1)\|_{r+1}=\|w(1)\|_{r+1}\le C\|u^\Delta\|_r.$$

The obtained estimates for ξ_1 and ξ_2 prove the proposition. □

Certainly the proposition above and its proof remain true for any mapping $\bar S_\tau : (u_0,\eta)\mapsto\Psi(u_0,\eta)\,|_{t=\tau}= u(\tau,x)$ with $\tau>0$. That is, the mapping

$$\bar S_\tau : H^r\times L_2(0,\tau;H^r)\to H^{r+1}\quad\text{where } r\in\mathbf{N}\text{ and }\tau>0$$

is locally Lipschitz. Let the force $\eta(t,x)$ be a C^∞–smooth function. Since for any N the operator $\bar S_t$ may be written as $\bar S_t(u_0,\eta)=(\bar S_{t/N})^N(u_0)$, where for any $j\ge1$ the j–th operator $\bar S_{t/N}$ corresponds to the force $\eta\,|_{[(j-1)t/N,jt/N]}$, then by the above and (1.2.15) each solution $u(t,x)$ is smooth in x for $t>0$. Expressing the time-derivatives of u using the equation, we get

COROLLARY 1.8.10. *If the function $\eta(t,x)$ is C^∞–smooth, then for any $u_0 \in H^1$ a solution $u(t,x)$ for* (B), (1.1.11) *is C^∞–smooth for $t > 0$.*

If η is C^∞–smooth only in x, i.e. if all derivatives $\partial_x^k \eta(t,x)$ are continuous functions of (t,x), then the argument above shows that the solution $u(t,x)$ is smooth in x for $t > 0$.

CHAPTER 2

Asymptotically sharp estimates for Sobolev norms of solutions

In this section we derive for moments of Sobolev norms of solutions for (1.1.13) estimates which are asymptotically sharp as ν converges to 0. They are crucially based on Oleinik–type estimates for solutions of the Burgers equation (1.1.13) with a deterministic r.h.s. $\partial_t \xi(t,x)$, and we start with these inequalities.

2.1. Oleinik's estimate.

Here we are concerned with estimates for u, u_x and $u_x^+ := \max(0, u_x)$, where $u = u(t,x)$, $0 \le t \le T < \infty$, solves (1.1.13), obtained using the celebrated Oleinik's argument (see [**95, 129**] and Problem 2.1.3 below). Firstly we state an obvious lemma.

LEMMA 2.1.1. *Consider a function $f \in C^1(S^1;\mathbb{R})$ satisfying $\int_{S^1} f(x)dx = 0$ such that $f'(x) \le c^*$, $x \in S^1$. Then we have $|f|_\infty \le c^*$ and $|f'|_1 \le 2c^*$.*

The following theorem gives us estimates for the norms of $u_x^+(t)$, $u(t)$ in L_∞ and for the norm of $u_x(t)$ in L_1, uniform in u_0 and ν.

THEOREM 2.1.2. *Assume that $u(t,x)$ solves (1.1.13) with $\xi \in \dot{X}_T^4$ and $u_0 \in H^1$. Then for any $T > 0$, $\theta \in (0,T]$ and $0 < \nu \le 1$,*

$$\sup_{\theta \le t \le T} |u_x^+(t,\cdot)|_\infty \le 4\theta^{-1}(1 + T\|\xi\|_{X_T^4}^{1/2} + T\|\xi\|_{X_T^4}), \tag{2.1.1}$$

$$\sup_{\theta \le t \le T} (|u(t,\cdot)|_\infty + |u_x(t,\cdot)|_1) \le 12\theta^{-1}(1 + T\|\xi\|_{X_T^4}^{1/2} + T\|\xi\|_{X_T^4}). \tag{2.1.2}$$

PROOF. Due to Lemma 2.1.1 estimate (2.1.1) implies (2.1.2), so it suffices to prove the former one. Suppose firstly that $u_0 \in H^4$ and $\xi \in \dot{X}_T^5$. Then Theorem 1.3.6 implies that $u = \mathcal{M}(u(0), \xi) \in X_T^4$. Decompose the solution u as

$$u = \xi + v.$$

Then v solves

$$v_t + uu_x - \nu v_{xx} = \nu \xi_{xx}. \tag{2.1.3}$$

Taking a derivative with respect to x and multiplying by t, we get

$$tv_{tx} + t(u_x^2 + uu_{xx}) = \nu t v_{xxx} + \nu t \xi_{xxx}. \tag{2.1.4}$$

We define

$$w := tv_x$$

and rewrite (2.1.4) as

$$(w_t - v_x) + tu_x^2 + (tu\xi_{xx} + uw_x) = \nu w_{xx} + \nu t \xi_{xxx}. \tag{2.1.5}$$

Now we study the function $w(t,x)$ on the cylinder $Q=[0,T]\times S^1$. Since $w_{|t=0}=0$ and for every t, $\int w(t,x)dx=0$, then either w vanishes identically or it attains its maximum $M>0$ at some point $(t_1,x_1)\in Q$, where $t_1>0$. In the former case $v_x(t,x)\equiv 0$ and $v(t,x)=0$ on Q, which implies (2.1.1). Now assume that w attains its maximum $M>0$ at $(t_1,x_1)\in Q$, $t_1>0$, and denote

$$K:=\max_{0\le t\le T}\|\xi(t,\cdot)\|_{C^3(S^1)}. \tag{2.1.6}$$

Then by (0.0.14), $K\le\|\xi(t,\cdot)\|_{X_T^4}<\infty$.

We want to prove that

$$M\le 3(1+TK+TK^{1/2}). \tag{2.1.7}$$

If $M\le TK$, then the estimate holds trivially. So we may assume that

$$M>TK. \tag{2.1.8}$$

Since $\xi,u,v\in X_T^4$, then $v,w\in X_T^3$ and we see from eq. (2.1.5) that $w_t\in X_T^1$. So w_t,w_x and w_{xx} are continuous functions and by Taylor's formula at (t_1,x_1) we get

$$w_t\ge 0,\ w_x=0,\ w_{xx}\le 0, \tag{2.1.9}$$

and therefore by (2.1.5) we have:

$$-v_x+tu_x^2+tu\xi_{xx}\le \nu t\xi_{xxx}.$$

Multiplying this inequality by t, we get that at the point (t_1,x_1)

$$-M+(M+t\xi_x)^2+t^2u\xi_{xx}-\nu t^2\xi_{xxx}\le 0. \tag{2.1.10}$$

As $t\le T$, than by (2.1.6)

$$|t\xi_x|\le TK. \tag{2.1.11}$$

Therefore in view of (2.1.8)

$$(M+t\xi_x)^2\ge (M-TK)^2. \tag{2.1.12}$$

Since for every t, $\int_{S^1}tv(t,x)dx=0$ and $tv_x=w\le M$, Lemma 2.1.1 implies that $|tv|\le M$, which yields that for $(t,x)\in Q$,

$$|tu|\le|t\xi|+M\le TK+M,\quad \text{so } |t^2u\xi_{xx}|=|(t\xi_{xx})(tu)|\le TK(TK+M). \tag{2.1.13}$$

Since $\nu t^2\xi_{xxx}\le T^2K$, then by (2.1.10) and (2.1.13) we have:

$$-M+(M-TK)^2-TK(TK+M)-T^2K\le 0.$$

The expression in the left-hand side is a quadratic polynomial in M. Computing its unique positive root, we get (2.1.7).

Since $tu_x=w+t\xi_x$ and $M\le 3(1+TK+TK^{1/2})$, for all $(t,x)\in Q$ we have $tu_x\le 4(1+TK+TK^{1/2})$. Therefore for $t\ge\theta$ and any x, $u_x(t,x)$ is bounded by the r.h.s. of (2.1.1) if $u(0)\in H^4$. So u satisfies (2.1.1) if $u(0)\in H^4$ and $\xi\in\dot X_T^5$.

Now we assume that $u_0\in H^1$ and $\xi\in\dot X_T^4$, so $u\in X_T^1$. Our goal is to prove that u still satisfies (2.1.1). For $j=1,2,\dots$ consider $u_0^j=\Pi_j u_0$ and $\xi^j=\Pi_j\xi$. Then $H^4\ni u_0^j\to u_0$ in H^1 and $X_T^5\ni\xi^j\to\xi$ in X_T^4, see (1.3.27) and its proof. For each $j\ge 1$, $u^j(0)$ and ξ^j define a solution $u^j\in X_T^4$ for which (2.1.1) holds. Take any $t\ge 0$. By Theorem 1.3.11 with $m=1$,

$$u_x^j(t,\cdot)\to u_x(t,\cdot)\text{ in }L_2. \tag{2.1.14}$$

So a suitable subsequence $\{u_x^{j_n}(t,x),\ n\ge 1\}$ converges to $u_x(t,x)$ a.s. in x. Since for $\theta\le t\le T$ the functions $u_x^{j_n}(t,x)$ are bounded by the r.h.s. of (2.1.1), then estimate (2.1.1) is valid for the solution u. □

PROBLEM 2.1.3. Repeating the proof of Theorem 2.1.2 establish *Oleinik's estimate*, which says that if $u(t,x)$ is a solution of (B) with $\xi=0$ such that $u(0,x)\in H^1$, then $tu_x(t,x)\le 1$ for any $t>0$ and every x.

PROBLEM 2.1.4. Verify that Theorem 2.1.2 remains true for solutions $u(t,x)$ of eq. (0.0.12), (1.1.5), where f satisfies (0.0.13).

By Theorem 1.2.3 and Theorem 2.1.2, we obtain:

COROLLARY 2.1.5. *If $B_4<\infty$, u_0 is a r.v. in H^1 and $u(t)=u(t;u_0)$, then for every $1\le p\le\infty$, $0<\theta\le 1$ and $\sigma>0$ there exists $C(p,\sigma,B_4)>0$ such that for any $T\ge\theta$ we have*

$$\mathbb{E}[\sup_{T\le t\le T+\sigma}|u_x^+(t,\cdot)|_\infty^p]\le C\theta^{-p}, \tag{2.1.15}$$

$$\mathbb{E}[\sup_{T\le t\le T+\sigma}(|u(t,\cdot)|_\infty^p+|u_x(t,\cdot)|_1^p)]\le C\theta^{-p}. \tag{2.1.16}$$

PROOF. We restrict ourselves to proving the first estimate; the second one follows from Lemma 2.1.1.

Firstly let $T=\theta$. Then by (2.1.1) with $T{=}\theta+\sigma$ (where $0<\theta\le 1$), for each ω we have that

$$\Phi(u^\omega):=\sup_{\theta\le t\le\theta+\sigma}|(u^\omega)_x^+(t,\cdot)|_\infty^p\le C'(p,\sigma)\theta^{-p}(1+\|\xi^\omega\|_{X_T^4}^p).$$

Taking the expectation and using (1.2.19) we get the required inequality.

Now let $u_0\in H^1$ be a random variable, $\mathcal{D}(u_0)=:\mu$, and $u=u(t;u_0)$. Considering the operator $\mathcal{M}^{T'}$ as in (1.3.34) with $T'=\theta+\sigma$, we use the inequality above jointly with Theorem 1.2.3 and get:

(2.1.17)

$$\begin{aligned}\mathbb{E}\Phi(u)=&\mathbb{E}\Phi\circ\mathcal{M}(u_0,\xi)=\int_{H^1}\mu(dv)\Big(\int_{\dot X_{T'}^4}\pi_{T'}^4(d\xi)\Phi\circ\mathcal{M}(v,\xi)\Big)\\ &\le C'(p,\sigma)\theta^{-p}\int_{H^1}\mu(dv)\int_{\dot X_{T'}^4}\pi_{T'}^4(d\xi)(1+\|\xi\|_{\dot X_{T'}^4}^p)=C(P,\sigma,B_4)\theta^{-p}.\end{aligned}$$

Finally let $T>\theta$. We set $T'=T-\theta>0$ and denote $v^\omega(\tau)=u(\tau+T')$. This is a weak solution of (1.1.13) with the random initial data $v(0)=u^\omega(T')$, and the left-hand term in the inequality which we are proving is $\mathbb{E}\Phi(u^\omega(\cdot+T'))=\mathbb{E}\Phi(v^\omega(\cdot))$. So the required estimate follows from (2.1.17) with $u_0=u(T')$. □

Note that since by writing any function $u(x)\in W_1^1$ as Fourier series $u=\sum\hat u_k e^{2i\pi kx}$ and integrating by parts in the formula $\hat u_k=\int u(x)e^{-2i\pi kx}$ we get that

$$|\hat u_k|\le\frac{1}{2\pi|k|}|u|_{1,1},\qquad |k|\ge 1, \tag{2.1.18}$$

then relations (2.1.2) and (2.1.16) provide an efficient tool to estimate Fourier coefficients of solutions for eq. (B).

PROBLEM 2.1.6. In the proof of Corollary 2.1.5 we implicitly used that if $u(t)$ solves (1.1.13) with $u_0 \in H^1$, then for $\tau > 0$, $|u_x^{\omega+}(\tau)|_\infty$ is a r.v. To justify that, show that for any $u_0 \in H^1$, $s < 2$ and $\tau > 0$, the mapping

$$(0, \infty) \times \dot{X}_T^4 \to H^s, \quad (\tau, \xi) \mapsto u(\tau; u_0, \xi)$$

is continuous (and next use that $H^{5/3} \subset C^1$).

A version of Oleinik's estimate.

Here we discuss a version of Theorem 2.1.2 which holds if $u_0 \in H^2$ and is valid for all $t \in [0, T]$. The price to pay is that the corresponding estimates are not anymore u_0–independent. The obtained results play a crucial role in Chapter 8.

THEOREM 2.1.7. *Assume that* $u(t, x)$ *solves* (1.1.13) *with* $\xi \in \dot{X}_T^4$ *and* $u_0 \in H^2$. *Then for any* $T > 0$*:*

$$\sup_{0 \le t \le T} |u_x^+(t, \cdot)|_\infty \le B, \tag{2.1.19}$$

$$\sup_{0 \le t \le T} |u(t, \cdot)|_\infty \le B, \qquad \sup_{0 \le t \le T} |u_x(t, \cdot)|_1 \le 2B \tag{2.1.20}$$

where $B = B_T(u_0, \xi) := \max(|u_{0x}^+|_\infty + \|\xi\|_{X_T^4},\ 4\|\xi\|_{X_T^4} + \|\xi\|_{X_T^4}^{1/2})$.

We emphasize that estimates (2.1.19) and (2.1.20) hold uniformly in ν.

PROOF. The proof is a simplified version of that of Theorem 2.1.2. As there, we firstly assume that

$$u_0 \in H^4, \qquad \xi \in X_T^5; \tag{2.1.21}$$

then $u \in X_T^4$. In view of Lemma 2.1.1 it suffices to establish (2.1.19).

As above, we use the decomposition $u = \xi + v$ and define K as in (2.1.6). Taking the derivative in x of (B) and considering the function $h = v_x$, we get:

$$h_t + u_x^2 + u u_{xx} = \nu h_{xx} + \nu \xi_{xxx}. \tag{2.1.22}$$

Consequently,

$$h_t + (h + \xi_x)^2 + u h_x + \xi \xi_{xx} + v \xi_{xx} - \nu \xi_{xxx} = \nu h_{xx}. \tag{2.1.23}$$

Now we examine the function $h(t, x)$ on the cylinder $Q = [0, T] \times S^1$. We want to prove that its maximum N on Q satisfies

$$N \le \max(|u_x^+(0, \cdot)|_\infty,\ 3K + K^{1/2}). \tag{2.1.24}$$

Since $\int h(t, x) dx \equiv 0$, then $N \ge 0$. If $N = 0$, then h vanishes identically and we clearly have (2.1.24). If h attains the maximal value N at some point $(0, x_1)$, $x_1 \in S^1$, then (2.1.24) holds because $u_x(0, \cdot) = h(0, \cdot)$ since $\xi(0) = 0$. Therefore it remains to consider the case of h attaining its maximum $N > 0$ at a point $(t_1, x_1) \in (0, T] \times S^1$.

We argue by contradiction, assuming that

$$N > \max(|u_x^+(0, \cdot)|_\infty,\ 3K + K^{1/2}). \tag{2.1.25}$$

By Theorem 1.3.6 with $m = 4$ the function $h = u_x - \xi_x$ satisfies $h, h_x, h_{xx} \in C(Q)$, and from (2.1.22) we see that $h_t \in C(Q)$. Consequently by Taylor's formula, at the point (t_1, x_1) we have

$$h = N,\ h_t \ge 0,\ h_x = 0,\ h_{xx} \le 0,$$

and therefore by (2.1.23) applied at this point,

$$(N+\xi_x)^2+\xi\xi_{xx}+v\xi_{xx}-\nu\xi_{xxx}\le 0. \tag{2.1.26}$$

In view of (2.1.25) $N+\xi_x\ge N-K\ge 0$, so

$$(N+\xi_x)^2\ge (N-K)^2 \quad \text{at } (t_1,x_1). \tag{2.1.27}$$

Since $v_x(t_1,x)=h(t_1,x)\le N$ for every x, then $|v(t_1,x_1)|\le N$ by Lemma 2.1.1. Using this and (2.1.27) we derive from (2.1.26) that

$$(N-K)^2-K^2-NK-K\le 0.$$

The expression in the left-hand side is a quadratic polynomial in N. Computing its unique positive root, we get that $N<3K+K^{1/2}$. This contradicts (2.1.25) and proves (2.1.24). Since $u_x=h+\xi_x$, then (2.1.19) is established if (2.1.21) holds. The general case follows from this result as in the proof of Theorem 2.1.2. □

Note that if $u(t,x)$ is a solution of (1.1.13) with $u_0\in H^2$, then Theorems 2.1.7 and 1.2.3 with $m=4$ immediately imply uniform in $\nu>0$ estimates on the l.h.s.'s in (2.1.15), (2.1.16) with $T=0$ in terms of σ,p and $\|u_0\|_2$.

2.2. Upper bounds for moments of Sobolev norms of solutions.

From now on, till the end of Chapter 2, we assume that

$$B_4<\infty \tag{2.2.1}$$

(we recall the agreement after Corollary 1.2.4). Then by Theorem 1.4.2 eq. (1.1.12) is well posed in the space H^3 (and is well posed in H^4 if $B_{4_*}<\infty$ for some $4_*>4$). Corollary 2.1.5 applies to solutions of (1.1.13). The aim of this section is to estimate from above the expected values $\mathbb{E}[\|u(t;u_0)\|_m^2]$, $m\ge 1$, for $t\ge\theta>0$, uniformly in $u_0\in H^1$ and with explicit control for the dependence of the estimates on $\nu>0$.

THEOREM 2.2.1. *Let $m\in\mathbf{N}$ and $B_{\max(4,m)}<\infty$. Then for each $\theta>0$ there exists $C_m=C_m(\theta,B_{\max(4,m)})>0$ such that for every non-random $u_0\in H^1$ we have:*

$$\mathbb{E}[\|u(t;u_0)\|_m^2]\le C_m\nu^{-(2m-1)}, \tag{2.2.2}$$

for all $t\ge\theta$.

PROOF. Below we denote by β_0 and C_m, C_{mj}, C'_{mj}, etc. various positive constants which only depend on m, θ and $B_{\max(4,m)}$.

i) Assume firstly that

$$u_0\in C^\infty, \qquad B_M<\infty \quad \forall M. \tag{2.2.3}$$

Then by Theorem 1.4.4 and Proposition 1.4.6 estimate (1.4.10) holds for every m, the solution $u(t;u_0)=:u(t)$ satisfies (1.4.22) and $\mathbb{E}\|u(t)\|_m^2$ is a C^1-function of t. Applying (1.3.14) to the second term in the r.h.s. of (1.4.22) we get that

$$\frac{d}{dt}\mathbb{E}\|u(t)\|_m^2\le -2\nu\mathbb{E}\|u(t)\|_{m+1}^2+C_m\mathbb{E}\Big(|u(t)|_{1,1}^{\frac{2m+3}{2m+1}}\|u(t)\|_{m+1}^{\frac{4m}{2m+1}}\Big)+B_m \quad \text{for } t\ge 0. \tag{2.2.4}$$

Let $t\ge\theta'$, where $\theta'=\theta/3$. Then by Hölder's inequality with $p=2m+1, q=\frac{2m+1}{2m}$ and Corollary 2.1.5, $u=u(t)$ satisfies

$$\mathbb{E}\Big(|u|_{1,1}^{\frac{2m+3}{2m+1}}\|u\|_{m+1}^{\frac{4m}{2m+1}}\Big)\le\big(\mathbb{E}[|u|_{1,1}^{2m+3}]\big)^{\frac{1}{2m+1}}\big(\mathbb{E}[\|u\|_{m+1}^2]\big)^{\frac{2m}{2m+1}}\le C'_m\big(\mathbb{E}[\|u\|_{m+1}^2]\big)^{\frac{2m}{2m+1}}.$$

Denote

$$X_j(t) = \mathbb{E}[\|u(t)\|_j^2], \qquad j \in \mathbf{N}.$$

By the previous estimates, (2.2.4) implies that:

$$\begin{aligned}\frac{d}{dt}X_m(t) &\le B_m - 2\nu X_{m+1}(t) + C_m X_{m+1}(t)^{\frac{2m}{2m+1}} \\ &= B_m - X_{m+1}(t)^{\frac{2m}{2m+1}}\left(2\nu X_{m+1}(t)^{\frac{1}{2m+1}} - C_m\right), \quad t \ge \theta'.\end{aligned} \tag{2.2.5}$$

Using inequality (1.3.11), Hölder's inequality and Corollary 2.1.5 we get that

$$X_m(t) \le C_{m1} X_{m+1}(t)^{\frac{2m-1}{2m+1}} \left(\mathbb{E}[|u(t)|_{1,1}^a]\right)^b \le C_{m2} X_{m+1}(t)^{\frac{2m-1}{2m+1}}, \quad t \ge \theta',$$

for suitable constants $a, b > 0$. Therefore:

$$X_{m+1}(t) \ge C_{m3} X_m(t)^{\frac{2m+1}{2m-1}}, \quad t \ge \theta'. \tag{2.2.6}$$

To establish (2.2.2) we fix $\beta \ge 1$ and assume that

$$\exists\, t_* \ge 2\theta' \text{ such that } X_m(t_*) > \beta\nu^{-(2m-1)} =: Y. \tag{2.2.7}$$

Below we derive a contradiction from (2.2.7), if $\beta \gg 1$. This will prove the required estimate.

Let us denote $s = t_* - t$. The calculations below are made for $s \in [0, \theta']$, so that accordingly $t \ge \theta'$, and (2.2.5) yields:

$$\frac{d}{ds}X_m(s) \ge -B_m + X_{m+1}(s)^{\frac{2m}{2m+1}}\left(2\nu X_{m+1}(s)^{\frac{1}{2m+1}} - C_m\right). \tag{2.2.8}$$

We claim that

$$X_m(s) > Y \quad \text{for} \quad s \in [0, \theta']. \tag{2.2.9}$$

Indeed, assuming that this is not true we consider the smallest $s \in (0, \theta']$ such that $X_m(s) = Y$. Denote this s as $\tilde\theta$. Then for any $0 < \tilde\theta_- < \tilde\theta$ we have that $X_m(s) > Y$ on the interval $[0, \tilde\theta_-]$, so by (2.2.6)

$$2\nu X_{m+1}^{\frac{1}{2m+1}} - C_m \ge 2\nu\left(C_{m3} X_m^{\frac{2m+1}{2m-1}}\right)^{\frac{1}{2m+1}} - C_m \ge 2C_{m3}^{\frac{1}{2m+1}}\beta^{\frac{1}{2m-1}} - C_m =: K_m(\beta),$$

$$s \in [0, \tilde\theta_-].$$

Choose $\beta_0 \gg 1$ large enough so that $K_m(\beta) > 1$ if $\beta \ge \beta_0$. Then by (2.2.8) and (2.2.6), for $s \in [0, \tilde\theta_-]$ we have that

$$\frac{d}{ds}X_m \ge -B_m + X_{m+1}^{\frac{2m}{2m+1}} K_m(\beta) \ge -B_m + C_{m3}^{\frac{2m}{2m+1}} X_m^{\frac{2m}{2m-1}} K_m(\beta) > 0, \tag{2.2.10}$$

where the last inequality is valid if $\beta \ge \beta_0 \gg 1$. Since $X_m(s)\mid_{s=0} = X_m(t)\mid_{t=t_*}$, then $X_m(\tilde\theta_-) \ge X_m(t_*)$. So also $X_m(\tilde\theta) \ge X_m(t_*) > Y$ (see (2.2.7)). This contradiction proves (2.2.9).

Repeating the calculation above for $0 \le s \le \theta'$ using (2.2.9), we get that estimate (2.2.10) holds for such s, and so

$$\frac{d}{ds}X_m \ge -B_m + C_{m3}^{\frac{2m}{2m+1}} X_m^{\frac{2m}{2m-1}} K_m(\beta) \ge \frac{1}{2} C_{m3}^{\frac{2m}{2m+1}} X_m^{\frac{2m}{2m-1}} K_m(\beta), \quad \forall\, s \in [0, \theta'],$$

if $\beta_0 \gg 1$. Consequently,

$$\frac{d}{ds}\left(X_m^{\frac{-1}{2m-1}}\right) \le -\frac{C_{m3}^{\frac{2m}{2m+1}}}{2(2m-1)} K_m(\beta) =: -C_{m4} K_m(\beta).$$

Integrating in time over the interval $[0,\theta']$, using that $X_m(s)|_{s=0}> Y$, we obtain:

$$X_m(s)^{\frac{-1}{2m-1}} \le -C_{m4}K_m(\beta)s + X_m(0)^{\frac{-1}{2m-1}} \le -C_{m4}K_m(\beta)s + \beta^{\frac{-1}{2m-1}}\nu, \quad s\in[0,\theta'].$$

Since $\nu\le 1$, then $X_m(\theta')^{\frac{-1}{2m-1}}<0$, provided that $\beta\ge\beta_0\gg 1$. This is a contradiction as X_m is a positive function. Thus (2.2.7) does not hold for $\beta=\beta(m,\theta,B_m)$ large enough. We have proved (2.2.2) with $C_m=\beta$ under the restriction (2.2.3).

ii) (removing the restriction (2.2.3)*).* Now we assume that $u_0\in H^1$ and $B_m<\infty$ (so $\xi^\omega\in\dot X^m_\infty$ for each ω by the agreement after Corollary 1.2.4). Let us take $u_0^N=\Pi_N u_0$ and $\xi^N=\Pi_N\xi$ (see (1.3.23)). Then u_0^N,ξ^N satisfy (2.2.3), $u_0^N\to u_0$ in H^1 and $\xi^{N,\omega}\to\xi^\omega$ in $\dot X^m_t$ for every ω, for the same reason as in the proof of (1.3.27). Then the solution $u^N(t)$, corresponding to $u_0:=u_0^N$, $\xi:=\xi^N$ satisfies (2.2.2), and $u^N(t)\to u(t)$ in H^1 for every ω. So by Theorem 11.1.4 u also meets (2.2.2). □

If $u_0\in H^1$ is a random variable, independent from the force ξ, then by (1.4.5) for any $t\ge\theta$,

$$\mathbb{E}\|u(t;u_0)\|_m^2=\int\mathbb{E}\|u(t;v)\|_m^2\mathcal{D}(u_0)(dv)\le C_m\nu^{-(2m-1)}.$$

So we have

COROLLARY 2.2.2. *If $B_{\max(4,m)}<\infty$, then the estimate* (2.2.2) *remains true for random initial data $u_0\in H^1$.*

In particular it means that if $B_{\max(4,m)}<\infty$, then the semigroup of measures $\{S_t^*\}$ is "immediately smoothing": if $\nu\in\mathcal{P}(H^1)$, then $S_t^*\nu\in\mathcal{P}(H^m)$ for any $t>0$ (see Proposition 11.3.3).

The simple observation that a validity of an estimate for deterministic initial data implies its validity for random initial data via the Fubini formula (1.4.5) remains true in similar situations below, where we usually do not repeat it.

REMARK 2.2.3. Theorem 2.2.1 applies when $B_4=0$, i.e. when $\eta=0$. Then (B) is the free Burgers equation and the expectation sign may be dropped in (2.2.2). See more in Section 10.1.1.

PROBLEM 2.2.4. Prove that a solution $u^f(t;u_0)$ of equation (0.0.12), (0.0.13), (1.1.5), (1.1.6), equal to $u_0\in H^1$ at $t=0$, satisfies (2.2.2) with a suitable $C_m(f,\theta,B_{\max(4,m)})$.
Hint: Repeat the proof of Theorem 2.2.1, using the assertions of Problem 1.3.5 and Problem 1.4.5.

By Theorem 2.2.1, if $B_{m_*}<\infty$ for some $m_*>m\ge 4$, then for $u_0\in H^1$ and $\tau>0$ there exists a null set $Q_{u_0}(\tau)$ such that $u(\tau;u_0)\in H^m$ if $\omega\notin Q_{u_0}(\tau)$. By the cocycle property (1.3.37) and (1.2.17), $u(t;u_0)$ stays in H^m for $t\ge\tau$ if $\omega\notin Q_{u_0}(\tau)$. Setting $Q_{u_0}=\cup_{n\in\mathbf{N}}Q_{u_0}(1/n)$ we get

COROLLARY 2.2.5. *If $B_{m_*}<\infty$ for some $m_*>m\ge 4$, then for any $u_0\in H^1$ there exists a null set Q_{u_0} such that $u(t;u_0)\in H^m$ for $t>0$ if $\omega\notin Q_{u_0}$.*

Assuming that the random force ξ is smoother than in Theorem 2.2.1 we estimate higher moments of Sobolev norms of solutions for (B), as well as moments of their norms in the spaces W_p^m with $p>1$:

COROLLARY 2.2.6. *Let $\theta > 0$, $m \in \mathbf{N}$ and $\alpha \geq 1$. Assume that $B_\mu < \infty$ for*

$$\mu = \mu_{m,\alpha} := \lceil \max\big(m, \frac{(2m-1)\alpha+2}{4}\big)\rceil$$

(see (0.0.16)*). Then there exists $C = C(\alpha, m, \theta, B_{\max(4,\mu)}) > 0$ such that for every $u_0 \in H^1$,*

$$\mathbb{E}\,[\|u(t;u_0)\|_m^\alpha] \leq C\nu^{-\frac{\alpha}{2}(2m-1)}, \quad \forall t \geq \theta. \tag{2.2.11}$$

PROOF. The case $\alpha \leq 2$ follows immediately from Theorem 2.2.1 by Hölder's inequality. So we may assume that $\alpha > 2$. Then $\mu \geq m+1$. By the Gagliardo–Nirenberg inequality (1.3.9), applied to $w = u_x$, we have:

$$\mathbb{E}\,[\|u(t)\|_m^\alpha] \leq C\mathbb{E}\left[\|u(t)\|_\mu^{\alpha\lambda}|u(t)|_{1,1}^{\alpha(1-\lambda)}\right], \quad \lambda = \frac{2m-1}{2\mu-1},$$

where $C = C(m, \alpha, \mu) > 0$. By our assumption on μ, $\alpha\lambda \leq 2$. Then by Hölder's inequality and Corollary 2.1.5,

$$\mathbb{E}\,[\|u(t)\|_m^\alpha] \leq C'\mathbb{E}\left[\|u(t)\|_\mu^2\right]^{\frac{\alpha\lambda}{2}}\mathbb{E}\left[|u(t)|_{1,1}^a\right]^b \leq C''\mathbb{E}\left[\|u(t)\|_\mu^2\right]^{\frac{\alpha\lambda}{2}},$$

where C', C'', a and b depend on α, m, θ. Thence, by the estimate (2.2.2), we obtain (2.2.11). □

COROLLARY 2.2.7. *Let $\theta > 0$, $m \in \mathbf{N}$, $1 < p \leq \infty$ and $\alpha \geq 1$. Assume that $B_M < \infty$ for all M*[1]*. Then there exists $C = C(\alpha, m, p, \theta, \vec{B}) > 0$ (we recall* (1.1.8)*) such that for every $u_0 \in H^1$,*

$$\mathbb{E}\left[|u(t;u_0)|_{m,p}^\alpha\right] \leq C\nu^{-\alpha(m-1/p)}, \quad \forall t \geq \theta. \tag{2.2.12}$$

PROOF. Let us take $M = m+1$ if $p \geq 2$, and if $1 < p < 2$ choose M large enough so that

$$\lambda' = \frac{m-1/p}{M-1/2} \geq \frac{m-1}{M-1}.$$

Then applying (1.3.11) we obtain:

$$\mathbb{E}\left[|u(t)|_{m,p}^\alpha\right] \leq C\mathbb{E}\left[\|u(t)\|_M^{\alpha\lambda'}|u(t)|_{1,1}^{\alpha(1-\lambda')}\right],$$

where $C = C(m, \alpha, p, M) > 0$. So by the Cauchy-Schwarz inequality and Corollary 2.1.5,

$$\mathbb{E}\left[|u(t)|_{m,p}^\alpha\right] \leq C'\left[\mathbb{E}\|u(t)\|_M^{2\alpha\lambda'}\right]^{1/2}\left[\mathbb{E}|u(t)|_{1,1}^{2\alpha(1-\lambda')}\right]^{1/2} \leq C''\left[\mathbb{E}\|u(t)\|_M^{2\alpha\lambda'}\right]^{1/2},$$

where C' and C'' depend on α, m, p, M, θ. Thence by the estimate (2.2.11) we obtain (2.2.12), since $\lambda'(2M-1) = 2(m-1/p)$. □

2.3. Energy balance and lower bounds for moments of norms of solutions.

In all assertions in this section we assume that

$$u = u(t,x) \text{ is a solution } u^\omega(t;u_0) \text{ with some } u_0 \in H^1,$$

unless otherwise stated.

[1] Cf. the discussion which follows Problem 1.3.2

Bounds for moments of H^m norms, $m \geq 1$.

Let $B_4 < \infty$. Applying (1.4.9) and Theorem 2.2.1 to $u(t) = u(t; u_0)$ we find that

$$\mathbb{E}e^{\sigma' \|u(t')\|^2} < \infty, \quad \mathbb{E}\|u(t')\|_4^2 < \infty \qquad \forall t' > 0.$$

So for any $t' > 0$, $u_0' := u(t')$ satisfies the assumptions for Part 2 of Proposition 1.4.6 with $M = 2$, and in view of the cocycle property (1.4.8) relation (1.4.11) is valid for the solution $u(t)$ for $t > 0$. So the latter satisfies the balance of energy relation

$$\frac{1}{2}\mathbb{E}[\|u(T+\sigma)\|^2] - \frac{1}{2}\mathbb{E}[\|u(T)\|^2] + \nu \int_T^{T+\sigma} \mathbb{E}[\|u(s)\|_1^2]ds = \frac{\sigma}{2} B_0 , \tag{2.3.1}$$

for any $T, \sigma > 0$.

The following theorem gives upper and lower bounds for the energy dissipation rate $\nu\mathbb{E}[\|u(t)\|_1^2]$.

THEOREM 2.3.1. *Assume that $B_0 > 0$ and $B_4 < \infty$. Then there exists $\sigma_0(B_0, B_4) > 0$ such that for any $\sigma \geq \sigma_0$ and $T \geq 1$ the solution $u(t)$ satisfies:*

$$\frac{1}{4}B_0 \leq \frac{\nu}{\sigma} \int_T^{T+\sigma} \mathbb{E}[\|u(s)\|_1^2]ds \leq \frac{3}{4}B_0 . \tag{2.3.2}$$

PROOF. Corollary 2.1.5 with $p = 2$ implies that $\mathbb{E}(\|u(t)\|^2) \leq C(B_4)$ if $t \geq 1$. For

$$\sigma \geq \frac{2C(B_4)}{B_0} =: \sigma_0, \tag{2.3.3}$$

this estimate yields:

$$\frac{1}{2\sigma}\mathbb{E}[\|u(T)\|^2] \leq \frac{1}{4}B_0, \qquad \frac{1}{2\sigma}\mathbb{E}[\|u(T+\sigma)\|^2] \leq \frac{1}{4}B_0.$$

Now (2.3.1) implies the theorem's statement. □

Everywhere below in this section σ_0 is defined by relation (2.3.3). The theorem above does not apply to the free Burgers equation $(\mathrm{B})|_{\eta=0}$ since then $B_0 = 0$. But a similar result is true:

PROBLEM 2.3.2. Prove that if $u(t,x)$ is a solution of the free Burgers equation $(\mathrm{B})|_{\eta=0}$ (such that $u(0,x) = u_0(x) \in H^1$, as always in this section) and $\|u_0\|^2 = \kappa^2 > 0$, then for any $\sigma \geq \sqrt{2}/\kappa$ we have

$$\frac{1}{4\sigma}\kappa^2 \leq \frac{\nu}{\sigma} \int_0^{\sigma} \|u(s)\|_1^2\, ds \leq \frac{1}{2\sigma}\kappa^2.$$

Hint: use that $u(t,x)$ satisfies the balance relation (1.3.33) and that $\|u(t)\|^2 \leq t^{-2}$ in view of Oleinik's inequality (see Problem 2.1.3) and Lemma 2.1.1.

Fix any $T \geq 1$, $\sigma \geq \sigma_0$ and consider the probability space

$$(Q,\ \mathcal{T},\ \rho) := \left([T, T+\sigma] \times \Omega,\ \mathcal{L} \otimes \mathcal{F},\ \frac{dt}{\sigma} \times \mathbb{P} \right), \tag{2.3.4}$$

where $\mathcal{L}$ is the Borel σ–algebra on $[T, T+\sigma]$.

NOTATION 2.3.3. For a random process $\zeta^\omega(t)$ we denote by $\langle\langle \zeta \rangle\rangle$ its expectation with respect to the measure ρ:

$$\langle\langle \zeta \rangle\rangle = \langle\langle \zeta^\omega \rangle\rangle_{T,\sigma} = \frac{1}{\sigma} \int_T^{T+\sigma} \mathbb{E}[\zeta^\omega(t)]dt. \tag{2.3.5}$$

Naturally, if $\zeta(t)$ is a stationary process, then $\langle\langle\zeta\rangle\rangle = \mathbb{E}\zeta(t)$, for any t. Using this notation, the inequality in Theorem 2.3.1 reads:

$$\frac{1}{4}B_0\nu^{-1} \le \langle\langle\|u\|_1^2\rangle\rangle \le \frac{3}{4}B_0\nu^{-1}.$$

The following theorem gives upper and lower bounds for $\langle\langle\|u\|_m^2\rangle\rangle$.

THEOREM 2.3.4. *Assume that $B_0 > 0$ and that for some $m \in \mathbf{N}$ we have $B_{\max(4,m)} < \infty$. Then the solution $u = u(t;u_0)$ of (1.1.13) satisfies*

$$C_m^{-1}\nu^{-(2m-1)} \le \langle\langle\|u\|_m^2\rangle\rangle \le C_m\nu^{-(2m-1)}, \tag{2.3.6}$$

where $C_m = C_m(B_0, B_{\max(4,m)}) > 1$.

We emphasize that the constant C_m does not depend on $T \ge 1$ and on $\sigma \ge \sigma_0$ in the definition of the averaging $\langle\langle\cdot\rangle\rangle$, and that due to (1.4.5) the estimates (2.3.6) remain true if $u_0 \in H^1$ is a r.v., independent from η, cf. Corollary 2.2.2.

PROOF. The constants C and C_m in this proof depend only on m, B_0 and $B_{\max(4,m)}$.

The right inequality in (2.3.6) follows from Theorem 2.2.1, and we have already established the left inequality with $m = 1$ in Theorem 2.3.1. By the Gagliardo–Nirenberg inequality (1.3.9)

$$\|u\|_1 = \|u_x\| \le C_m\|u_x\|_{m-1}^{\frac{1}{2m-1}}|u_x|_1^{\frac{2m-2}{2m-1}}.$$

Since Corollary 2.1.5 with $\theta = \sigma = 1$ implies that $\mathbb{E}|u_x(t)|_1^p \le C_p$ for $t \ge 1$, then applying Hölder's inequality to the integral $\langle\langle\cdot\rangle\rangle$ we derive from the last displayed formula that

$$\langle\langle\|u\|_1^2\rangle\rangle \le C_m'\langle\langle\|u\|_m^2\rangle\rangle^{\frac{1}{2m-1}}\langle\langle|u_x|_1^2\rangle\rangle^{\frac{2m-2}{2m-1}} \le C_m\langle\langle\|u\|_m^2\rangle\rangle^{\frac{1}{2m-1}},$$

i.e. $\langle\langle\|u\|_m^2\rangle\rangle \ge C_m^{1-2m}\langle\langle\|u\|_1^2\rangle\rangle^{2m-1}$. Now evoking Theorem 2.3.1 we obtain the left inequality in (2.3.6). □

PROBLEM 2.3.5. Using Remark 2.2.3 and Problem 2.3.2, prove that if $u_0 \in C^\infty$ and $u(t,x)$ is a solution of the deterministic equation (1.1.13) with $\xi = 0$, then for $m \ge 2$,

$$C_m^{-1}\nu^{-(2m-1)} \le \frac{1}{\sigma}\int_0^\sigma \|u(t)\|_m^2 dt \le C_m\nu^{-(2m-1)},$$

if $\sigma \ge \sqrt{2}/\|u_0\|$. On which quantities does the constant C_m now depend?

REMARK 2.3.6. If $b_s \equiv b_{-s}$ and $u_0^\omega(x)$ is a space-homogeneous random field, then by Theorem 1.5.6 the solution $u(t,x) = u(t;u_0)$ also is space-homogeneous. In this case, for any $t \ge 0$ and $x \in S^1$,

$$\mathbb{E}\|u(t)\|_m^2 = \mathbb{E}\int |\partial_x^m u(t,x)|^2\,dx = \int \mathbb{E}\,|\partial_x^m u(t,x)|^2\,dx = \mathbb{E}|\partial_x^m u(t,x)|^2,$$

since by homogeneity the quantity $\mathbb{E}\,|\partial_x^m u(t,x)|^2$ is x-independent. By Theorem 2.3.4 it is finite, and its time-average $\frac{1}{\sigma}\int_T^{T+\sigma}\mathbb{E}|\partial_x^m u(t,x)|^2\,dt$ equals $\langle\langle\|u\|_m^2\rangle\rangle$ and satisfies (2.3.6).

COROLLARY 2.3.7. *If $B_0 > 0$ and $B_M < \infty$ for all M, then for every $m \in \mathbf{N}$ and $\alpha \geq 1$, there exists a constant $C_m(\alpha) = C_m(\alpha, \vec{B}) > 0$ (we recall (1.1.8)) such that the solution u satisfies*

$$C_m^{-1}(\alpha)\nu^{-m+\frac{1}{2}} \leq \langle\langle \|u\|_m^\alpha \rangle\rangle^{\frac{1}{\alpha}} \leq C_m(\alpha)\nu^{-m+\frac{1}{2}}. \tag{2.3.7}$$

PROOF. The right inequality follows immediately by averaging in time in (2.2.11). For $\alpha \geq 2$, Hölder's inequality and (2.3.6) yield the left inequality. Using again Hölder's inequality, we get:

$$\langle\langle \|u\|_m^2 \rangle\rangle = \langle\langle \|u\|_m^{\frac{2}{3}} \|u\|_m^{\frac{4}{3}} \rangle\rangle \leq \langle\langle \|u\|_m \rangle\rangle^{\frac{2}{3}} \langle\langle \|u\|_m^4 \rangle\rangle^{\frac{1}{3}}. \tag{2.3.8}$$

Therefore by (2.3.7) with $\alpha = 2$ and $\alpha = 4$, the quantity $\langle\langle \|u\|_m \rangle\rangle$ is bounded from below by

$$\langle\langle \|u\|_m^2 \rangle\rangle^{\frac{3}{2}} \langle\langle \|u\|_m^4 \rangle\rangle^{-\frac{1}{2}} \geq \left(\frac{\nu^{-2m+1}}{C_m(2)} \right)^{3/2} \left(C_m(4)\nu^{-4m+2} \right)^{-1/2} =: C^{-1}\nu^{-m+\frac{1}{2}},$$

which yields (2.3.7) for $\alpha = 1$. Finally, for $\alpha \in (1,2)$, the left inequality in (2.3.7) follows from the case $\alpha = 1$ and Hölder's inequality. □

The result below is used in Part II to analyse small-scale quantities, related to the 1d turbulence (we recall the notation (2.3.4)):

COROLLARY 2.3.8. *Let $B_0 > 0$ and $B_M < \infty$ for each M. Then for every $m \in \mathbf{N}$ there exist $\alpha(m) = \alpha(m, \vec{B}) \in (0,1)$ and $C(m) = C(m, \vec{B}) > 0$ such that the solution $u(t)$ satisfies*

$$\rho_{\nu,m} := \rho\left(\alpha\nu^{-m+\frac{1}{2}} \leq \|u^\omega(t)\|_m \leq \alpha^{-1}\nu^{-m+\frac{1}{2}} \right) \geq C(m). \tag{2.3.9}$$

PROOF. For any $a > 0$ consider the set

$$Y_a = \{(t,\omega) \in Q : \|u^\omega(t)\|_m \leq a\}.$$

Using Corollary 2.3.7 firstly with $\alpha = 1$ and next with $\alpha = 2$, we get

$$\begin{aligned} C_m(1)^{-1}\nu^{-m+\frac{1}{2}} &\leq \int_{Y_a} \|u^\omega(t)\|_m d\rho + \int_{Q\backslash Y_a} \|u^\omega(t)\|_m d\rho \\ &\leq a + \left(\int_Q \|u^\omega(t)\|_m^2 d\rho \right)^{\frac{1}{2}} (\rho(Q\backslash Y_a))^{\frac{1}{2}} \leq a + C_m(2)\nu^{-m+\frac{1}{2}} (\rho(Q\backslash Y_a))^{\frac{1}{2}}. \end{aligned}$$

We choose $a = a_\nu = \frac{1}{2}C_m(1)^{-1}\nu^{-m+\frac{1}{2}}$ and denote by Y the set Y_{a_ν}. Then, by the last inequality,

$$\rho\left(\|u^\omega(t)\|_m \geq a_\nu\right) \geq \rho(Q\backslash Y) \geq \left(\frac{1}{2} \left(C_m(1)C_m(2)\right)^{-1} \right)^2 =: C'_m.$$

Let $\alpha < \max\left(1, C_m(1)^{-1}/2\right)$. By the last inequality, (2.3.7) (with $k = \alpha = 1$) and Markov's inequality, we have:

$$\rho_{\nu,m} = \rho\left(\|u^\omega(t)\|_m \geq \alpha\nu^{-m+\frac{1}{2}} \right) - \rho\left(\|u^\omega(t)\|_m > \alpha^{-1}\nu^{-m+\frac{1}{2}} \right) \geq C'_m - \alpha C_m(1).$$

Now choose $\alpha > 0$ so small that the term in the r.h.s. is strictly positive. This implies (2.3.9). □

Bounds for moments of W^m_∞ norms, $m \ge 0$.

PROPOSITION 2.3.9. *Consider $m \in \mathbf{N}_0$. Assume that $B_0 > 0$ and that $B_{\max(4,m)} < \infty$. Then there exists $C_m(B_0, B_{\max(4,m)}) > 1$ such that*

$$C_m^{-1}\nu^{-2m} \le \langle\langle |u|^2_{m,\infty}\rangle\rangle \le C_m \nu^{-2m}. \tag{2.3.10}$$

PROOF. *Case $m \ge 1$.* To prove the right inequality we observe that by the Gagliardo–Nirenberg inequality (1.3.9),

$$|u|^2_{m,\infty} \le C\|u\|_m \|u\|_{m+1}.$$

Applying next next the Cauchy-Schwarz inequality and (2.3.6) we obtain the required estimate:

$$\langle\langle |u|^2_{m,\infty}\rangle\rangle \le C\langle\langle \|u\|^2_m\rangle\rangle^{1/2}\langle\langle \|u\|^2_{m+1}\rangle\rangle^{1/2} \le C_m\nu^{-2m}.$$

To get the left inequality we proceed in the same way as when proving the lower bound for $\langle\langle \|u\|^2_m\rangle\rangle$:

$$\|u\|_1 = \|u_x\| \le |u_x|_\infty^{\frac{1}{2}} |u_x|_1^{\frac{1}{2}} \le C_m |u_x|_{m-1,\infty}^{\frac{1}{2m}} |u|_\infty^{\frac{m-1}{2m}} |u_x|_1^{\frac{1}{2}},$$

where the second upper estimate follows from the Landau–Kolmogorov inequality (1.3.12). This implies the desired lower bound in (2.3.10).

Case $m = 0$: The upper estimate follows from Corollary 2.1.5. To obtain the lower bound we observe that by (1.3.12) we have

$$\langle\langle |u|^2_{1,\infty}\rangle\rangle \le C\langle\langle |u|_\infty |u|_{2,\infty}\rangle\rangle \le C\langle\langle |u|^2_\infty\rangle\rangle^{1/2}\langle\langle |u|^2_{2,\infty}\rangle\rangle^{1/2}.$$

Now using (2.3.10) with $m = 1, 2$, we get that

$$\langle\langle |u|^2_\infty\rangle\rangle \ge C\langle\langle |u|^2_{1,\infty}\rangle\rangle^2\langle\langle |u|^2_{2,\infty}\rangle\rangle^{-1} \ge C.$$

This completes the proof of the proposition. □

COROLLARY 2.3.10. *If $B_0 > 0$ and $B_M < \infty$ for each M, then for $m \in \mathbf{N}_0$ and $\alpha \ge 1$, there exists a constant $C'_m(\alpha) = C'_m(\alpha, \vec{B}) > 0$ such that*

$$(C'_m(\alpha))^{-1}\,\nu^{-m} \le \langle\langle |u|^\alpha_{m,\infty}\rangle\rangle^{\frac{1}{\alpha}} \le C'_m(\alpha)\,\nu^{-m}.$$

PROOF. To prove the right inequality it suffices to average in time in (2.2.12) if $m \ge 1$, and in (2.1.16) if $m = 0$.

The left inequality with $\alpha \ge 2$ follows from the left one in (2.3.10) and Hölder's inequality. To prove the left inequality with $\alpha = 1$ we replace in (2.3.8) the norm $\|u\|_m$ with $|u|_{m,\infty}$ and then argue as when proving Corollary 2.3.7. Finally, the left inequality with $\alpha \in (1, 2)$ follows from that with $\alpha = 1$ and Hölder's inequality. □

Bounds for moments of W^m_p norms.

LEMMA 2.3.11. *If $B_0 > 0$ and $B_M < \infty$ for each M, then for every $m \in \mathbf{N}$ and $1 < p < \infty$, and also for $m = p = 1$, there exists a constant $C_{m,p} = C_{m,p}(\vec{B}) > 0$ such that*

$$(C_{m,p})^{-1}\,\nu^{-2(m-1/p)} \le \langle\langle |u|^2_{m,p}\rangle\rangle \le C_{m,p}\,\nu^{-2(m-1/p)}. \tag{2.3.11}$$

PROOF. The scheme of the proof is similar to that of Theorem 2.3.4.

The right inequality follows immediately from Corollary 2.1.5 if $m = p = 1$ and from Corollary 2.2.7 if $p > 1$.

To get the left inequality we consider two cases:
Case 1: $m = 1$, $1 \le p \le 2$. The case $p = 2$ has already been treated in (2.3.6). For $p < 2$ we have

$$\langle\langle |u_x|_2^2 \rangle\rangle \le \langle\langle |u_x|_p^p |u_x|_\infty^{(2-p)} \rangle\rangle \le \langle\langle |u_x|_p^2 \rangle\rangle^{p/2} \langle\langle |u_x|_\infty^2 \rangle\rangle^{(2-p)/2}.$$

Consequently, by Proposition 2.3.9 and Theorem 2.3.4 we get

$$\langle\langle |u_x|_p^2 \rangle\rangle \ge \langle\langle |u_x|_\infty^2 \rangle\rangle^{-(2-p)/p} \langle\langle |u_x|_2^2 \rangle\rangle^{2/p} \ge C(p)\nu^{2(2-p)/p}\nu^{-2/p} = C(p)\nu^{-2(1-1/p)},$$

which proves the desired estimate.
Case 2: $m = 1, p > 2$. The scheme of the proof is the came as in Case 1. First, using Theorem 11.1.13 and the Cauchy-Schwarz inequality we have

$$\langle\langle |u_x|_2^2 \rangle\rangle \le \langle\langle |u_x|_p^{p/(p-1)} |u_x|_1^{(p-2)/(p-1)} \rangle\rangle \le \langle\langle |u_x|_p^2 \rangle\rangle^{p/2(p-1)} \langle\langle |u_x|_1^2 \rangle\rangle^{(p-2)/2(p-1)}.$$

Then by Corollary 2.1.5 and Theorem 2.3.4 we get

$$\langle\langle |u_x|_p^2 \rangle\rangle \ge \langle\langle |u_x|_1^2 \rangle\rangle^{-(p-2)/p} \langle\langle |u_x|_2^2 \rangle\rangle^{2(p-1)/p} \ge C(p)\nu^{-2(1-1/p)},$$

which proves the required estimate.
Case 3: $m \ge 2$. We proceed as when proving the left inequality (2.3.6). Namely, by the Gagliardo–Nirenberg (1.3.9) and Cauchy-Schwarz inequalities we have

$$\langle\langle \|u\|_1^2 \rangle\rangle = \langle\langle |u_x|_2 \rangle\rangle \le C_{m,p} \langle\langle |u_x|_{m-1,p}^2 \rangle\rangle^\lambda \langle\langle |u_x|_1^2 \rangle\rangle^{1-\lambda}, \qquad \lambda = \frac{p}{2(pm-1)}.$$

From here and Corollary 2.1.5 we deduce that

$$\langle\langle |u|_{m,p}^2 \rangle\rangle \ge C_{m,p} \langle\langle \|u\|_1^2 \rangle\rangle^{\lambda^{-1}} \ge C_{m,p}\nu^{-\lambda^{-1}},$$

which implies the desired estimate since $\lambda^{-1} = 2(m - 1/p)$. □

We skip the derivation of the following result, which is similar to that of Corollary 2.3.7 and Corollary 2.3.10.

COROLLARY 2.3.12. *Assume that $B_0 > 0$ and $B_M < \infty$ for each M. The for any $\alpha \ge 1$ and for every $m \in \mathbf{N}$ and $1 < p < \infty$, or for $m = p = 1$, there exists a constant $C'_{m,p}(\alpha) = C'_{m,p}(\alpha, \vec{B}) > 0$ such that*

$$(C'_{m,p}(\alpha))^{-1}\ \nu^{-(m-1/p)} \le \langle\langle |u|_{m,p}^\alpha \rangle\rangle^{\frac{1}{\alpha}} \le C'_{m,p}(\alpha)\ \nu^{-(m-1/p)}.$$

EXAMPLE 2.3.13. Let us consider the balance of energy (1.4.25) for the 1d vorticity $\zeta = u_x$ and average it in time from T to $T + \sigma$:

$$\begin{aligned} &\frac{1}{\sigma}\mathbb{E}\Big(\tfrac{1}{2}\int \zeta(T+\sigma, x)^2 dx\Big) - \frac{1}{\sigma}\mathbb{E}\Big(\tfrac{1}{2}\int \zeta(T, x)^2 dx\Big) \\ &= -\nu\langle\langle \int \zeta_x^2(t,x)dx \rangle\rangle - \tfrac{1}{2}\langle\langle \int \zeta^3(t,x)dx \rangle\rangle + \tfrac{1}{2}B_1. \end{aligned} \tag{2.3.12}$$

The term $-\frac{1}{2}\langle\langle \int \zeta^3 dx \rangle\rangle$ is the (averaged) vorticity stretching. If $B_M < \infty$ for all M and $B_0 > 0$, then due to the results above in this section the l.h.s. is bounded from above by $C\nu^{-1}$, while $\nu\langle\langle \int \zeta_x^2 dx \rangle\rangle \sim \nu^{-2}$. So we get from (2.3.12) that

$$-\tfrac{1}{2}\langle\langle \int \zeta^3(t,x)dx \rangle\rangle \sim \nu^{-2}.$$

I.e. the vorticity stretching is a large positive quantity.

In view of the estimates (2.3.11), (2.3.6) and (2.3.10), solutions of the Burgers equation (1.1.12) satisfy

$$\langle\langle |u_x|_1^2\rangle\rangle \sim 1, \quad \langle\langle |u_x|_2^2\rangle\rangle \sim \nu^{-1}, \quad \langle\langle |u_x|_\infty^2\rangle\rangle \sim \nu^{-2}. \tag{2.3.13}$$

In the language of the theory of turbulence, it means that for small ν the function $u_x(t,x)$ is highly intermittent. Below in this section we will show that unlike the derivatives u_x, solutions u of (B) are functions of order 1 uniformly in $\nu \in (0,1]$, in the sense that all moments of all their L_p-norms admit lower and upper bounds by ν-independent constants. Cf. Fig. 6.0.2, representing a typical solution of the free Burgers equation at a fixed time. We will return to the intermittent behaviour of derivatives $u_x(t,x)$ below in Section 7.2, where it will be discussed in the context of 1d turbulence.

Bounds for moments of L_p norms.

Here we obtain ν–independent upper and lower bounds for moments of the L_p norms of solutions. We start with an auxiliary lemma.

LEMMA 2.3.14. *Let $f \in C^1(S^1)$ be a zero-average function. Then we have the estimate*

$$|f|_\infty^3 \le 3|f_x^+|_\infty |f|_2^2.$$

PROOF. If $f \equiv 0$ the result is obvious. Otherwise let us define three positive numbers:

$$F{=}|f|_\infty; \quad G{=}|f_x^+|_\infty; \quad l{=}F/G.$$

By Lemma 2.1.1, $l \le 1$. Now let x_f be any point on S^1 where $|f|$ reaches its maximum. Without loss of generality we may assume that $f(x_f) < 0$ since if $f(x_f) > 0$, then we may replace $f(x)$ by the function $-f(-x)$, without modifying F, G and the L_2-norm. Denote by I the interval $[x_f, x_f + l]$. For $x \in I$, we have

$$f(x) \le -F + (x - x_f)G \le 0.$$

Consequently,

$$|f|_2^2 \ge \int_I f^2 \ge \int_x^{x+l} (F - yG)^2\, dy \ge lF^2 - l^2FG + l^3G^2/3 = F^3/3G.$$

This implies the assertion if $f \not\equiv 0$. □

THEOREM 2.3.15. *Assume that $B_0 > 0$ and $B_4 < \infty$. Then there exists a constant $C(B_0, B_4) > 0$ such that*

$$C^{-1} \le \langle\langle |u|_2^2\rangle\rangle \le C.$$

PROOF. The upper estimate follows immediately from Corollary 2.1.5.

To prove the lower bound, we observe that by Lemma 2.3.14 and Hölder's inequality it follows that

$$\langle\langle |u|_\infty^2\rangle\rangle \le 3^{2/3}\langle\langle |u|_2^{4/3}|u_x^+|_\infty^{2/3}\rangle\rangle \le 3^{2/3}\langle\langle |u|_2^2\rangle\rangle^{2/3}\langle\langle |u_x^+|_\infty^2\rangle\rangle^{1/3}.$$

Now (2.3.10) and Corollary 2.1.5 imply that

$$\langle\langle |u|_2^2\rangle\rangle \ge \frac{1}{3}\langle\langle |u|_\infty^2\rangle\rangle^{3/2}\langle\langle |u_x^+|_\infty^2\rangle\rangle^{-1/2} \ge C(B_0, B_4).$$

□

PROBLEM 2.3.16. Prove that if $B_0 > 0$ and $B_4 < \infty$, then for each $\mu \leq 0$ there exists $C_\mu \geq 1$ such that every solution $u = u(t; u_0)$ with $u_0 \in H^1$ meets the estimates

$$C_\mu^{-1} \leq \langle\langle \|u\|_\mu^2 \rangle\rangle \leq C_\mu \qquad \forall\, 0 < \nu \leq 1, \ \ \forall\, \mu \leq 0. \tag{2.3.14}$$

Hint: For the lower bound use the theorem above jointly with (2.1.18) and (2.1.16) (cf. Section 3.4, where estimates similar to (2.3.14) are established for second moments of the stationary measure for equation (B)). The upper bound follows from Theorem 2.3.15 .

COROLLARY 2.3.17. *Assume that $B_0 > 0$ and $B_4 < \infty$. Then for $\alpha \geq 1$ there exists a constant $C_\alpha(B_0, B_4) \geq 1$ such that*

$$C_\alpha^{-1} \leq \langle\langle |u|_p^\alpha \rangle\rangle \leq C_\alpha \quad \textit{for all} \quad p \in [1, \infty]. \tag{2.3.15}$$

PROOF. The upper estimate follows by averaging relation (2.1.16) with respect to $\langle\langle\cdot\rangle\rangle$. To prove the lower one it suffices to establish it for $p = \alpha = 1$. We have:

$$\langle\langle |u|_2^2 \rangle\rangle \leq \langle\langle |u|_1 |u|_\infty \rangle\rangle \leq \langle\langle |u|_1^2 \rangle\rangle^{1/2} \langle\langle |u|_\infty^2 \rangle\rangle^{1/2} \leq C_3 \langle\langle |u|_1^2 \rangle\rangle^{1/2},$$

where we used the already established upper bound in (2.3.15) with $p = \infty, \alpha = 2$. Now the required estimate follows from Theorem 2.3.15. $\square$

CHAPTER 3

Mixing in the stochastic Burgers equation

In this chapter we study the long-time behaviour of solutions for equation (B). In Sections 3.1.1 and 3.2 we do not use the results of Chapter 2, and there we only assume that

$$B_{1_*} < \infty$$

for some real number $1_* > 1$. In Sections 3.1.1, 3.2 and 3.3 we do not control the dependence of the involved constants on ν, while in Section 3.4 we examine how the objects we study depend on ν. Accordingly there the constants C are independent from ν.

3.1. Stationary measure and the Bogolyubov-Krylov method.

DEFINITION 3.1.1. A measure $\mu \in \mathcal{P}(H^1)$ is said to be *stationary for eq.* (1.1.12) if $S_t^* \mu = \mu$ for all $t \geq 0$. In other words, if $\mathcal{D}(u_0) = \mu$, then $\mathcal{D}(u(t; u_0)) \equiv \mu$.

A solution $u(t)$ of (B) such that $\mathcal{D}(u(t)) \equiv \mu$ is said to be a *stationary solution.*

THEOREM 3.1.2. *There exists a stationary measure $\mu \in \mathcal{P}(H^1)$ for eq.* (1.1.12). *Moreover, $\langle \mu, \|u\|_2^2 \rangle < \infty$, so the measure is supported by the space H^2.*

PROOF. Let us denote $u(t) := u(t; 0)$, set $\mu(t) = \mathcal{D}(u(t)) \in \mathcal{P}(H^1)$ and note that $S_\theta^*(\mu(t)) = \mathcal{D}(u(t+\theta)) = \mu(t+\theta)$ for $t, \theta \geq 0$. Consider *the Bogolyubov-Krylov ansatz.* Namely, for $T \in \mathbf{N}$, define the measure

$$m_T = \frac{1}{T} \int_0^T \mu(s) ds \in \mathcal{P}(H^1).$$

For $R > 0$ consider the complement $\big(\overline{B_2^R}\big)^c$ to the closed H^2-ball $\overline{B_2^R}$ in H^1. Then

$$\begin{aligned} m_T\Big(\big(\overline{B_2^R}\big)^c\Big) &= \frac{1}{T} \int_0^T \mu(s)\Big(\big(\overline{B_2^R}\big)^c\Big) ds = \frac{1}{T} \int_0^T \mathbb{P}\big(\|u(s)\|_2 > R\big) ds \\ &\leq \frac{1}{R^2 T} \int_0^T \mathbb{E}\|u(s)\|_2^2 \, ds \leq \frac{C}{\nu R^2}, \end{aligned}$$

where we used (1.4.11) with $m = 1$ and Markov's inequality. Thence,

$$m_T\big(\overline{B_2^R}\big) \geq 1 - \frac{C}{\nu R^2}, \qquad \forall R > 0.$$

Since by the Rellich–Kondrachov theorem [**59**] the closed balls $\overline{B_2^R}$ are compact sets in H^1, then the family of measures $M = \{m_T\}_{T \in \mathbf{N}}$ is tight in $\mathcal{P}(H^1)$. So M is weakly relatively compact in $\mathcal{P}(H^1)$ by Theorem 1.7.10. Consequently,

$$m_{j_k} \rightharpoonup \mu \in \mathcal{P}(H^1) \tag{3.1.1}$$

for a suitable sequence $j_k \to \infty$.

Now we show that μ is a stationary measure. Since S_t^* is linear and $S_t^*(\mu(s)) = \mu(s+t)$ for $s, t \geq 0$, then for $\theta > 0$ we get:

$$S_\theta^*(m_{j_k}) = \frac{1}{j_k}\int_1^{1+j_k} S_\theta^*(\mu(t))dt = \frac{1}{j_k}\int_1^{1+j_k}\mu(t+\theta)dt = \frac{1}{j_k}\int_{1+\theta}^{1+j_k+\theta}\mu(t)dt$$
$$= \underbrace{-\frac{1}{j_k}\int_1^{1+\theta}\mu(t)dt}_{=:I} + \underbrace{\frac{1}{j_k}\int_1^{1+j_k}\mu(t)dt}_{=m_{j_k}} + \underbrace{\frac{1}{j_k}\int_{1+j_k}^{1+j_k+\theta}\mu(t)dt.}_{=:K}$$

For any $f \in C_b(H^1)$ the terms $|\langle K, f\rangle|$ and $|\langle I, f\rangle|$ are bounded from above by $\frac{\theta}{j_k}|f|_\infty$, so they converge to zero when $k \to \infty$. Since $m_{j_k} \rightharpoonup \mu$ and the operator S_θ^* is weakly continuous (see Theorem 1.7.11), then passing to the limit in the equality above, we get $S_\theta^*\mu = \mu$.

To prove the second assertion of the theorem we note that by (1.4.11) with $m = 1$, $\langle m_T, \|u\|_2^2\rangle \leq C$ for any $T \in \mathbf{N}$. So $\langle \mu, \|u\|_2^2\rangle \leq C$ in view of convergence (3.1.1) and Corollary 11.1.5. □

The stationary measure obtained in Theorem 3.1.2 is smoother by one then the force η^ω in eq. (B):

PROPOSITION 3.1.3. *Assume that $B_m < \infty$ for some integer $m \geq 2$. Then the stationary measure μ constructed in Theorem 3.1.2 is supported by the space H^{m+1} and $\langle \mu, \|u\|_{m+1}^2\rangle \leq C_{m+1} < \infty$.*

PROOF. Using Theorem 1.4.4 and repeating the argument used to prove Theorem 3.1.2 with the space H^1 replaced by H^m, we get that the the family of measures $M = \{m_T\}$ is tight in $\mathcal{P}(H^m)$. So the sequence $\{j_k\}$ as in (3.1.1) has a subsequence $\{j_k'\}$ such that the measures $m_{j_k'}$ converge weakly in $\mathcal{P}(H^m)$ to some measure μ_m, and $\langle \mu_m, \|u\|_{m+1}^2\rangle \leq C_{m+1} < \infty$. The limit μ_m must coincide with μ, which implies the assertions. □

In view of (1.2.15) we get

COROLLARY 3.1.4. *If $B_m < \infty$ for all $m \geq 1$, then $\mu(C^\infty) = 1$.*

3.2. Recurrence lemmas and L_1-nonexpansion.

LEMMA 3.2.1. *Assume that $B_{m+1} < \infty$ for some $m \geq 0$. Then for every $\varepsilon, T > 0$ the probability $\gamma_{\varepsilon,T}^m := \mathbf{P}\{\|\xi\|_{X_T^m} < \varepsilon\}$ satisfies*

$$\gamma_{\varepsilon,T}^m \geq f_m(\varepsilon/\sqrt{T}), \tag{3.2.1}$$

where

$$f_m(x) = \frac{1}{2}\exp\Big(-\frac{B_m\pi^2}{2x^2}\Big\lceil\frac{2\sqrt{B_{m+1}}}{\pi x}\Big\rceil\Big) \geq \frac{1}{2}e^{-\kappa_m(x^{-3}+x^{-2})} =: f_m^0(x)$$

(see (0.0.16)). The constant $\kappa_m > 0$ depends on B_{m+1}.

PROOF. We may limit ourselves to the case $T = 1$ since the r.v.'s $\xi(t)$ and $\sqrt{T}\xi(t/T)$ have the same law (see Problem 11.5.1), and therefore $\gamma_{\varepsilon,T}^m \equiv \gamma_{\varepsilon/\sqrt{T},1}^m$.

For an $N \geq 1$ let us define the $2N$-dimensional process ξ^N as in (1.2.4) and set ${}^N\xi = \xi - \xi^N$. Since ξ^N and ${}^N\xi$ are independent, then

$$\gamma_{\varepsilon,1}^m \geq \mathbf{P}\{\|\xi^N\|_{X_1^m} < \varepsilon/\sqrt{2}\}\cdot\mathbf{P}\{\|{}^N\xi\|_{X_1^m} < \varepsilon/\sqrt{2}\}.$$

i) Process ${}^N\xi$. We have

$$
\begin{aligned}
{}^NP := \mathbf{P}\{\|{}^N\xi\|_{X_1^m} < \varepsilon/\sqrt{2}\} &\geq 1 - 2\varepsilon^{-2}\mathbb{E}\|{}^N\xi\|^2_{X_1^m} \\
&\geq 1 - 8\varepsilon^{-2}\mathbb{E}\|{}^N\xi(1)\|^2_m \geq 1 - 8\varepsilon^{-2}(2\pi N)^{-2}B_{m+1},
\end{aligned}
$$

where to get the first inequality we use Markov's one, while to get the second we evoke Doob's inequality (11.5.2) with $p = 2$, applied to

$$
|\mathbf{w}(t)|^2 = \sum_{|s|=N+1}^{M} |2\pi s|^{2m} b_s^2 w_s(t)^2, M = \infty
$$

(the result with $M = \infty$ follows from that with finite M by passing to the limit). The third inequality follows from (1.2.18) with $m := m + 1$ and the definitions of B_m, B_{m+1}. Choosing $N = N_\varepsilon = \left\lceil \frac{2\sqrt{B_{m+1}}}{\pi\varepsilon} \right\rceil$ we achieve that ${}^NP \geq \frac{1}{2}$.

ii) Process ξ^N. Since for $\xi = \sum b_s\xi_s e_s$ relations $|\xi_s| < \frac{1}{\sqrt{2}}\varepsilon B_m^{-1/2}$ for $|s| \leq N$ imply that

$$
\|\xi^N\|_m^2 \leq \frac{\varepsilon^2}{2} B_m^{-1} \sum_{|s|\leq N} b_s^2 (2\pi s)^{2m} < \frac{\varepsilon^2}{2},
$$

then

$$
P^N := \mathbf{P}\{\|\xi^N\|_{X_1^m} < \tfrac{1}{\sqrt{2}}\varepsilon\} \geq \prod_{|s|\leq N} \mathbf{P}\{\sup_{0\leq t\leq 1} |\beta_s(t)| < \tfrac{1}{\sqrt{2}}\varepsilon B_m^{-1/2}\} \geq \rho^{2N},
$$

where

$$
\rho = \rho(\varepsilon') = \mathbf{P}\{\sup_{0\leq t\leq 1} |\beta_s(t)| < \varepsilon'\}, \quad \varepsilon' = \tfrac{1}{\sqrt{2}}\varepsilon B_m^{-1/2}.
$$

But it is well known in the probability theory that

$$
\rho(\varepsilon') \geq e^{-\pi^2/8\varepsilon'^2} \tag{3.2.2}
$$

(e.g. see [**23**, Section 4.8], or Problem 3.2.3 below). Thence,

$$
P^N > \exp\Big(-2N_\varepsilon \frac{B_m\pi^2}{2\varepsilon^2}\Big) = \exp\Big(-\frac{B_m\pi^2}{2\varepsilon^2}\Big\lceil \frac{2\sqrt{B_{m+1}}}{\pi\varepsilon} \Big\rceil\Big).
$$

iii) We have seen that

$$
\gamma^m_{\varepsilon,1} \geq {}^NP \cdot P_N \geq \frac{1}{2}\exp\Big(-\frac{B_m\pi^2}{2\varepsilon^2}\Big\lceil \frac{2\sqrt{B_{m+1}}}{\pi\varepsilon} \Big\rceil\Big) \geq \frac{1}{2}e^{-\kappa_m(\varepsilon^{-3}+\varepsilon^{-2})}.
$$

The lemma is proved. □

REMARK 3.2.2. It is easy to see from the prof that if only $B_m < \infty$, then still $\gamma^m_{\varepsilon,T} \geq g(\varepsilon/\sqrt{T})$ for some continuous increasing function $g(x)$, positive for $x > 0$ and vanishing for $x = 0$, depending on the sequence $\{b_s\}$. If $B_{m_*} < \infty$ for some $m_* > m$, then the function $g(x)$ may be chosen to depend only on m_* and B_{m_*}.

PROBLEM* 3.2.3. In the notation of the proof of Lemma 3.2.1 derive estimate (3.2.2), using that $\rho = u(0,0)$, where $u(t,x)$, $(t,x) \in [0,1] \times (-\varepsilon', \varepsilon') =: \Pi$ satisfies the diffusion equation with absorption

$$
u_t + \tfrac{1}{2}u_{xx} = 0 \quad \text{for } (t,x) \in \Pi, \quad u(1,x) \equiv 1, \; u(t,\pm\varepsilon') = 0 \;\; \forall t < 1. \tag{3.2.3}
$$

See [**61**, Section X.5], [**84**, Section III.6] and [**69**, Section VIII.5].

LEMMA 3.2.4. *For all $\varepsilon > 0$ and $R > 0$ there exists $T_\varepsilon = T_\varepsilon(R)$ and $\delta_\varepsilon = \delta_\varepsilon(R, B_{1_*}) > 0$ such that if $u_0 \in \bar{B}_1^R = \{\|u\|_1 \le R\}$, then*

$$\mathbb{P}(\|u(T_\varepsilon; u_0)\| < \varepsilon) \ge \delta_\varepsilon. \tag{3.2.4}$$

PROOF. Consider a solution of the free equation $u^e(t) = u(t; u_0, 0)$. Taking the scalar product in L_2 of u^e with the equation and integrating by parts in space, we get:

$$\frac{1}{2}\frac{d}{dt}\|u^e(t)\|^2 = -\nu\|u^e_x(t)\|^2 \le -\nu\|u^e(t)\|^2.$$

By Gronwall's lemma, we obtain that $\|u^e(t)\|^2 \le e^{-2\nu t}\|u_0\|^2$. Consequently, choosing $T_\varepsilon = \nu^{-1}\ln(2R/\varepsilon)$ we get $\|u^e(T_\varepsilon)\| \le \frac{\varepsilon}{2}$. By Theorem 1.3.11 the mapping $\mathcal{M}_{T_\varepsilon} : H^1 \times X^1_{T_\varepsilon} \to H^1$ is locally Lipschitz. Therefore there exists $\gamma_\varepsilon = \gamma_\varepsilon(R) > 0$ such that if $\|\xi\|_{X^1_T} \le \gamma_\varepsilon$, then $\|u^e(T_\varepsilon) - u(T_\varepsilon; u_0, \xi)\|_1 \le \frac{\varepsilon}{2}$. Thence, if $\|\xi\|_{X^1_T} \le \gamma_\varepsilon$, then $\|u(T_\varepsilon; u_0, \xi)\| < \varepsilon$. By Lemma 3.2.1 and Remark 3.2.2 with $m = 1$, $\mathbb{P}(\|\xi\|_{X^1_T} \le \gamma_\varepsilon) =: \delta_\varepsilon > 0$. This yields (3.2.4). □

LEMMA 3.2.5. *(Recurrence.) Assume that $B_4 < \infty$. Then for $u_0 \in H^1$ and $\varepsilon > 0$, we have:*

$$\pi(T, \varepsilon) := \mathbb{P}(\inf_{t \in [0,T]} \|u(t; u_0)\| \ge \varepsilon) \underset{T \to +\infty}{\longrightarrow} 0, \tag{3.2.5}$$

the convergence being uniform in u_0.

PROOF. By Markov's inequality and Corollary 2.1.5, there exists a constant R such that for all $u' \in H^1$ we have $\mathbb{P}(\|u(1; u')\| \ge R) \le \frac{1}{2}$. Using this fact, the Kolmogorov-Chapman equality (1.6.8) applied to the characteristic function of the ball $B_0^\varepsilon = \{\|u\| < \varepsilon\}$, and (3.2.4) with $T_\varepsilon = T_\varepsilon(R)$, we get:

$$\mathbb{P}(\|u(1 + T_\varepsilon; u')\| < \varepsilon) \ge \int_{B_0^R} \mathbb{P}(\|u(T_\varepsilon; v)\| < \varepsilon)\Sigma_1(u')(dv) \ge \frac{1}{2}\delta_\varepsilon, \quad \forall u' \in H^1.$$

Consequently,

$$\mathbb{P}(\|u(1 + T_\varepsilon; u')\| \ge \varepsilon) \le 1 - \frac{1}{2}\delta_\varepsilon, \quad \forall u' \in H^1. \tag{3.2.6}$$

Setting $T_j = j(1 + T_\varepsilon)$, denoting by Q_j the event $\{\|u(T_j)\| \ge \varepsilon\}$ and by Q^N the event $\bigcap_{i=1}^N Q_i$, we have $Q^N = \{\omega \in Q^{N-1} : \|u(T_N)\| \ge \varepsilon\}$. Using the Kolmogorov-Chapman equality in the form (1.6.10) jointly with (3.2.6), we get:

$$\begin{aligned} \mathbb{P}(Q^N) &= \mathbb{E}\big(\chi_{\{u(\tau;u_0)_{[0,T_{N-1}]} \in Q^{N-1}\}} \mathbb{P}\{\|u(1 + T_\varepsilon; u(T_{N-1}; u_0))\| \ge \varepsilon\}\big) \\ &\le (1 - \frac{1}{2}\delta_\varepsilon)\mathbb{E}\big(\chi_{\{u(\tau;u_0)_{[0,T_{N-1}]} \in Q^{N-1}\}}\big) = (1 - \frac{1}{2}\delta_\varepsilon)\mathbb{P}(Q^{N-1}). \end{aligned}$$

By induction, it follows that $\mathbb{P}(Q^N) \le (1 - \frac{1}{2}\delta_\varepsilon)^N$. Finally, since $\pi(T, \varepsilon) \le \mathbb{P}(Q^N)$ for $T \ge T^N$, then we get (3.2.5). □

LEMMA 3.2.6. *(L_1-nonexpansion.) For $j = 1, 2$, fix $u_j(0) \in H^1$. Then for each $\xi \in X_T^{1_*}$, where 1_* is any number larger than one, solutions $u_j(t) = u(t; u_j(0), \xi) \in H^1$ satisfy:*

$$|u_1(T) - u_2(T)|_1 \le |u_1(0) - u_2(0)|_1. \tag{3.2.7}$$

PROOF. Since $u_j(T)$ $(j = 1,2)$ continuously depends on $u_j(0) \in H^1$ and $\xi \in X_T^{1*}$, then firstly approximating $u_j(0)$ and ξ by $\Pi_N u_j(0)$ and $\Pi_N \xi$, $N = 1,2,\dots$, and next approximating $\Pi_N \xi$ by functions $\sum_{|s|\le N} g_s(t)e_s(x)$ whose Fourier coefficients $g_s(t)$ are C^∞–smooth, we reduce the problem to the case when $u_1(0), u_2(0)$ and ξ are C^∞–smooth functions. Then by Theorem 1.8.6, the solutions u_1 and u_2 belong to the space X_T^m, for any m. From the equation we get that the same is true for u_{1t} and u_{2t}.

Now let us set $w = u_1 - u_2$. Due to what was said, w is a classical solution of the problem

$$\left\{ \begin{array}{c} w_t + \frac{1}{2}(w(u_1+u_2))_x - \nu w_{xx} = 0, \\ w(0) = u_1(0) - u_2(0) =: w_0. \end{array} \right. \tag{3.2.8}$$

Consider the dual Cauchy problem:

$$\left\{ \begin{array}{c} \phi_t + \frac{1}{2}(u_1+u_2)\phi_x + \nu\phi_{xx} = 0, \\ \phi(T,x) = \phi_T(x), \end{array} \right. \tag{3.2.9}$$

where $t \in [0,T]$, $\phi_T(x) \in C^\infty$ and $|\phi_T|_\infty = 1$. The theory of parabolic equations (e.g. see [**59, 112**]) implies the existence of a unique classical solution ϕ to the problem (3.2.9),[1] and that this solution satisfies the maximum principle: $|\phi(t)|_\infty \le 1$ for all $t \in [0,T]$.

Considering the relation

$$0 = \int_0^T \int_{S^1} \phi\big(w_t + \frac{1}{2}(w(u_1+u_2))_x - \nu w_{xx}\big)\, dx\, dt,$$

where w solves (3.2.8) and ϕ solves (3.2.9), and integrating by parts in dx and dt, we get that $\langle w(T), \phi_T\rangle = \langle w_0, \phi(0)\rangle$. Since $|\phi(0)|_\infty \le 1$, then

$$|\langle w(T), \phi_T\rangle| \le |w_0|_1, \tag{3.2.10}$$

for any ϕ_T as above.

Let $\chi(t)$ be an odd smooth function $\mathbb{R} \to \mathbb{R}$, equal to one for $t \ge 1$ and monotonically increasing for $0 \le t \le 1$, so $\chi(0) = 0$ and $0 \le \chi \le 1$ for $0 \le t \le 1$. For $\varepsilon > 0$ we set $\chi_\varepsilon(t) = \chi(t/\varepsilon)$. Then $\chi_\varepsilon(r) \to \mathrm{sgn}(r)$ as $\varepsilon \to 0$, for each r. For $\varepsilon > 0$ and $T \ge 0$, the function $\phi_T^\varepsilon = \chi_\varepsilon(w(T,\cdot))$ is smooth, $|\phi_T^\varepsilon|_\infty = 1$ and $\phi_T^\varepsilon(x) \to \mathrm{sgn}(w(T,x))$ as $\varepsilon \to 0$, for all x. Thus, using the dominated convergence theorem and (3.2.10) with $\phi_T = \phi_T^\varepsilon$, we get:

$$|w(T)|_1 = \Big|\int w(T,x)\,\mathrm{sgn}(w(T,x))dx\Big| = \lim_{\varepsilon\to 0}\Big|\int w(T,x)\phi_T^\varepsilon(x)dx\Big| \le |w_0|_1.$$

□

COROLLARY 3.2.7. *For any $u_0 \in H^1$, $\xi \in \dot{X}^4$ and $t \ge 0$,*

$$|u(t; u_0, \xi)|_1 \le |u_0|_1 + 4\|\xi\|_{X_t^4} + \|\xi\|_{X_t^4}^{1/2}. \tag{3.2.11}$$

PROOF. By the lemma above and Theorem 2.1.7 with $T = t$,

$$|u(t;u_0,\xi)|_1 - |u(t;0,\xi)|_1 \le |u_0|_1, \quad |u(t;0,\xi)|_1 \le 4\|\xi\|_{X_t^4} + \|\xi\|_{X_t^4}^{1/2},$$

so the result follows. □

[1] Alternatively the unique solvability of the problem easily follows by arguing as when proving Proposition 1.3.14 (which deals with an equation of the same type as (3.2.8)).

3.3. Uniqueness of the stationary measure and the mixing property.

In this section we fix any $m \in \mathbf{N}$ and assume that

$$B_4,\ B_{m+1} < \infty.$$

Our goal is to prove that the stationary measure μ for the stochastic Burgers equation is unique and mixing: every solution of the equation converges to μ in distribution. The proof uses the Döblin method of two equations and is non-technical because of the L_1–nonexpansion property (3.2.7). We note that for more complicated stochastic partial differential equations it is much more difficult to prove the mixing, e.g. see in [**103**, Section 3].

Consider two copies of the Burgers equation (B) with the same forcing and different initial conditions:

$$\begin{cases} (u_1)_t + u_1(u_1)_x - \nu(u_1)_{xx} = \eta, & u_1(0) = u_{10}, \\ (u_2)_t + u_2(u_2)_x - \nu(u_2)_{xx} = \eta, & u_2(0) = u_{20}, \end{cases} \tag{3.3.1}$$

where u_{10} and u_{20} are random variables in H^1, independent from η. Let $u_1(t) = u_1(t; u_{10})$ and $u_2(t) = u_2(t; u_{20})$. Then

$$U(t) = (u_1(t), u_2(t))$$

is a solution of the system (3.3.1) for the initial condition

$$U(0) = (u_{10}, u_{20}) =: U_0.$$

In the same way as in the proof of Lemma 3.2.5, we get

LEMMA 3.3.1. *For any* $\varepsilon > 0$ *and* $U_0 \in H^1 \times H^1$, $\mathbb{P}\left(\inf_{t\in[0,T]} \|U(t)\| \geq \varepsilon\right) \underset{T\to\infty}{\to} 0$, *uniformly in* U_0.

We recall that the Lipschitz-dual distance $\|\mu_1 - \mu_2\|^*_{L(H^m)}$ between measures $\mu_1, \mu_2 \in \mathcal{P}(H^m)$ was defined in (1.7.4).

LEMMA 3.3.2. *Let* $r \geq 1$. *There exists* $C(\nu) > 0$ *such that if* $u_{10}, u_{20} \in H^m$ *and* $|u_{10}|_1, |u_{20}|_1 \leq \frac{1}{r}$, *then for any* $t \geq 1$, *we have:*

$$\|\Sigma_t(u_{10}) - \Sigma_t(u_{20})\|^*_{L(H^m)} \leq C r^{-\frac{2}{2m+3}}.$$

PROOF. By the Gagliardo–Nirenberg inequality (1.3.9) we have

$$\|u_1(t) - u_2(t)\|_m \leq C' |u_1(t) - u_2(t)|_1^{\theta} \|u_1(t) - u_2(t)\|_{m+1}^{1-\theta}, \quad \theta = \tfrac{2}{2m+3}, \quad C' > 0.$$

Now consider $f \in C_b(H^m)$ such that $\|f\|_{L(H^m)} \leq 1$. Then

$$\begin{aligned} |\langle \Sigma_t(u_{10}), f\rangle - \langle \Sigma_t(u_{20}), f\rangle| &= |\mathbb{E}(f(u_1(t)) - f(u_2(t)))| \leq \mathbb{E}[\|u_1(t) - u_2(t)\|_m] \\ &\leq C'\ \mathbb{E}[|u_1(t) - u_2(t)|_1^{\theta} \|u_1(t) - u_2(t)\|_{m+1}^{1-\theta}] \\ &\leq C'\ \mathbb{E}[\ |u_1(t) - u_2(t)|_1^{\theta}\ (\|u_1(t)\|_{m+1}^{1-\theta} + \|u_2(t)\|_{m+1}^{1-\theta})]. \end{aligned}$$

Since $|u_1(0) - u_2(0)|_1 \leq \frac{2}{r}$, we have $|u_1^{\omega}(t) - u_2^{\omega}(t)|_1 \leq \frac{2}{r}$ for every ω (see Lemma 3.2.6). Consequently, the r.h.s. in the estimate above is bounded by

$$C' \left(\frac{2}{r}\right)^{\theta} (\mathbb{E}[\|u_1(t)\|_{m+1}^{1-\theta}] + \mathbb{E}[\|u_2(t)\|_{m+1}^{1-\theta}]).$$

By Theorem 2.2.1 and Hölder's inequality, for $j = 1, 2$,

$$\mathbb{E}\|u_j(t)\|_{m+1}^{1-\theta} \leq C_2 \nu^{-(m+1/2)(1-\theta)}.$$

Consequently,

$$|\langle \Sigma_t(u_{10}), f\rangle - \langle \Sigma_t(u_{20}), f\rangle| \le C(\nu) r^{-\theta}$$

if $f \in \bar{B}^1_{L(H^m)}$, which proves the lemma. ☐

PROPOSITION 3.3.3. *For any $u_{10}, u_{20} \in H^m$,*

$$d_t(u_{10}, u_{20}) := \|\Sigma_t(u_{10}) - \Sigma_t(u_{20})\|^*_{L(H^m)} \le d_t^0, \quad \forall t \ge 0, \tag{3.3.2}$$

where $t \mapsto d_t^0$ is a positive function, independent from u_{10} and u_{20}, and converging to zero as $t \to +\infty$.

PROOF. *Step 1:* For any $f \in \bar{B}^1_{L(H^m)}$ we have

$$\langle \Sigma_t(u_{10}), f\rangle - \langle \Sigma_t(u_{20}), f\rangle = \mathbb{E}[f(u_1(t)) - f(u_2(t))] = \mathbb{E}[F(U(t))] =: g_f(t; U_0);$$
$$U_0 = (u_{10}, u_{20}),$$

where $u_j(t) = u(t; u_j)$ and for $U = (u_1, u_2)$ we set $F(U) = f(u_1) - f(u_2)$. Since the supremum over $f \in \bar{B}^1_{L(H^m)}$ in the l.h.s. above equals $d_t(u_{10}, u_{20})$, then

$$g_f(t; U_0) \le \sup_{\|f\|_{L(H^m)} \le 1} g_f(t; U_0) = d_t(u_{10}, u_{20}), \quad t \ge 0. \tag{3.3.3}$$

Step 2: For $r \ge 1$ denote by O_r the closed set $O_r = \{u \in H^m : |u|_1 \le \frac{1}{r}\}$ and consider the *hitting time* for a trajectory $U^\omega(t) = (u_1(t), u_2(t)) \in H^m \times H^m$ and the set $O_r \times O_r$ (see Appendix 11.7):

$$\tau^\omega_{r,t} = \inf\{s \in [0, t] : U^\omega(s) \in O_r \times O_r\},$$

where $\tau^\omega_{r,t} = t$ if the set under the inf-sign is empty.

Observe that:

i) Since $|u_j^0|_1 \le \|u_j^0\|$ for $j = 1, 2$, Lemma 3.3.1 implies that the probability $\mathbb{P}(\tau_{r,t} \ge t - 1)$ tends to 0 uniformly in U^0, as t tends to $+\infty$.

ii) If $\tau_{r,t} < t$, then $U(\tau_{r,t}) \in O_r \times O_r$.

iii) For $U^0 \in O_r \times O_r$ and $t \ge 1$ we get from (3.3.3) and Lemma 3.3.2 that $g_f(t; U^0) \le C r^{-\frac{2}{2m+3}}$.

Step 3: By the strong Markov property (11.7.1), $\mathbb{E}[F(U(t))] = \mathbb{E}[g_f(t - \tau_{r,t}; U(\tau_{r,t}))]$. Consequently, for any $r \ge 1$,

$$\mathbb{E}[F(U(t))] = \mathbb{E}[\chi_{\{\tau_{r,t} \le t-1\}} g_f(t - \tau_{r,t}; U(\tau_{r,t}))] + \mathbb{E}[\chi_{\{\tau_{r,t} > t-1\}} g_f(t - \tau_{r,t}; U(\tau_{r,t}))], \tag{3.3.4}$$

where $\chi_\cdot$ is the characteristic function. By ii) and iii), if $\tau_{r,t} \le t - 1$, then $U(\tau_{r,t}) \in O_r \times O_r$ and $g_f(t - \tau_{r,t}; U(\tau_{r,t})) \le C r^{-\frac{2}{2m+3}}$. Since $|F(U(t))| \le 2$, then (3.3.4) implies that:

$$|\mathbb{E}[F(U(t))]| \le C r^{-\frac{2}{2m+3}} + 2\mathbb{P}\{\tau_{r,t} > t - 1\}.$$

Let $\varepsilon > 0$. We choose $r = r_\varepsilon$ in such a way that $C r_\varepsilon^{-\frac{2}{2m+3}} \le \frac{1}{2}\varepsilon$. Next we use i) to find $t(\varepsilon)$, satisfying

$$\mathbb{P}\{\tau_{r_\varepsilon,t} > t - 1\} \le \frac{1}{4}\varepsilon \quad \text{if } t \ge t(\varepsilon).$$

Then $\mathbb{E}[F(U(t))] \le \varepsilon$ for all $t \ge t(\varepsilon)$, uniformly in $f \in \bar{B}^1_{L(H^m)}$. We have seen that $d_t(u_{10}, u_{20}) \le \varepsilon$ for $t \ge t(\varepsilon)$, uniformly in (u_{10}, u_{20}). This proves (3.3.2). ☐

THEOREM 3.3.4. *The stationary measure* μ *given by Theorem* 3.1.2 *is unique. Moreover, for any* $\lambda \in \mathcal{P}(H^1)$ *we have:*

$$\lim_{t\to+\infty} \|S_t^*\lambda - \mu\|_{L(H^m)}^* = 0, \tag{3.3.5}$$

uniformly in λ. *In particular, for any* $u_0 \in H^1$ *we have* $\lim_{t\to+\infty} \|\Sigma_t(u_0) - \mu\|_{L(H^m)}^* = 0$, *uniformly in* u_0.

PROOF. Let μ be the stationary measure built in Theorem 3.1.2 and μ_1 be another stationary measure. If (3.3.5) holds, then

$$\|\mu_1 - \mu\|_{L(H^m)}^* = \|S_t^*\mu_1 - \mu\|_{L(H^m)}^* \underset{t\to+\infty}{\longrightarrow} 0,$$

and consequently $\mu_1 = \mu$.

Now we prove (3.3.5). Assume first that $\lambda \in \mathcal{P}(H^m)$, consider any $f \in \bar{B}^1_{L(H^m)}$ and denote

$$X_t^f := |\langle f, S_t^*\lambda\rangle - \langle f, \mu\rangle| = |\langle f, S_t^*\lambda\rangle - \langle f, S_t^*\mu\rangle|.$$

We have:

$$\langle f, S_t^*\lambda\rangle = \langle S_t f, \lambda\rangle = \int_{H^m} S_t f(u_1)\lambda(du_1) = \int_{H^m}\int_{H^m} S_t f(u_1)\lambda(du_1)\mu(du_2).$$

Similarly,

$$\langle f, S_t^*\mu\rangle = \int_{H^m}\int_{H^m} S_t f(u_2)\mu(du_2)\lambda(du_1).$$

Therefore

$$X_t^f \le \int_{H^m}\int_{H^m} |S_t f(u_1) - S_t f(u_2)|\lambda(du_1)\mu(du_2).$$

By Proposition 3.3.3, the integrand $|S_t f(u_1) - S_t f(u_2)| = |\langle \Sigma_t(u_1) - \Sigma_t(u_2), f\rangle|$ is bounded by d_t^0. Consequently $X_t^f \le d_t^0$ for any f as above, so $\|S_t^*\lambda - \mu\|_{L(H^m)}^* \le d_t^0$. Since $d_t^0 \to 0$ as $t \to \infty$, we recover (3.3.5).

Now let $\lambda \in \mathcal{P}(H^1)$. Since $S_t^*\lambda = S_{t-1}^*(S_1^*\lambda)$ and $S_1^*\lambda \in \mathcal{P}(H^m)$ by Corollary 2.2.2, then (3.3.5) also holds for this choice of λ. □

Assume that in (1.1.6) $b_s \equiv b_{-s}$. Then by Theorem 1.5.6 the random field $\xi(t,x)$ is space-homogeneous. Let $u^{st}(t,x)$ be a stationary solution of (B), $\mathcal{D}u^{st}(t) \equiv \mu$. Then for any $y \in \mathbb{R}$ the function $u^{st}(t, x+y)$ satisfies eq. (B) with $\xi(t,x)$ replaced by the shifted random field $\xi(t, x+y)$, distributed as $\xi(t,x)$. So $u^{st}(t,x+y)$ is a weak solution of (B), and for any $\tau \ge 0$

$$S_\tau^*\mathcal{D}u^{st}(t, \cdot + y) = \mathcal{D}u^{st}(t+\tau, \cdot + y) = \mathcal{D}u^{st}(t, \cdot + y).$$

Thus, $\mathcal{D}u^{st}(t)$ is a stationary measure, and by the theorem above it equals μ. We have proved

COROLLARY 3.3.5. *If* $b_s \equiv b_{-s}$, *then the stationary measure* μ *is space-homogeneous.*

In view of (1.6.5) Theorem 3.3.4 implies that

$$S_t f(v) = \langle f, \Sigma_t(v)\rangle \to \langle f, \mu\rangle \quad \text{as} \quad t \to \infty, \quad \forall\, v \in H^1, \tag{3.3.6}$$

for each $f \in C^b(H^m)$. This convergence still holds for functionals f with at most a polynomial growth at infinity, see Problem 3.4.2.

The property (3.3.5) is called *the mixing*, and a Markov process $u(t;u_0)$ generating a Markov semigroup $\{S_t^*\}$ satisfying (3.3.5) is said to be *mixing* in the space H^m.

Below in Theorem 4.2.5 we prove that the speed of convergence $S_t^*\lambda \rightharpoonup \mu$ with respect to the norms $\|\cdot\|_{L(L_p)}$, $p<\infty$, is uniform in ν, and in Chapter 10 we briefly consider the Burgers equation with other random forces and discuss for them mixing, its rate and its consequences.

3.4. Moments of the stationary measure.

By Proposition 3.1.3 if $B_{m-1}<\infty$ for some integer $m\ge 3$, then the second moment $\langle \mu^\nu, \|u\|_m^2\rangle$ of the stationary measure μ^ν is finite. Now we will show that if the force η in (B) is smooth in x and $B_0>0$, then all moments of $\|u\|_m$ are bounded from above and from below, and the corresponding estimates are asymptotically sharp as $\nu\to 0$.

THEOREM 3.4.1. *Assume that $B_0>0$.*

i) If $B_M<\infty$ for all M, then for any $m\in\mathbf{N}$ and each $\alpha\in(0,\infty)$ the stationary measure μ^ν satisfies

$$\int_{H^m}\|u\|_m^\alpha \mu^\nu(du)\le C_{m,\alpha}\nu^{-\frac{\alpha}{2}(2m-1)}, \tag{3.4.1}$$

$$\int_{H^m}\|u\|_m^\alpha \mu^\nu(du)\ge c_{m,\alpha}\nu^{-\frac{\alpha}{2}(2m-1)}, \tag{3.4.2}$$

where $C_{m,\alpha}$ and $c_{m,\alpha}$ depend on $\vec{B}$.

ii) If we only know that $B_4<\infty$, then for any $1\le p\le\infty$ and $\alpha\ge 1$,

$$(C'_\alpha)^{-1}\le\int |u|_p^\alpha\mu^\nu(du)\le C'_\alpha, \tag{3.4.3}$$

and

$$\int |u|_{1,1}^\alpha\mu^\nu(du)\le C'_\alpha, \tag{3.4.4}$$

for a suitable $C'_\alpha(B_0,B_4)\ge 1$.

PROOF. On the probability space $(Q,\mathcal{T},\rho)$ as in (2.3.4) with $T=1$ consider the r.v.'s $U_n^{t,\omega}=u^\omega(n+t)$, $n=1,2,\dots$, where $u^\omega(\tau)=u^\omega(\tau;0)$. Convergence (3.3.5) implies that

$$\mathcal{D}(U_n)\rightharpoonup\mu^\nu \quad \text{as } n\to\infty, \quad \text{in } H^m. \tag{3.4.5}$$

On the other hand, Corollary 2.3.7 with $T{=}n+1$ tell us that for any $m\in\mathbf{N}$ and any $\alpha\ge 1$,

$$c_{m,\alpha}\nu^{-\frac{\alpha}{2}(2m-1)}\le\mathbb{E}^Q\|U_n\|_m^\alpha\le C_{m,\alpha}\nu^{-\frac{\alpha}{2}(2m-1)}, \tag{3.4.6}$$

$$\mathbb{E}^Q\|U_n\|_m^{\alpha+1}\le C_{m,\alpha}\nu^{-\frac{\alpha+1}{2}(2m-1)}, \tag{3.4.7}$$

uniformly in n (we recall that the constants in the corollary do not depend on $T\ge 1$ in the definition of the probability space Q).

Due to (3.4.5) and Skorokhod's representation theorem (see Theorem 1.7.2), there exist random variables $\{\xi_n, n\in\mathbf{N}_0\}$, such that $\mathcal{D}(\xi_n)=\mathcal{D}(U_n)$ for $n\ge 1$, $\mathcal{D}(\xi_0)=\mu^\nu$, and

$$\xi_n\to\xi_0 \quad \text{in } \ H^m, \text{ a.s.}$$

By (3.4.7) the sequence of r.v.'s $\{\|\xi_n\|_m^\alpha, n \geq 1\}$ is uniformly integrable, so by Theorem 11.1.8

$$\mathbb{E}^Q \|U_n\|_m^\alpha = \mathbb{E}\|\xi_n\|_m^\alpha \to \mathbb{E}\|\xi_0\|_m^\alpha = \langle \|u\|_m^\alpha, \mu^\nu \rangle.$$

Now (3.4.1) and (3.4.2) follow from (3.4.6).

To prove (3.4.3) and (3.4.4) we repeat the argument above, replacing there the norm $\|u\|_m$ with $|u|_p$ or with $|u|_{1,1}$ and using (2.3.15) or (2.1.16). □

PROBLEM 3.4.2. Let $B_M < \infty$ for all M and $f \in C(H^m)$, $m \in \mathbf{N}$, satisfying $|f(u)| \leq C(1 + \|u\|_m^N)$, for some $C, N > 0$. Then (3.3.6) still holds.
Hint: Multiplying $f(u)$ by $\chi^R(\|u\|_m)$, where $\chi^R(t)$ is a continuous function, equal to 1 for $t \leq R \in \mathbf{N}$, vanishing for $t \geq R+1$ and monotone between R and $R+1$, we decompose f as $f_R + f^R$, where $f_R = f(u)\chi^R(\|u\|_m) \in C^b(H^m)$ and $f^R = f(u)(1 - \chi^R(\|u\|_m))$ satisfies $|f^R(u)| \leq CR^{-1}\|u\|_m^{N+1}$. Applying (3.3.6) to f_R and estimating $S_t f^R$ and $\langle f^R, \mu \rangle$ using Corollary 2.3.7 and the theorem above, we get the result.

Assuming that $B_M < \infty$ for all M and $B_0 > 0$, let us consider the second moments of the stationary measure μ^ν in the spaces H^m, $m \in \mathbb{R}$,

$$M_m^2(\nu) = \int_{H^m} \|u\|_m^2 \mu^\nu(du).$$

For $m \in \mathbf{N}_0$ their asymptotic behaviour as $\nu \to 0$ is described by Theorem 3.4.1. For $m \leq 0$ we have

$$M_m^2(\nu) \leq M_0^2(\nu) \leq C_{0,2}. \tag{3.4.8}$$

Using (2.1.18) and (3.4.4), we get

$$\langle |\hat{u}_k|^2, \mu^\nu \rangle \leq (2\pi k)^{-2} \langle |u|_{1,1}^2, \mu^\nu \rangle \leq C_2'(2\pi k)^{-2}.$$

Due to this estimate and (3.4.3), for any $m \leq 0$ and $M \in \mathbf{N}$ we have:

$$\begin{aligned} M_m^2(\nu) &= \langle \|u\|_m^2, \mu^\nu \rangle \geq (2\pi M)^{2m} \sum_{k=-M}^{M} \langle |\hat{u}_k|^2, \mu^\nu \rangle \\ &= (2\pi M)^{2m} \big(M_0^2(\nu) - \sum_{|k|>M} \langle |\hat{u}_k|^2, \mu^\nu \rangle \big) \geq (2\pi M)^{2m} \big(c_{0,2} - CM^{-1} \big). \end{aligned}$$

Choosing M large enough so that $2CM^{-1} \leq c_{0,2}$ we achieve that

$$M_m^2(\nu) \geq C^m \qquad \forall m \leq 0, \ \forall \nu \in (0,1], \tag{3.4.9}$$

for some $C > 0$. We see that asymptotically, as $\nu \to 0$, the moments $M_m^2(\nu)$ with $m \leq 0$ behave as $M_0^2(\nu)$.

Now let us consider $M_m^2(\nu)$ as a function of ν in the log-log scale and evaluate its slope at $\nu = 0$:

$$e_m = \lim_{\nu \to 0} \frac{\ln M_m^2(\nu)}{\ln \nu^{-1}}.$$

By the above, the function $\mathbb{Z} \ni m \mapsto e_m$ has the graph given in Figure 3.4.1. The quantity e_m is the *scaling exponent* of $M_m(\nu)$ in ν as $\nu \to 0$. Due to the theorem above and (3.4.8), (3.4.9), the function $m \mapsto e_m$ has the graph given in Figure 3.4.1. This is a piece-wise linear function, broken between $m = 0$ and $m = 1$.

In view of Theorem 2.3.4 and estimate (2.3.14), for any solution $u = u(t; u_0)$ with $u_0 \in H^1$ its second moments $\langle\langle \|u\|_m^2 \rangle\rangle$, $m \in \mathbb{Z}$, obey the same estimates as

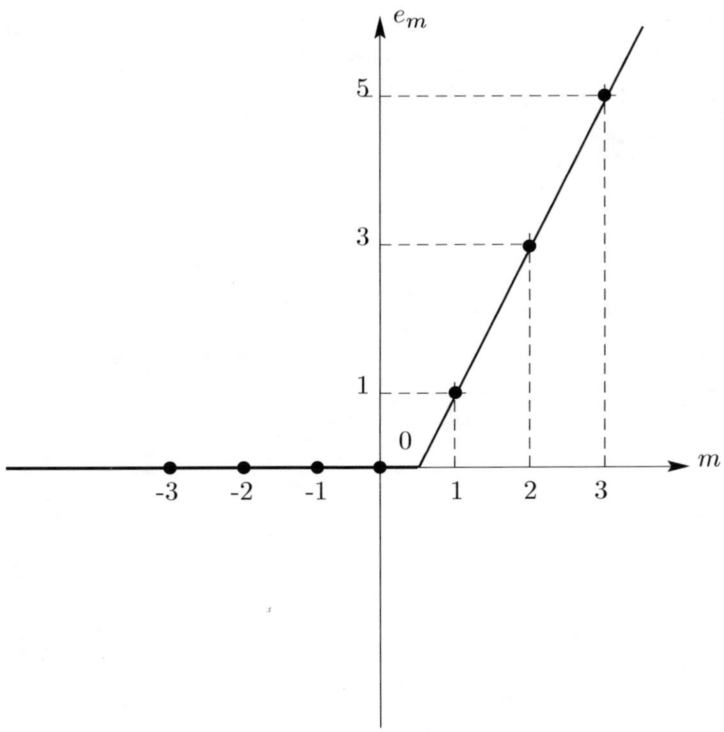

FIGURE 3.4.1. The quantity e_m as a function of $m \in \mathbb{Z}$.

M_m^2 with modified multiplicative constants. So the scaling exponent for $\langle\langle \|u\|_m^2 \rangle\rangle$ equals e_m as above, for any u_0.

Other scalings for the force η in equation (B).

We recall that the quantity

$$\tfrac{1}{2} M_0^2(\nu) = \int_{L_2} \tfrac{1}{2} |u|_2^2 \mu^\nu(du)$$

is the averaged kinetic energy of a stationary solution for eq. (B). The results above show that it remains of order 1 as $\nu \to 0$, which makes the equation with this scaling of the force natural for interpretation in terms of the one-dimensional hydrodynamics. Now we will show that – as it should be – if we scale the force η by an additional factor ν^a with $a \neq 0$, then for small ν the averaged kinetic energy will be either very small, or very large.

So let us consider the stochastic Burgers equation with a rescaled force:

$$\begin{aligned} &u_t(t,x) + u(t,x)u_x(t,x) - \nu u_{xx}(t,x) = \nu^a \eta(t,x), \\ &t \geq 0, \quad x \in S^1 = \mathbb{R}/\mathbb{Z}, \end{aligned} \tag{3.4.10}$$

where $a \in \mathbb{R}$ is a constant and η is defined as in (1.1.5). Let μ^ν be its stationary measure, which exists and is unique by Theorem 3.3.4.

PROBLEM 3.4.3. Assume that $B_4 < \infty$. Then

(i) if $a > 0$, then $\int |u|_2^2 d\mu_\nu(u) \underset{\nu \to 0}{\longrightarrow} 0$;

(ii) if $a < 0$, then $\int |u|_2^2 d\mu_\nu(u) \underset{\nu \to 0}{\longrightarrow} +\infty$.

CHAPTER 4

Stochastic Burgers equation in the space L_1

In this chapter, we study generalised strong solutions of the problem (1.1.13), where $u_0 \in L_1$ (which is the space of L_1–functions on S^1 with zero mean–value). They are obtained as limits of regular solutions with $u_0 \in H^1$.

4.1. Setting and definitions.

Let

$$B_{1_*} < \infty$$

for some number $1_* > 1$. For any $T > 0$ consider the mappings defined in (1.3.34) and (1.3.35):

$$\mathcal{M} : H^1 \times \dot{X}_T^{1_*} \to X_T^1, \qquad \mathcal{M}_t : H^1 \times \dot{X}_T^{1_*} \to H^1, \;\; 0 \le t \le T.$$

Applying (3.2.7) with T replaced by any $t \in [0, T]$ we get that the mapping $\mathcal{M}$ satisfies

(4.1.1)

$$\|\mathcal{M}(u_{01}, \xi) - \mathcal{M}(u_{02}, \xi)\|_{C(0,T;L_1)} \le |u_{01} - u_{02}|_1 \qquad \forall \xi \in \dot{X}_T^{1_*}, \;\forall u_{01}, u_{02} \in H^1.$$

For $n \ge 2$ denote by $\Phi_n(x)$ Fejér's kernel and consider the convolution operators

$$L_1 \to H^1, \;\; u \mapsto u_n = u \star \Phi_n.$$

We recall that for a function $u(x)$, written as Fourier series (1.1.3), the convolution $u \star \Phi_n$ equals $(S_1(x) + \dots S_{n-1}(x))/n$, where $S_k(x) = \sum_{|s| \le k} u_s e_s(x)$. See [**91**, Section VIII]. Since $u_n \to u$ in L_1 as $n \to \infty$ for each $u \in L_1$ by Fejér's theorem (see [**91**], Section VIII.2.3), then by (4.1.1) for every $u_0 \in L_1$ and $\xi \in \dot{X}_T^{1_*}$ the curves $\mathcal{M}(u \star \Phi_n, \xi) \in X_T^1$ converge in the space $C(0, T; L_1)$, as $n \to \infty$, to a limit which we still denote $\mathcal{M}(u, \xi)$. The constructed extension of the operator $\mathcal{M}$ to the space $L_1 \times \dot{X}_T^{1_*}$ is a pointwise limit of continuous mappings

$$L_1 \times \dot{X}_T^{1_*} \to C(0, T; L_1), \quad (u, \xi) \mapsto \mathcal{M}(u \star \Phi_n, \xi),$$

so it is Borel–measurable, see [**51**], Theorem 4.2.2. Since by (4.1.1) for any $u, v \in L_1$ we have

$$\begin{aligned}\|\mathcal{M}(u, \xi) - \mathcal{M}(v, \xi)\|_{C(0,T;L_1)} &= \lim \|\mathcal{M}(u \star \Phi_n, \xi) - \mathcal{M}(v \star \Phi_n, \xi)\|_{C(0,T;L_1)} \\ &\le \lim |u \star \Phi_n - v \star \Phi_n|_1 = |u - v|_1,\end{aligned}$$

then the extended mapping $\mathcal{M}(u, \xi)$ is 1-Lipschitz in u. The same argument applies to $\mathcal{M}_t$, and we get

PROPOSITION 4.1.1. *For any $T > 0$ the mappings $\mathcal{M}^T$ and $\mathcal{M}_t^T$, $0 \le t \le T$, extend by continuity to Borel–measurable maps*

(4.1.2) $$\mathcal{M}^T : L_1 \times \dot{X}_T^{1_*} \to C(0, T; L_1), \qquad \mathcal{M}_t^T : L_1 \times \dot{X}_T^{1_*} \to L_1,$$

which are 1–Lipschitz in the first argument.

This result allows to define a *generalised solution* $u(t,x)$ of the Cauchy problem (B), (1.1.5), (1.1.11) with $\xi \in \dot X_T^{1_*}$ and $u_0 \in L_1$ as $u(t,x;u_0,\xi) = \mathcal{M}^T(u_0,\xi) \in C(0,T;L_1)$.

For any $u_0 \in L_1$ let us approximate it by H^1-functions $u_j = u_0 \star \Phi_j$, $j \geq 1$. Considering the cocycle relation (1.3.37) with u_0 replaced by u_j and passing to the limit we see that (1.3.37) remains true for $u_0 \in L_1$. Using this relation and the regularity (1.3.34), we get:

LEMMA 4.1.2. *Let $u_0 \in L_1$ and $u(t) = u(t;u_0,\xi)$ be a generalised solution, where $\xi \in \dot X_T^{m_*}$ for some $m_* > m \in \mathbf{N}$. Assume that $u(\tau) \in H^m$ for some $0 \leq \tau < T$. Then $u(t) \in H^m$ for all $t \geq \tau$.*

That is, if a generalised solution became regular at some moment of time, it remains regular for larger times.

Naturally, a *(strong) generalised solution* of the stochastic Cauchy problem (1.1.13) with $u_0 \in H^1$ is the random process in H^1

$$u^\omega(t;u_0) = \mathcal{M}(u_0;\xi^\omega)(t).$$

Accordingly, we define the transition probability for the process of generalised solutions for (B) as the mapping

$$\Sigma : \mathbb{R}_+ \times L_1 \to \mathcal{P}(L_1), \qquad (t,v) \mapsto \mathcal{M}_t^T \circ (\delta_v \times \pi_T^{1_*}) \tag{4.1.3}$$

for some $T \geq t$, where $\pi_T^{1_*} \in \mathcal{P}(\dot X_T^{1_*})$ is the law of the force $\xi\mid_{[0,T]}$ (we recall that $B_{1_*} < \infty$). By the construction of the maps (4.1.2), for any function $f \in C_b(L_1)$ we have

$$\langle f, \Sigma_t(u)\rangle = \lim_{n\to\infty} \langle f, \Sigma_t(u\star\Phi_n)\rangle, \tag{4.1.4}$$

where Σ_t in the r.h.s. is the regular transition probability from Section 1.6. Since for any n the function

$$\mathbb{R}_+ \times L_1 \ni (t,u) \mapsto \langle f, \Sigma_t(u\star\Phi_n)\rangle$$

is continuous, then the limiting function of (t,u) is measurable. The 1–Lipschitz continuity of $\mathcal{M}_t^T$ in $u \in L_1$ immediately implies that the l.h.s. of (4.1.4) is Lipschitz-continuous in $u \in L_1$ for any $f \in B^1_{L_1}$. These properties jointly mean that the mapping (4.1.3) indeed is a transition probability (see Section 11.3). So the generalised solutions of (B) define a family of Markov processes in L_1, see Section 11.6.

In the same way as in the regular case (when $u_0 \in H^1$) we define operators S_t^* by:

$$S_t^* : \mathcal{P}(L_1) \to \mathcal{P}(L_1), \quad \mu \mapsto \mathcal{M}_t \circ (\mu \times \pi^{1_*}) := \mathcal{D}(u^\omega(t;u_0^\omega)), \tag{4.1.5}$$

where $\mathcal{D}u_0 = \mu$.

PROBLEM 4.1.3. Prove that the operators S_t^* in (4.1.5) are continuous.
Hint: If $\mu_n \rightharpoonup \mu_0$ in $\mathcal{P}(L_1)$, then by Skorokhod's theorem we can find r.v.'s $u_0^n, n \geq 0$, such that $\mathcal{D}(u_0^n) = \mu_n$ and $u_0^n \to u_0^0$, a.s. From here and Proposition 4.1.1 it follows that $\mathbb{E}f(u(t;u_0^n)) \to \mathbb{E}f(u(t;u_0^0))$, for each $f \in C_b(L_1)$.

Repeating the argument in the proof of Theorem 1.6.1 (or using general properties of Markov processes, see [**69, 80, 139**]) we see that the maps S_t^* in (4.1.4) make a semigroup.

If μ is a Borel measure on H^1, then via the continuous embedding $i : H^1 \to L_1$ it defines a Borel measure $i \circ \mu$ on L_1. Since H^1 is a Borel subset of L_1,[1] then for any Borel set Q in L_1, $i \circ \mu(Q)$ equals $\mu(Q \cap H^1)$. So for $\mu \in \mathcal{P}(H^1)$,

$$i \circ S_t^* \mu = S_t^*(i \circ \mu). \tag{4.1.6}$$

That is, the constructed semigroup in $\mathcal{P}(L_1)$ agrees with the regular semigroup in $\mathcal{P}(H^1)$, defined in Section 11.3.2.

A remarkable property of the operators (4.1.5) is that they are non-expanding:

PROPOSITION 4.1.4. *For any measures $\mu_1, \mu_2 \in \mathcal{P}(L_1)$ and any $t \geq 0$ we have*

$$\|S_t^* \mu_1 - S_t^* \mu_2\|_{L(L_1)}^* \leq \|\mu_1 - \mu_2\|_{L(L_1)}^*.$$

PROOF. We may assume that $\mu_j = \mathcal{D}(u_{0j}^{\omega_1})$ and $S_t^* \mu_j = \mathcal{D}(\mathcal{M}_t(u_{0j}^{\omega_1}, \xi^{\omega_2}))$, for $j = 1, 2$ (that is, we choose as the probability space $\Omega = \Omega_1 \times \Omega_2 = \{(\omega_1, \omega_2)\}$). Then for any function $f \in \bar{B}^1(L_1)$ we have

$$\begin{aligned}\langle f, S_t^* \mu_1 \rangle - \langle f, S_t^* \mu_2 \rangle =& \mathbb{E}^{\omega_2} \mathbb{E}^{\omega_1} \big(f(\mathcal{M}_t(u_{01}^{\omega_1}, \xi^{\omega_2})) - f(\mathcal{M}_t(u_{02}^{\omega_1}, \xi^{\omega_2})) \big) \\ &\leq \mathbb{E}^{\omega_2} \|\mu_1 - \mu_2\|_{L(L_1)}^* = \|\mu_1 - \mu_2\|_{L(L_1)}^*,\end{aligned}$$

where we use that for any ξ the function $L_1 \ni u \mapsto f(\mathcal{M}_t(u, \xi))$ is bounded by 1 and 1-Lipschitz (see Proposition 4.1.1). So the required estimate follows. □

The operators (4.1.5) are instantly smoothing. To state and to prove this property we start with another proposition:

PROPOSITION 4.1.5. *If $B_m < \infty$ for some $m \geq 4$, then for any random variable $u_0 \in L_1$ the generalised solution $u(t; u_0)$ satisfies*

$$\mathbb{E}\|u(t; u_0)\|_m^2 \leq C(m, B_m, t)\nu^{-(2m-1)} \quad \forall t > 0. \tag{4.1.7}$$

PROBLEM 4.1.6. Prove the proposition.

Hint: Define $u_0^n = \Phi_n \star u_0$, where Φ_n is Fejér's kernel. Then $u(t; u_0^n) \to u(t; u_0)$ in L_1, a.s. For u_0 replaced by u_0^n the estimate (4.1.7) follows from Theorem 2.2.1. Now deduce the required result from Theorem 11.1.4, noting that the L_1-convergence above implies the convergence in H^{-1}.

Since for $\nu \in \mathcal{P}(L_1)$ we have $S_t^* \nu = \mathcal{D}u(t; u_0)$, where $\mathcal{D}(u_0) = \nu$, then by Proposition 4.1.5 for any $t > 0$ the measure $S_t^* \nu$ is supported by H^m. Evoking Proposition 11.3.3 (and regarding $S_t^* \nu$ as a measure on H^{-1}) we have that the mappings S_t^* smooth measures immediately: if $B_m < \infty$ for some natural number m, then

$$S_t^* \nu \in \mathcal{P}(H^m) \quad \forall \nu \in \mathcal{P}(L_1), \ \forall t > 0.$$

In view of Proposition 4.1.5 for any $u_0 \in L_1$ and $t > 0$ there exists a null set $Q_{u_0}(t)$ such that $u^\omega(t; u_0) \in H^m$ if $\omega \notin Q_{u_0}(t)$. Assume that $B_{m_*} < \infty$ for some $m_* > m$. Then by Lemma 4.1.2 a solution $u^\omega(\tau; u_0)$ with $\omega \notin Q_{u_0}(t)$ stays in H^m for $\tau \geq t$, so considering the null set $\cup_{n \in \mathbf{N}} Q_{u_0}(1/n)$ we get

COROLLARY 4.1.7. *If $B_{m_*} < \infty$ for some $m_* > m \geq 4$, then for any fixed $u_0 \in L_1$ the generalised solution $u^\omega(\tau; u_0)$ almost surely belongs to the space H^m for all $t > 0$.*

[1] Indeed, the function $u \mapsto \|u\|_1^2 \leq \infty$ is measurable on L_1 as a limit of obviously measurable functions $u \mapsto \sum_{|s| \leq N} |u_s|^2$. So $H^1 = \{u : \|u\|_1^2 < \infty\}$ is a measurable set.

4.2. Mixing for L_1 solutions.

Our goal in the rest of this chapter is to obtain a ν-independent estimate for the rate of mixing for the semigroup (4.1.5). Accordingly, the constants in this section do not depend on ν. From now on, we consider the problem (1.1.13), where

$$B_5 < \infty, \qquad u_0 \in L_1,$$

and study its generalised solution $u(t) = \mathcal{M}_t(u_0, \xi)$, defined as in Section 4.1. In this section the letters k, l and N stand for various natural numbers, so always

$$k, l, N \in \mathbf{N}.$$

A *stationary measure in* $\mathcal{P}(L_1)$ is defined in the same way as in Definition 3.1.1: this is a measure which is invariant for all maps S_t^* as in (4.1.4).

By (4.1.6), every measure which is stationary for equation (B) in H^1 is stationary for the flow maps of generalised solutions in L_1. Next, if $\mu \in \mathcal{P}(L_1)$ is stationary for (B) and $u(t)$ is a corresponding stationary generalised solution, then $u(t)$ satisfies (4.1.7), so μ is a measure in H^1, stationary for the flow of regular solutions. Accordingly, there exists a unique stationary measure μ for the semigroup $\{S_t^*\}$ in $\mathcal{P}(L_1)$, and since $B_4 < \infty$, then it is supported by the space H^4.

Below we study the mixing properties of μ in L_1, following the lines of Chapter 3.

Firstly, we take any $k, N \in \mathbf{N}$ and note that for $\tau \in [(k-1)N, kN]$ a solution $u^\omega(\tau) = u^\omega(\tau; u_0, \xi^\omega)$ may be written as

$$u^\omega((k-1)N + t) = \mathcal{M}^N\big(u^\omega((k-1)N), \xi_{k-1}^\omega\big)(t), \quad 0 \le t \le N,$$

where $\xi_{k-1}^\omega(t) = \xi^\omega((k-1)N + t) - \xi^\omega((k-1)N)$.

So in view of Theorem 2.1.2 with $\theta = T = N$, if $u_0 \in H^1$ then the solution u^ω obeys the following property:

(4.2.1)
$$\text{if} \sup_{(k-1)N \le t \le kN} \|\xi^\omega(t) - \xi^\omega((k-1)N)\|_4 \le N^{-2}/4, \text{ then } |u^\omega(kN; u_0)|_1 \le 24N^{-1}$$

(for a fixed ω). This statement remains true for all $u_0 \in L_1$. Indeed, if for a fixed ω we approximate $u_0 \in L_1$ by H^1-functions, then (4.2.1) holds for the approximating solutions, and in view of (4.1.2) with $t = kN$ the estimate remains true for the original generalised solution $u(\tau; u_0)$.

Secondly, in the two lemmas below we establish that for any $N \in \mathbf{N}$, there exists $G(N) \in \mathbf{N}$ such that the L_1-norm of $u^\omega(t)$ with high probability becomes smaller than N^{-1} at a random point t from the time interval $[0, NG(N)]$, uniformly in u_0.

LEMMA 4.2.1. *There exist constants $C, \kappa > 0$, depending only on B_5, such that for any N the event*

$$\Gamma_N = \{\omega : \sup_{(k-1)N \le t \le kN} \|\xi^\omega(t) - \xi^\omega((k-1)N)\|_4 > N^{-2}/4 \quad \forall k \le G(N)\},$$

where $G(N) = Ce^{N^8}$, satisfies

$$\mathbf{P}(\Gamma_N) \le N^{-1}. \tag{4.2.2}$$

PROOF. Since ξ has i.i.d. increments on disjoint intervals, then in view of Lemma 3.2.1

$$\mathbf{P}(\Gamma_N) \le \big(1 - f_4^0(1/4N^{5/2})\big)^G, \quad G := G(N). \tag{4.2.3}$$

i) If N is sufficiently large (in terms of B_5), then $f_4^0 := f_4^0(1/4N^{5/2}) \geq \frac{1}{2}e^{-2\kappa_4(4N^{5/2})^3}$, and $\ln(1-f_4^0) \leq -\frac{1}{2}f_4^0$. So

$$\mathbf{P}(\Gamma_N) \leq e^{G\ln(1-f_4^0)} \leq e^{-\frac{1}{2}Gf_4^0}$$

and (4.2.2) holds if $G \geq (2\ln N)/f_4^0$. Since for N as above there exists C such that

$$(2\ln N)/f_4^0 \leq 4\ln N\, e^{2\kappa_4(4N^{5/2})^3} \leq Ce^{N^8},$$

then (4.2.2) is proved if $N \geq N_0(B_5)$.

ii) In view of (4.2.3), for $N < N_0$ estimate (4.2.2) obviously is valid if $C \geq 1$ is large enough.

The lemma is proved. □

For $t \geq 0$ consider the function

$$U^\omega(t) := \sup_{u_0 \in L_1} |u^\omega(t;u_0)|_1.$$

It is measurable in (t,ω). Indeed, since the function $|u^\omega(t;u_0)|_1$ is continuous in u_0 and the space L_1 is separable, then $U^\omega(t)$ is measurable as the supremum of a countable family of measurable functions (e.g. see [**130**], Section 23).

LEMMA 4.2.2. *For G as in Lemma* 4.2.1 *and every N we have*

$$\mathbf{P}(\inf_{k\leq NG(N)} U^\omega(kN) > 24N^{-1}) \leq N^{-1}. \tag{4.2.4}$$

PROOF. By Lemma 4.2.1, for any N,

$$\mathbf{P}(\sup_{(k-1)N\leq t\leq kN} \|\xi(t)-\xi((k-1)N)\|_4 \leq N^{-2}/4 \text{ for some } k \leq G(N)) \geq 1-N^{-1}.$$

Consequently, by (4.2.1) we get

$$\mathbf{P}(\,|u^\omega(l;u_0)|_1 \leq 24N^{-1} \text{ for some } l \in [1, NG(N)] \text{ and all } u_0 \in L_1) \geq 1-N^{-1}.$$

This proves (4.2.4). □

Thirdly, arguing as in Section 3.3 we show that for large t the measures $\Sigma_t(u_{10})$ and $\Sigma_t(u_{20})$ are close in the space $\mathcal{P}(L_1)$, uniformly in ν, u_{10} and u_{20}:

PROPOSITION 4.2.3. *There exists a constant $\kappa(B_5) > 0$ such that for any $u_{10}, u_{20} \in L_1$ and $t \geq 3$,*

$$d_t(u_{10},u_{20}) := \|\Sigma_t(u_{10}) - \Sigma_t(u_{20})\|^*_{L(L_1)} \leq \kappa(\ln t)^{-1/8}.$$

PROOF. Similarly to Section 3.3 we consider the curve

$$U(t) = (u_1(t), u_2(t)) \in L_1 \times L_1, \quad u_j(t) = u(t;u_{j0}),\ j=1,2,$$

and the closed sets $O_N = \bar{B}_{L_1}^{24N^{-1}}$, $N \in \mathbf{N}$. Next we define the hitting times

$$\tau_N^\omega := \min\{l \in \mathbf{N},\ l \leq NG(N):\ U^\omega(l) \in O_N \times O_N\}$$

($\tau_N^\omega = \infty$ if the set is empty). Applying Lemma 4.2.2 to solutions u_1 and u_2 we find that

$$\mathbf{P}(\tau_N^\omega = \infty) \leq N^{-1}.$$

If $\tau_N^\omega < \infty$, then $U^\omega(t') \in O_N \times O_N$ for some $t' \leq NG(N)$. Since $U^\omega(t') \in O_N \times O_N$ implies that $|u_1^\omega(t') - u_2^\omega(t')|_1 \leq 48N^{-1}$, then by Lemma 3.2.6 $|u_1^\omega(t) - u_2^\omega(t)|_1 \leq 48/N$ for $t \geq \tau_N^\omega$. So for $t \geq NG(N)$ we have

$$d_t(u_{10}, u_{20}) = \sup_{\|f\|_{L(L_1)} \leq 1} \mathbb{E}\big(f(u_1(t)) - f(u_2(t))\big) \leq N^{-1} \cdot 2 + 48N^{-1},$$

where the first term in the r.h.s. comes from the integration over the event $\{\tau_N^\omega = \infty\}$ and the second – over its complement. Therefore $d_t(u_{10}, u_{20}) \leq 50N^{-1}$ if $t \geq NG(N)$ for any $N \in \mathbf{N}$. As $G(N) = Ce^{N^8}$, then for a suitable $t_0(B_5) \geq 3$ relation $t \geq \max(t_0, e^{2N^8})$ implies that $d_t(u_{10}, u_{20}) \leq 50N^{-1}$. That is,

$$d_t(u_{10}, u_{20}) \leq 50\Big(\frac{\ln t}{2}\Big)^{-1/8} < 100(\ln t)^{-1/8} \quad \text{if} \quad t \geq t_0.$$

Since $d_t(u_{10}, u_{20}) \leq 2$ for all t, then the assertion follows if we choose

$$\kappa = \max(100, 2(\ln t_0)^{1/8}).$$

□

Finally, repeating the proof of Theorem 3.3.4 we derive from Proposition 4.2.3 an estimate of the rate of mixing for the semigroup $\{S_t^*\}$ in $\mathcal{P}(L_1)$:

THEOREM 4.2.4. *For any probability measure $\mu_0 \in \mathcal{P}(L_1)$ and $t \geq 3$ we have*

$$\|S_t^*\mu_0 - \mu\|_{L(L_1)}^* \leq \kappa(\ln t)^{-1/8}, \tag{4.2.5}$$

where μ is the unique stationary measure for eq. (B), *and κ the same positive constant as in Proposition* 4.2.3, *depending only on B_5.*

Note that by Proposition 4.1.4 the distance in the l.h.s. of (4.2.5) decreases to zero monotonically when time grows.

Using Remark 3.2.2 and repeating the proof of Theorem 4.2.4 we see that if we weaken the assumption $B_5 < \infty$ to $B_{4_*} < \infty$ for some number $4_* > 4$, then

$$\|S_t^*\mu_0 - \mu\|_{L(L_1)}^* \leq g(t), \tag{4.2.6}$$

for a suitable ν-independent function $g(t)$, going to zero when $t \to \infty$, which depends only on 4_* and B_{4_*}. While if only $B_4 < \infty$, then (4.2.6) holds with a function g, depending on the sequence $\{b_s\}$.

Finally we show that Theorem 4.2.4 implies that, in fact, the semigroup (4.1.5) is uniformly in ν mixing with respect to any L_p-norm with finite p:

THEOREM 4.2.5. *For any $\mu_0 \in \mathcal{P}(L_1)$, $t \geq 3$ and $1 \leq p < \infty$,*

$$\|S_t^*\mu_0 - \mu\|_{L(L_p)}^* \leq C_p(\ln t)^{-\kappa_p}, \tag{4.2.7}$$

where C_p and $\kappa_p > 0$ depend on p and B_5.

PROOF. We may assume that $1 < p < \infty$ since for $p = 1$ the result is proved.

i) Let us abbreviate $S_t^*\mu_0 =: \mu_t$. Since by (2.1.16) and (3.4.3) $\langle |u|_1^\gamma, \mu_t\rangle \leq C(B_4, \gamma)$ and $\langle |u|_1^\gamma, \mu\rangle \leq C(B_4, \gamma)$ for each $t \geq 1$ and $\gamma > 0$, then in view of (4.2.5) and Proposition 11.4.2 we have the following estimate for the Kantorovich distance between the measures μ_t and μ (see (11.4.1)):

$$\|\mu_t - \mu\|_{Kant, L_1} \leq C_\gamma(B_5)(\ln t)^{-\frac{1-1/\gamma}{8}}, \quad t \geq 3,$$

for any $\gamma > 1$. Choosing γ large enough we get that

$$\|\mu_t - \mu\|_{Kant, L_1} \leq C(B_5)(\ln t)^{-1/9}, \quad t \geq 3. \tag{4.2.8}$$

ii) Now estimate (4.2.8) and the Kantorovich–Rubinstein Theorem (see Theorem 11.4.3) imply that for $t \geq 3$ there are r.v.'s U_t and $\tilde{U}_t \in L_1$ such that $\mathcal{D}U_t = \mu_t$, $\mathcal{D}\tilde{U}_t = \mu$ and

$$\mathbb{E}|U_t - \tilde{U}_t|_1 \leq C(\ln t)^{-1/9}. \tag{4.2.9}$$

Moreover, by (2.1.16) and (3.4.3), $U_t, \tilde{U}_t \in L_p$, a.s. For any $f \in \bar{B}^1_{L(L_p)}$ we have

(4.2.10)

$$\begin{aligned}\mathbb{E}|\langle f, \mu_t\rangle - \langle f, \tilde{\mu}\rangle| &= \mathbb{E}|f \circ U_t - f \circ \tilde{U}_t| \leq \mathbb{E}|U_t - \tilde{U}_t|_p \leq \mathbb{E}|U_t - \tilde{U}_t|_1^{1-\theta}|U_t - \tilde{U}_t|_{2p}^{\theta} \\ &\leq \left(\mathbb{E}|U_t - \tilde{U}_t|_1\right)^{1-\theta}\left(\mathbb{E}(|U_t|_{2p} + |\tilde{U}_t|_{2p})\right)^{\theta}, \quad \theta = (2p-2)/(2p-1),\end{aligned}$$

where the second upper estimate is the Riesz–Thorin inequality (11.1.5). Since

$$\mathbb{E}(|U_t|_{2p} + |\tilde{U}_t|_{2p}) = \langle |u|_{2p}, \mu_t\rangle + \langle |u|_{2p}, \mu\rangle \leq C_{2p}(B_5)$$

in view of (2.1.16) and (3.4.3), then by (4.2.9) the r.h.s. of (4.2.10) is bounded by $C_p(\ln t)^{-\kappa_p}$ with $\kappa_p = 1/9(2p-1)$, uniformly in $f \in \bar{B}^1_{L(L_p)}$. So (4.2.7) follows. $\square$

CHAPTER 5

Notes and comments, I

Results of Chapter 1 are well known. The method used by us to construct solutions of the stochastic Burgers equation is, in our opinion, the easiest and the most elementary since it does not involve the martingale technique and stochastic integrals (to avoid the latter we use Ito's formula only in the integrated form (11.8.11)). Certainly our treatment of the equation is not unique. In particular, another way to solve it may be found in Section 14 of [**44**] (also see the paper [**42**]). Our approach to estimate high Sobolev norms of solutions, based on a priori estimates for their L_∞ norms, is also sufficiently traditional; e.g. see [**6, 99**]. Since Oleinik's estimate has no analogue for the Galerkin approximations, then the final upper estimates (2.2.2) are obtained in two steps: firstly, in Theorem 1.4.4 we use the Galerkin method to establish preliminary results which imply that for smooth initial data the function $t \mapsto \mathbb{E}\|u(t)\|_m^2$ is C^1-smooth and satisfies the differential inequality (2.2.4). Secondly, in Theorem 2.2.1 we analyse this differential inequality using Oleinik's estimate for solutions of (B), and obtain (2.2.2).

The method of bounding from below the first Sobolev norm using the energy balance and Oleinik's maximum principle is due to Biryuk [**17, 18**]. Obtaining lower bounds for higher Sobolev norms by interpolating them with the lower bound for some low Sobolev norm and the upper estimate for the L_∞ norm was suggested in [**99, 107**] for the complex Ginzburg–Landau equations (see Section 10.4.3), and then was used in [**17, 18, 26, 28**] for the Burgers equation and its stochastic version. The fact that the lower and upper estimates for Sobolev norms of solutions are asymptotically sharp (i.e. involve ν to the same negative degree) is one more remarkable feature of the Burgers equation.

Results of the first two sections of Chapter 3 are well known. The proof of the mixing in Section 3.3 is inspired by the corresponding results for the 2d Navier–Stokes system (see in Chapter 6) and is based on the very strong assertion of Lemma 3.2.6. It was suggested in [**26, 28**], where it was also observed that the rate of mixing with respect to the L_1 norm does not depend on $\nu > 0$.[1] We note that the first rigorous proof of the mixing for stochastic Burgers equation was given in Section 14 of [**44**]. There (following the work [**63**] on the stochastic 2d Navier–Stokes system) the random force is assumed to be rough: the coefficients b_s all equal 1.

The L_1-construction in Section 4.1 is routine. Section 4.2 is written for this book, and the L_p-result of Theorem 4.2.5 may be new.

[1] The ν-uniform rate of mixing for eq. (B) in L_1 was calculated in [**26, 28**] with a mistake, corrected in this book.

Part 2

One-Dimensional Turbulence

CHAPTER 6

Turbulence and Burgulence

The presentation in this chapter is made on the physical level of rigour.

3d hydrodynamical turbulence. For the purposes of this discussion we restrict ourselves to the hydrodynamical turbulence in space-periodic fluid flows. We choose the units in such a way that the period of the flow in all the three coordinates is one, its speed is a quantity of order one, and we assume that the mean-value of the velocity field over the cube of periods vanishes. After that we regard the involved quantities as dimensionless. In this way it would be impossible to explain the heuristic derivation of the laws of the theory of turbulence, based on a dimensional analysis, but by making the characteristics of fluid flows dimensionless we simplify presentation and discussion of the laws of the theory of turbulence relevant for our book.

Due to the agreement above the Reynolds number Re of a flow, defined as

$$Re = \frac{\text{(space-scale of the flow)(characteristic speed of the flow)}}{\text{(viscosity)}}$$

(e.g. see in [**68**, **114**]), is

$$Re \sim \nu^{-1}, \tag{6.0.1}$$

where ν is the viscosity of the fluid. If Re is large, then the velocity field $u(t,x)$ of the flow becomes very irregular, i.e. turbulent. A rigorous description of such flows is still missing, but because of the tremendous practical importance of the turbulent motion of fluids and gazes, hundreds of books and many thousands of papers have been dedicated to heuristic description of the phenomenon. These publications appeal to a number of related (non-rigorous) theories of turbulence. Among these, the one created by A. N. Kolmogorov in three articles published in the year 1941 [**88**–**90**] (and partially based on the previous work by Taylor and von Karman–Howard [**143**, **149**]) agrees the best with experiments and is the most popular. It is usually referred to as the *K41 theory*. It postulates that the velocity field $u(t,x)$ of a turbulent flow to which it applies is a stationary in time and homogeneous in space random field (i.e. translations of u in time and in space do not change its distribution). We give below a brief description of the basic claims of the theory, mostly inspired by [**114**, Section III], and [**68**]. More precisely, what we call the K41 theory is a rather rough presentation of its main statements, as they are understood now. In particular, in the works [**88**]-[**90**] A. N. Kolmogorov did not use the Fourier transform, while below we frequently employ it.[1]

[1] However, in the same year 1941 the Fourier method was applied to the problem of turbulence in the frame of the K41 theory by Obukhov, then a student of Kolmogorov. See below relation (6.0.10).

The K41 theory claims that although the large-scale characteristics of a turbulent flow $u(t,x)$ obviously depend on the features of the force applied to the fluid or gas to create turbulence (and this dependence is well visible), the small-scale behaviour [2] is universal and depends on the particularities of the system only through a few parameters. To make this point clearer, we will introduce several definitions which apply to a snapshot $u(t,\cdot)$ of the flow, considered at a fixed moment of time t. Moreover, we assume that u satisfies some smoothness assumptions so that the quantities treated below are well-defined. We will only consider the turbulent regime $0<\nu\ll 1$, corresponding to large values of the Reynolds number Re. The first relevant parameter of the theory is the *dissipation scale* l_d (also known as the *Kolmogorov inner scale*). To define it, let us decompose $u(t,x) =: u(x)$, $x\in\mathbb{R}^d$, in Fourier series:

$$u(x)=\sum_{\xi\in\mathbb{Z}^d}\hat{u}(\xi)e^{2\pi i\xi\cdot x},\quad \hat{u}(0)=0$$

(although Kolmogorov's theory deals with three-dimensional velocity-fields $u(x)$, for what follows it is convenient to develop a d-dimensional notation). Then $l_d=\nu^{-c_d}$, where $c_d\geq 0$ is the smallest number such that for every $c>c_d$ and each $N\geq 0$,

$$\langle|\hat{u}(\xi)|^2\rangle\leq C_{N,c}|\xi|^{-N}\qquad \forall\,|\xi|\geq\nu^{-c},\ \forall\nu\in(0,1], \tag{6.0.2}$$

with suitable ν–independent constants $C_{N,c}$, where here and below the brackets stand for some averaging. The K41 theory predicts that

$$l_d=\nu^{-3/4} \tag{6.0.3}$$

(we recall the agreement (6.0.1)), and that for $|x_1-x_2|\ll 1/l_d$ a suitably averaged increment of the velocity field $u(t,x_1)-u(t,x_2)$ is of order $|x_1-x_2|$, while for $|x_1-x_2|$ larger than $1/l_d$ it behaves differently.[3]

The interval $I_{diss}=(l_d,\infty)$ is called the *dissipation range* (in the Fourier presentation). The *energy range* is an interval of the form $I_{energ}=[1,l_e)$, $l_e=\nu^{-c_e}$, $0\leq c_e\leq c_d$, where c_e is the smallest non-negative exponent such that the zones $\{|\xi|\leq \text{Const }\nu^{-c_e}\}$ contain most of the energy of the flow in the Fourier presentation:

$$\sum_{|\xi|\leq C_\delta l_e,\,\xi\in\mathbb{Z}^d}\langle|\hat{u}(\xi)|^2\rangle\geq(1-\delta)\sum_{\xi\in\mathbb{Z}^d}\langle|\hat{u}(\xi)|^2\rangle, \tag{6.0.4}$$

for any $\delta>0$ with a suitable C_δ. K41 states that (in our setting) $c_e=0$, so $l_e=\nu^0$. The complement to the two ranges above, the interval $I_{inert}=[l_e,l_d]$, is the *inertial range*. This is the most interesting zone, where the flow exhibits a non-trivial universal small-scale behaviour described more precisely below, as opposed to the dissipation range (where damping plays the dominant role) and the energy range (where there is a heavy dependence on the large-scale features).

[2] Here and below, by a *small-scale* property we mean a feature which manifests itself on small scales in the physical space, which corresponds – as we will see later and which often is the case – to a certain behaviour of large Fourier coefficients.

[3] There is no rigorous mathematical result which would imply that the indicated behaviour of space-increments of a velocity-field $u(t,x)$ is a consequence of the decay (6.0.2) of its Fourier transform, so this property should be regarded as an assertion of the K41 theory. For the one-dimensional model of turbulence based on the Burgers equation, the property is proved below, see relation (6.0.21) and the related discussion.

In physics the quantities l_e and l_d are defined modulo multiplicative constants. So below in specific statements involving the three ranges, we may replace l_e and l_d by $C_1 l_e$ and $C_2 l_d$.

Discussing increments $u(x+r)-u(x)$ of a vector-field u of a turbulent flow we will usually impose a restriction that $|r|^{-1}$ belongs to I_{diss}, or I_{energ}, or I_{inert}. That is, $|r|$ belongs to $[0, l_d^{-1}) =: I^x_{diss}$, or $(l_e^{-1}, \infty) =: I^x_{energ}$, or $[l_d^{-1}, l_e^{-1}] =: I^x_{inert}$. Accordingly, we will call the interval I^x_{diss} the *dissipative range in the x-presentation*, and similarly for the two other intervals.

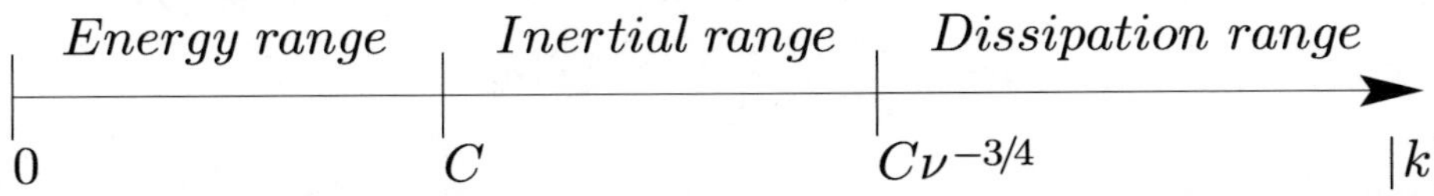

FIGURE 6.0.1. Kolmogorov scales

The two most important quantities used to describe the small-scale behaviour of a flow at a fixed time moment are:

- In the space-presentation, the *longitudinal structure function*

$$S^{\|}_{p,r} = \left\langle \left| \frac{(u(x+r)-u(x))\cdot r}{|r|} \right|^p \right\rangle, \qquad p \geq 0 \tag{6.0.5}$$

  (the function does not depend on x in view of the assumed spatial homogeneity of the random field $u(x)$).
- In the Fourier-presentation, the *energy spectrum* of a flow $u(x)$,

$$E_k = \langle e_k \rangle, \qquad e_k = |\Sigma_k|^{-1} \sum_{\xi \in \Sigma_k} \tfrac{1}{2} |\hat{u}(\xi)|^2, \qquad k > 0, \tag{6.0.6}$$

  where $\Sigma_k \subset \mathbb{Z}^d$ is a suitable layer around the sphere $S^{d-1}(k) = \{\xi \in \mathbb{Z}^d : |\xi| = k\}$ of a width significantly larger than 1. The width of Σ_k may grow indefinitely with k.[4]

The K41 theory predicts that if $\nu \ll 1$ (so the Reynolds number of a flow is large), then for $|r| \in I^x_{inert}$ we have:

$$S^{\|}_{2,r} \sim |r|^{2/3}. \tag{6.0.7}$$

This relation is called *Kolmogorov's 2/3-law.* The theory also states that the third moment of the speed's increments without the modulus sign behaves similarly:

$$\left\langle \left((u(x+r)-u(x)) \cdot r/|r| \right)^3 \right\rangle \sim -r. \tag{6.0.8}$$

The dimension argument used by Kolmogorov to derive (6.0.7) also implies that

$$S^{\|}_{p,r} \sim |r|^{p/3}, \quad p > 0. \tag{6.0.9}$$

This relation, although not claimed in the K41 papers, was frequently suggested in later works related to the Kolmogorov theory. More detailed predictions are also made concerning the second and third moments of the increments $u(x+r)-u(x)$ with $|r| \in I^x_{inert}$. The corresponding argument is based on the assumption that

[4] It is natural to take in (6.0.6) the averaging not over integer spheres $S^{d-1}(k)$ but over layers Σ_k since depending on $k \in \mathbf{N}$, the number of points and the geometrical configuration of the spheres $S^{d-1}(k)$ can change drastically.

the velocity field u is isotropic (i.e. is statistically invariant under rotations), and cannot coexist with the periodicity assumption $x \in \mathbf{T}^3$ which we have imposed. We do not touch those more delicate results in our book and do not discuss them here.

On the other hand, for $k \in I_{inert}$ K41 states that

$$E_k \sim k^{-5/3}. \tag{6.0.10}$$

This relation, called the *Kolmogorov–Obukhov law*, was published in the same year as the K41 papers in two articles of Obukhov [**126, 127**], then a student of Kolmogorov. Comparing (6.0.10) with (6.0.2) we see that the dissipation scale l_d separates the region where the energy spectrum E_k behaves algebraically from the one where it decays faster than any negative degree of k.

The K41 predictions are in good agreement with experimental and numerical data for the energy spectrum, for the function $S^{||}_{2,r}$ and for the third moment in (6.0.8). However, there are discrepancies between the prediction (6.0.9) and experiments when $p \geq 4$, which are commonly attributed to the *small-scale intermittency*, see [**68**, Section 8]. This rather involved feature of turbulent flows in particular implies that for $p \geq 4$ the function $\left(S^{||}_{p,r}\right)^{1/p}$ significantly depends on p. To measure the strength of the intermittency experts in turbulence usually use the *flatness*, which at a space-scale l is defined by:

$$F(l) = S^{||}_{4,l}/(S^{||}_{2,l})^2. \tag{6.0.11}$$

The larger F, the more intermittent is the velocity field. Relation (6.0.9) implies that if $\nu \ll 1$, then $F \sim 1$ for $|l|^{-1} \in I_{inert}$, which does not agree well with measurements of F. The "intermittency objection" does not change the fact that up until now K41 is the most consistent theory of turbulence.

It is a natural question, related to the above discussion, to know which mathematical model of fluid motion should be used to obtain a consistent rigorous theory to justify (at least partially) the K41. In his papers, Kolmogorov never mentioned the equations which should underly the theory. But starting from the work of Edwards in 1963 [**57**] it became a dominant opinion that this should be the Navier–Stokes system, perturbed by a smooth and homogeneous in x random force:

$$\dot{u}(t,x) + (u \cdot \nabla)u - \nu\Delta u + \nabla p = \eta^\omega(t,x), \qquad \operatorname{div} u = 0. \tag{6.0.12}$$

Since our discussion is restricted to space-periodic turbulence, we supplement the equations with the zero mean-value periodic boundary condition

$$x \in \mathbb{T}^3 = \mathbb{R}^3/\mathbb{Z}^3, \qquad \int_{\mathbb{T}^3} u(t,x)\,dx = \int_{\mathbb{T}^3} \eta(t,x)\,dx = 0.$$

The adequacy of (6.0.12) for the problem of turbulence was supported by Kolmogorov in his talk at a meeting of the Moscow Mathematical Society in 1978 [**92**]. See in [**66, 148**] for more discussion of (6.0.12) as an equation underlying statistical theories of turbulence.

In view of the well-known difficulties related to the 3d Navier–Stokes system, rigorous results concerning solutions of (6.0.12) are very far from what is needed to justify the K41 theory (or any other theory of turbulence). In particular, although it is known that (6.0.12) has stationary in time and homogeneous in space solutions (see in [**148**]), their properties are poorly understood and it is not clear if they play a special role among other solutions (we recall that the K41 theory insists upon the importance of stationary and homogeneous solutions). More importantly, it is not yet proved rigorously that the scaling in (6.0.12) is correct in the sense that the energy $\langle \frac{1}{2}|u(t)|^2_{L_2}\rangle$ of solutions for (6.0.12) remains of order one when $\nu \to 0$. If this is not the case, then the characteristic speed of the flow, equal to the square root of the energy, is not of order one, contrary to what we assume in our discussion.

Due to the equation the balance relation for the velocity u reads as

$$\frac{d}{dt}\mathbb{E}\frac{1}{2}|u(t)|^2_{L_2} + \nu\mathbb{E}|\nabla u(t)|^2_{L_2} = \dots,$$

where the dots correspond to terms of order one coming from the force (we keep the same agreement below). If we believe that $\frac{1}{2}\mathbb{E}|u(t)|^2_{L_2} \sim 1$, then in accordance with the balance relation the rate of dissipation of energy [5]

$$\epsilon := \nu\mathbb{E}|\nabla u|^2_{L_2}$$

also should be of order one. The fact that ϵ stays of order 1 as $Re \to \infty$ (or even converges to a limit) is, on the physical level of rigour, well established in the theory of turbulence, is older than the K41 theory and goes back at least to the work of G. I. Taylor in 1930's. It is well checked experimentally and numerically and is known as the *dissipation anomaly*, see in [**50, 146**].

Since $\epsilon \sim 1$, then $|\nabla u|^2_{L_2}$ is a large quantity of order ν^{-1}, although we stir the equation by a smooth force of order one, and $|u|^2_{L_2} \sim 1$. This is a manifestation of the *direct cascade of energy to high frequencies.* Indeed, writing $|u(t)|^2_{L_2}$ and $|\nabla u|^2_{L_2}$ as

$$|u(t)|^2_{L_2} = \sum |\hat{u}(t,\xi)|^2, \qquad |\nabla u(t)|^2_{L_2} = 4\pi^2 \sum |\xi|^2 |\hat{u}(t,\xi)|^2,$$

we derive from the relations $\mathbb{E}\sum|\hat{u}(t,\xi)|^2 \sim 1$ and $\mathbb{E}\sum|\xi|^2|\hat{u}(t,\xi)|^2 \sim \nu^{-1}$ that a substantial part of the energy $\mathbb{E}\sum \frac{1}{2}|\hat{u}(t,\xi)|^2$ is carried by high frequencies $|\xi| \gg 1$, even if originally at $t = 0$ all of it was concentrated in low modes and the high-frequency components of the force η are very small (or even vanish). When the energy is transferred to high modes, there it is dissipated by the Laplacian $\nu\Delta$. The cascade of energy agrees well with the Kolmogorov–Obukhov law (6.0.10), and was suggested by Kolmogorov as a mechanism responsible for the appearance of turbulence. As a manifestation of the direct cascade and of the growth of ∇u, experts in turbulence often consider the stretching of vorticity $\zeta(t,x) = \nabla_x \times u(t,x) \in \mathbb{R}^3$ (e.g. see in [**146**, (C.16)]). Namely, due to (6.0.12), the balance relation for the "energy of vorticity" is

$$\frac{d}{dt}\mathbb{E}\frac{1}{2}|\zeta(t)|^2_{L_2} + \nu\mathbb{E}|\nabla\zeta(t)|^2_{L_2} = \mathbb{E}\sum_{i,j=1}^{3}\int (\zeta_i\zeta_j s_{ij})dx + \dots, \tag{6.0.13}$$

where $s_{ij} = \frac{1}{2}(\partial u_i/\partial x_j + \partial u_j/\partial x_i)$ is the rate of strain, and as above the dots stand for terms of order one, coming from the force. The term $\sum_{i,j}\int(\zeta_i\zeta_j s_{ij})dx =: \mathrm{Str}$ is called the *vorticity stretching.* Numerical studies of eq. (6.0.12) clearly show that the expectation of Str is a large positive quantity. In view of the balance relation, to keep the energy of vorticity $\mathbb{E}\frac{1}{2}|\zeta(t)|^2_{L_2}$ from indefinite growth, the rate of dissipation of vorticity $\nu\mathbb{E}|\nabla\zeta(t)|^2_{L_2}$ should be large. So

$$\mathbb{E}|\nabla\zeta(t)|^2_{L_2} \gg \nu^{-1} \sim \mathbb{E}|\nabla u(t)|^2_{L_2} \gg \mathbb{E}|u(t)|^2_{L_2} \sim 1. \tag{6.0.14}$$

[5] The notation ϵ for the rate of dissipation of energy is due to Kolmogorov and is standard in the theory of turbulence. This is one of the relevant parameters in the Kolmogorov theory. In particular, the constants in (6.0.7) and (6.0.10) depend on it. We do not specify this dependence which can be found in [**68, 114**].

These relations convincingly show that a large part of the energy of the solutions u is contained in high frequencies. Unfortunately no rigorous theory in support of the numerical fact that $\mathbb{E}(\mathrm{Str}) \gg 1$ currently exists.

2d turbulence. A number of physical situations lead to turbulent flows $u = (u_1, u_2, u_3)(t, x)$, (approximately) stratified to horizontal motions $\{u_3 = \text{const}\}$ with (u_1, u_2) independent from x_3. In particular, this happens in the Earth atmosphere, where the vertical motion of air is much slower than the horizontal one. Completely stratified flows are described by solutions of the 2d Navier–Stokes system (6.0.12) and motivate its study. Again, discussing the 2d equations we assume the zero mean-value periodic boundary conditions, i.e. that

$$x \in \mathbb{T}^2, \qquad \int_{\mathbb{T}^2} u(t,x)\,dx = \int_{\mathbb{T}^2} \eta(t,x)\,dx = 0. \tag{6.0.15}$$

Unlike the 3d Navier–Stokes system, the 2d equations are well posed and their study is much more advanced (e.g. see in [**103**]). But the two systems differ significantly. A major difference comes from the fact that since for stratified 3d flows the vorticity $\zeta = \nabla \times u$ is a vertical vector, then the vorticity of a 2d vector field $u = (u_1, u_2)$ is a scalar function $\zeta = \partial u_2/\partial x_1 - \partial u_1/\partial x_2$ (the vertical component of the vector above). It turns out that under periodic boundary conditions, the balance relation for it is the same as for the velocity field:

$$\frac{1}{2}\frac{d}{dt}\mathbb{E}|\zeta(t)|_{L_2}^2 + \nu\mathbb{E}|\nabla\zeta(t)|_{L_2}^2 = \dots$$

(as before the dots stand for terms coming from the force). In other words, no vortex stretching takes place in 2d turbulence! Accordingly, the cascade to high frequencies, where the energy is dissipated by the Laplacian, is less vigorous. So to keep the energy of solutions of the 2d system of order one, the random force η should be not of order 1, but much smaller, proportional to $\sqrt{\nu}$:

$$\eta(t,x) = \sqrt{\nu}\,\eta'(t,x), \qquad \eta' \sim 1. \tag{6.0.16}$$

See [**100**, Section 10.3], and [**103**, Section 5.2.4]. With this scaling of the forcing higher norms of solutions also stay of order one as $\nu \to 0$; in particular $\mathbb{E}|\nabla u|_{L_2}^2 \le$ Const uniformly in ν. That is, no dissipation anomaly takes place in 2d, see [**100**, **103**].

For a number of classes of smooth in x random forces η' it was proved, starting with the paper [**102**], that the 2d equation (6.0.12) defines a mixing system: in a functional space $H = \{u(x)\}$, there is a unique measure μ_ν (a stationary measure for the equation, cf. Chapter 3), such that every solution u converges to μ_ν in distribution:

$$\mathcal{D}u(t,\cdot) \rightharpoonup \mu_\nu \quad \text{as} \quad t \to \infty;$$

see [**46**, **103**, **104**] for a discussion of the result and for a history of its proof. The equation has a stationary in time solution $u_\nu^{st}(t,x)$ satisfying $\mathcal{D}u_\nu^{st}(t,\cdot) \equiv \mu_\nu$, and if the force is periodic and homogeneous in x, this solution also is. This is in good agreement with the assertion postulated by K41 (as well as by other statistical theories, e.g. [**7**, **143**, **149**]) that turbulence prefers stationary and homogeneous flows. But in striking difference with the relations (6.0.14), predicted and numerically observed for 3d turbulence, for a stationary 2d turbulent flow $u_\nu^{st}(t,x)$ we have

$$\mathbb{E}|\nabla^2 u(t)|_{L_2}^2 \sim \mathbb{E}|\nabla u(t)|_{L_2}^2 \sim \mathbb{E}|u(t)|_{L_2}^2 \sim 1,$$

see [**103**, Section 3].

Although formally Kolmogorov's heuristic derivation of the laws of the K41 theory also applies to 2d turbulence, it was always clear for physicists that the main predictions of the theory, including the 2/3-law and the Kolmogorov–Obukhov law, should not hold in the 2d case. Still it was suggested by a number of researches that the spectral power law (6.0.10) holds true for 2d turbulence if we suitably adjust there the exponent 5/3, e.g. see [**93**]. Convincing experimental study of the energy spectrum E_k of turbulent 2d flows is hardly possible since existing techniques do not allow to obtain strongly stratified 3d turbulent flows, and the results of numerical study of the 2d energy spectrum are also not completely convincing when $\nu \ll 1$. Analytical study of 2d equations (6.0.12), (6.0.16) with small $\nu > 0$ revealed a number of interesting universal properties of their solutions and stationary measures (see [**100**, Section 11], [**103**, Section 5] and [**104**]). This, however, does not tell much about the asymptotic behaviour of the energy spectrum and structure function.

Since the lack of vortex stretching makes the 2d Navier–Stokes equations with small ν significantly different from the real 3d equations and since the former ones, being much simpler than the latter, still do not allow to understand deeply the properties of their solutions in the turbulent regime $\nu \ll 1$, the 2d Navier–Stokes system is not a satisfactory model for hydrodynamical turbulence. In search for instrumental models for the phenomenon, starting from 1930's physicists often have looked at 1d equations.

1d turbulence (burgulence). The Burgers equation

$$u_t(t,x) - \nu u_{xx} + u u_x = \eta(t,x) \tag{6.0.17}$$

owes its name to the Dutch physicist Jan Burgers who studied it starting from the late 1930's, see the book [**35**] (however, the equation was considered before him, for example by Forsyth [**67**] and Bateman [**8**], in 1906 and 1915 respectively). Burgers believed that the equation makes a simplified version of the Navier–Stokes system and keeps some of its crucial features. In particular, his opinion was that (6.0.17) with small ν makes a model for statistical theories of hydrodynamical turbulence and he performed a program of intensive study of the equation in this perspective, e.g. see [**34**]. This work was highly appreciated by Burgers' colleagues, including von Neumann, see [**150**, p. 437]. Since then physicists have kept a constant interest in the equation and its stochastic versions (with a random force η and/or random initial data $u(0,x)$), e.g. see in [**85**]. See the Introduction for a discussion on the history of the equation.

As before, discussing the equation we assume the zero mean-value periodic boundary conditions:

$$x \in S^1 = \mathbb{R}/\mathbb{Z}, \qquad \int u(t,x)\,dx = \int \eta(t,x)\,dx = 0.$$

Concerning the force η we will suppose that either it is zero, or it has the form (1.1.6) (then (6.0.17) is the stochastic Burgers equation). Similarly to the stochastic 3d Navier-Stokes system (6.0.12) [6] the scaling in eq. (6.0.17) is correct in the sense that the energy $\frac{1}{2}\mathbb{E}|u|^2_{L_2}$ of a solution u remains of order one as $\nu \to 0$, see Theorem 2.3.15 . Besides, by Theorem 2.3.1 for solutions of the equation the

[6] if we believe that indeed this system correctly describes 3d turbulence!

dissipation anomaly

$$\epsilon := \nu\langle\langle\|u_x\|^2\rangle\rangle \sim 1$$

is a proven fact. Moreover, naming the function $\zeta{=}u_x$ the vorticity of a solution u, we saw in Example 2.3.13 that its energy satisfies the balance relation

$$\frac{d}{dt}\mathbb{E}\Big(\tfrac{1}{2}|\zeta(t)|_{L_2}^2\Big) + \nu\mathbb{E}|\zeta_x(t)|_{L_2}^2 = -\frac{1}{2}\mathbb{E}\int \zeta^3(t,x)\,dx + \frac{1}{2}B_1$$

with the non-trivial vorticity stretching term $-\frac{1}{2}\mathbb{E}\int\zeta^3(t,x)\,dx$. Analysis of this relation, based on estimates from Part I, shows that there

$$\mathbb{E}\tfrac{1}{2}|\zeta(t)|_{L_2}^2 \sim \nu^{-1}, \quad \nu\mathbb{E}|\zeta_x(t)|_{L_2}^2 \sim \nu^{-2}, \quad \mathbb{E}\int \zeta^3(t,x)\,dx \sim -\nu^{-2}.$$

This is very similar to what is expected to hold in the hydrodynamical turbulence, see (6.0.14).

What has been just said convincingly supports Burgers' belief that eq. (6.0.17) makes a reasonable model for 3d turbulence.

Studying in [**34**] statistical properties of solutions for eq. (6.0.17) with $\eta = Cu$, Burgers discovered on the physical level of rigour a number of features of its solutions, similar to those in the K41 theory. In particular, he established that for initial data of order one the dissipation scale l_d is ν^{-1}, and that the energy spectrum is $E_k \sim k^{-2}$ for $k \lesssim \nu^{-1}$. In the late 90's U. Frisch and his collaborators started to re-examine statistical properties of solutions for the Burgers equation with random initial data, numerically and analytically (still on the physical level of rigour). Using an heuristic argument sharper and more convincing than that of Burgers they established (see [**1**, Section 7] and [**10**, Section 2.2]) that for small ν the dissipation space scale is

$$l_d = \nu^{-1}, \tag{6.0.18}$$

so the inertial range for burgulence is $I_{inert} = [C, \nu^{-1}]$. They also observed that the vorticity $\zeta = u_x$, as a function of x, is very intermittent; see Figure 6.0.2, which clearly shows that ζ is a function of order 1 on most of S^1 and is large on a small subset of S^1. Next they proved on the physical level of rigour that for small ν the structure function [7]

$$S_{p,r}(u) = \langle\int |u(x+r) - u(x)|^p dx\rangle$$

of a solution u for (6.0.17) behaves as

$$S_{p,r} \sim |r|^{\min(p,1)}, \qquad |r| \in I^x_{inert}, \tag{6.0.19}$$

and derived from this the spectral power law

$$E_k \sim k^{-2}, \qquad k \in I_{inert}, \tag{6.0.20}$$

predicted by Burgers. As we already mentioned in the introduction, based on obvious similarities between the laws (6.0.18)–(6.0.20) and (6.0.3), (6.0.7), (6.0.10), U. Frisch suggested to call the 1d turbulence in eq. (6.0.17) "burgulence".

The results of U. Frisch and others, presented above, inspired E, Khanin, Mazel and Sinai [**56**] to undertake a profound study of the stochastic Burgers equation (1.1.12) under the inviscid limit $\nu \to 0$. This work was developed in subsequent

[7] We added to the definition of the structure function the averaging in dx since solutions of the Burgers equation which we discuss are not assumed to be space-homogeneous.

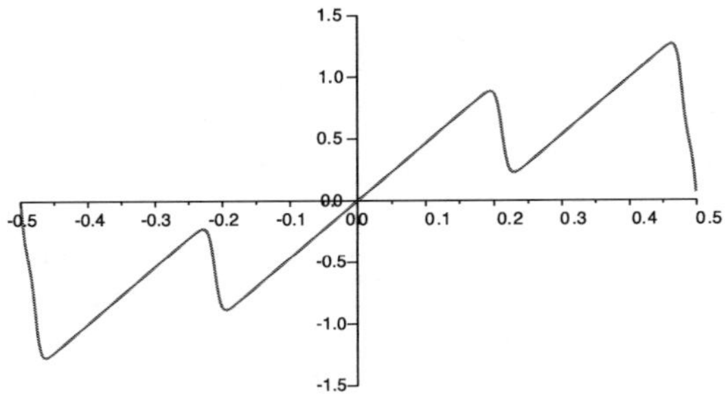

FIGURE 6.0.2. Typical solution of the Burgers equation

publications, including [**11, 70, 78**]. There, exploiting the Lax–Oleinik formula for inviscid solutions it was established that under the limit $\nu \to 0$ eq. (1.1.12) defines a Markov process in a suitable function space. Then, for this process, the authors of [**56**] obtained a presentation in terms of a stochastic Lagrangian, and studied properties of solutions. Nonetheless the approach of [**56**] has not yet made it possible to justify rigorously the laws of burgulence.

A rigorous derivation of the laws (6.0.18)–(6.0.20), based on the results of Part I, is the main goal of Part II (and of our book as a whole). We will justify them in three different senses:

Firstly, in Section 7.1 we understand the averaging $\langle\cdot\rangle$ which enters the definitions of the objects involved in the laws (6.0.18)–(6.0.20) as $\langle\langle\cdot\rangle\rangle$, i.e. as the local averaging in time and the averaging in ensemble (see (2.3.5)). We prove that for burgulence in the stochastic Burgers equation the three ranges are : $I_{diss} = (\nu^{-1}, \infty)$, $I_{energy} = [1, C_e)$ and $I_{inert} = [C_e, \nu^{-1}]$ (we recall that the edges of the ranges ν^{-1} and C_e are defined modulo multiplicative ν-independent constants).

Next in Section 7.2 we prove the law of averaged powers of increments (6.0.19) for solutions $u(t,x)$ of the equation, and in addition show that in the dissipation range the function $S_{p,r}$ behaves as $|r|^p$ times a non-positive degree of ν:

$$S_{p,r} \sim \nu^{1-\max(1,p)}|r|^p, \qquad |r| \in I^x_{diss}. \tag{6.0.21}$$

From this we obtain that the dissipation range in the x-presentation I^x_{diss} may be equivalently defined as the maximal interval of the form $(0, \nu^{c_d})$, such that for r from this interval the function $(S_{p,r})^{1/p}$ behaves linearly in $|r|$, for all $p \geq 1$.

In Section 7.3 we derive from (6.0.19) and some results from Part I the spectral power law (6.0.20), where the layer Σ_k in the definition of E_k is $\Sigma_k = \{\xi \in \mathbb{Z}^+ : M^{-1}k \leq |\xi| \leq Mk\}$, for a suitable $M > 1$.

Secondly, in Section 7.4, exploiting the mixing in the stochastic Burgers equation, established in Sections 3.3 and 4.2, we show that if we understand $\langle\cdot\rangle$ as the averaging only in ensemble, then the law of averaged powers of increments and the spectral power law hold for any solution $u(t,x)$, asymptotically as $t \to \infty$. Moreover, the rates of convergence for both laws admit ν-independent bounds. Now consider the stationary measure μ^ν, describing the limiting in time statistical properties of solutions. Then if in the formulas for $S_{p,r}(u)$ and $E_k(u)$ we understand the brackets $\langle\cdot\rangle$ as the averaging against $\mu^\nu(du)$, the two laws still hold. In other words, if $u^{st}(t,x)$ is a stationary solution, $\mathcal{D}(u^{st}(t,\cdot)) \equiv \mu^\nu$, and in the definitions of the structure function $S(u^{st})$ and of the energy spectrum $E_k(u^{st})$ we understand

$\langle\cdot\rangle$ as the averaging in ensemble for a fixed moment of time t, then the two laws (6.0.19) and (6.0.20) hold for u^{st}, for any t. Moreover, if in (1.1.6) $b_s \equiv b_{-s}$, then both the random force η and the measure μ^ν are space-homogeneous. This is in line with the K41 theory, dealing only with stationary in time and homogeneous in space flows. It also agrees well with the general belief in the theory of turbulence that when time grows, statistical characteristics of turbulent flows (which in our notation is the measure μ^ν) converge to a universal statistical equilibrium at a rate, independent from the viscosity. See in [**7**], e.g. pages 6-7 and 109.

Thirdly in Section 8 we start with establishing the well known fact that if $u^\nu(t)$ is a solution of the stochastic Burgers equation, then the inviscid limit $u^e(t) = \lim_{\nu\to 0} u^\nu(t)$ holds in any space L_p, $p < \infty$, and next show that the structure function of the inviscid limit $S_{p,r}(u^e)$, defined with $\langle\cdot\rangle := \langle\langle\cdot\rangle\rangle$, satisfies the law of averaged powers of increments for all $|r| \le c$ with a suitable $c > 0$, while the energy spectrum satisfies the power law for all $k \ge 1$. So for the inviscid solutions the dissipation range degenerates to the empty set. Moreover, the limiting inviscid dynamics is mixing with a unique stationary measure μ^e, and if in the formulas for $S_{p,r}$ and E_k the brackets $\langle\cdot\rangle$ stand for the averaging against μ^e, then the law (6.0.19) holds with $I^x_{inert} = (0, c']$, while the power law (6.0.20) holds for all k. It is natural to call this behaviour of the inviscid solutions *inviscid burgulence*, as opposed to *viscous burgulence* above, which describes solutions for the Burgers equation with small $\nu > 0$.[8] Our results on the inviscid burgulence have no analogy in the K41 theory since there the Reynolds number *Re* of the fluid flow is a fixed finite quantity, and since on the mathematical side of the question, the behaviour of solutions of the 3d hydrodynamical equations on time intervals of order 1, when $Re \to \infty$, is a completely open problem.

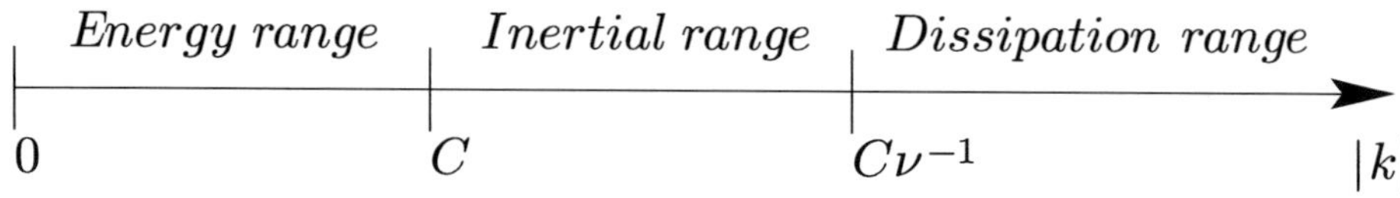

FIGURE 6.0.3. Burgulence scales

It is crucial for our argument that the averaging $\langle\cdot\rangle$, used to define the structure function and the energy spectrum involves the local in time averaging, and we do not know if the laws of burgulence (6.0.19) and (6.0.20) hold for solutions of the stochastic Burgers equation at fixed finite moments of time, under averaging in ensemble only. Nevertheless, in Section 7.5 we show that the laws (6.0.19) and (6.0.20) remain valid with high probability if $\langle\cdot\rangle$ is understood as the time-averaging along sufficiently long time intervals, without taking the mean-value in ensemble (in this case $S_{p,r}$ and E_k are random variables).

In Section 10.1.1 we review the results showing that the assertions of Chapter 7 concerning the behaviour of solutions for the stochastic Burgers equation on time intervals of order one essentially remain true for the free Burgers equation $(6.0.17)_{\eta=0}$ with initial data of order one. This group of results describes *decaying burgulence*. In Section 10.1.2 we discuss the Burgers equation (6.0.17), where η is a

[8] Probably it is surprising (if not disappointing) that two principal laws of the K41 theory, notoriously hard to justify rigorously, have in burgulence close analogues which admit relatively simple proofs (we owe this remark to Prof. L. Pastur).

random kick-force of period $\theta > 0$, and explain that the main results from Chapter 7 remain true for its solutions. Next in Section 10.2 we prove that when the period θ of a kick-force goes to zero and the force is scaled by the factor $\sqrt{\theta}$, statistical properties of the corresponding solutions $u^\theta(t;u_0)$ become close, in a number of senses, to those of the solutions $u(t;u_0)$ of the stochastic Burgers equation (1.1.12). In particular, we prove that $\mathcal{D}u^\theta(t;u_0) \rightharpoonup \mathcal{D}u(t;u_0)$ as $\theta \to 0$, uniformly in time. Section 10.1.3 contains a discussion of eq. (0.0.1), where the random force η is given by a bounded random Haar series. Finally, in Section 10.3 we discuss the restrictions on the random force η which guarantee that the mixing in (6.0.17) is exponential. Then solutions of the equation are ergodic in the sense that they satisfy the strong law of large numbers and the central limit theorem.

Space-homogeneous turbulence. In the discussion above we restricted ourselves to space-periodic flows, modeled by the stochastic Navier–Stokes system on $\mathbb{T}^2$ and $\mathbb{T}^3$, and to the stochastic Burgers equation on S^1. We saw that the latter makes an adequate model to rigorously derive three principal laws of 1d turbulence (burgulence). Certainly, it is possible that some advanced analysis of the 3d Navier–Stokes system on $\mathbb{T}^3$, which is beyond hope for the mathematics of today, will lead to a derivation of the K41 theory (rather of its improved version), based on those equations. But Kolmogorov, as well as Taylor and von Karman and Howard before him, reckoned that turbulent fields are stationary homogeneous (and isotropic). In response to the direct question whether space–periodic solutions of the 3d stochastic Navier–Stokes system may describe the K41 theory, Kolmogorov replied that he is not certain since "space-periodic solutions may be too regular".[9] So it may be that a justification of K41 will come through a deep examination of space-homogeneous solutions for the stochastic Navier–Stokes system in 3d. Existence of such solutions of low regularity is established in [**148**], but neither their uniqueness, nor their physically relevant properties, have been obtained so far.

However, significant progress in the study of space-homogeneous time-stationary solutions for the stochastic Burgers equation has been achieved in the last few years. A breakthrough was made by Bakhtin, Cator and Khanin [**2**] who considered the inviscid Burgers equation on the whole line with random forcing which is a Poisson process in space-time. Using heavy techniques from the percolation theory they proved uniqueness of its space-homogeneous time-stationary solution (up to an unavoidable additive constant). Their result was extended to the viscous case by Bakhtin and Li [**4**]. For more references and a discussion see the review [**3**].

Very recently, a setting much closer to that in our book was treated by Dunlap, Graham and Ryzhik [**54**]. Namely, they consider the viscous Burgers equation on the whole line with random forcing which is smooth in space and white in time, and prove uniqueness (again, up to an additive constant) of its space-homogeneous time-stationary solution. They use PDE methods, some of which are analogous to those in our book. See also the preprint [**53**] by Dunlap which is concerned with the multidimensional generalisation of their results for $d = 2, 3$ and proves the existence of space-homogeneous time-stationary solutions.

[9]The authors thank Prof. A. Fursikov for these reminiscences from the period when he and Prof. M. Vishik were writing the book [**148**] and a number of times discussed their work with Kolmogorov.

CHAPTER 7

Rigorous burgulence

In this chapter we assume that

$$B_0 > 0 \quad \text{and} \quad B_m < \infty \qquad \forall m \tag{7.0.1}$$

(and modify the random force ξ following the agreement after Corollary 1.2.4). Then by Corollary 2.2.5 solutions of eq. (B) with initial data in H^1 a.s. are smooth in x for $t > 0$. Our goal in this chapter is to study small scale properties of these solutions in their relation to burgulence and prove the results announced in Chapter 6. The proofs in Sections 7.1-7.3 are based on the theorems from Chapter 2, and those in Section 7.4 also use the results on mixing from Chapters 3, 4.

The theory built in Sections 7.1-7.3 only uses the assumption (7.0.1) and the assertions of Theorems 2.3.4 and 2.1.2 (as well as their corollaries). So it holds true for any family of random processes $u^\nu(t; u_0) \in H^1$, depending on the parameters $u_0 \in H^1$, $\nu \in (0, 1]$, such that:

(H0) $u^\nu(t; u_0) \in H^m$ for every m if $t > 0$;

(H1) Oleinik's estimates (2.1.1), (2.1.2) hold for any $0 < \nu \leq 1$, $T > 0$, $0 < \theta \leq T$ and any $u_0 \in H^1$;

(H2) for any $u_0 \in H^1$, u^ν satisfies (2.2.2) for each $m \in \mathbf{N}$.

(H3) for any $u_0 \in H^1$, u^ν satisfies (2.3.2).

As we have seen previously, hypotheses (H0)–(H3) jointly with the Hölder and Gagliardo-Nirenberg inequalities imply sharp estimates for moments of Sobolev norms of processes $u^\nu(t; u_0)$.

We could state the results in Sections 7.1-7.3 as properties of an abstract family of processes $u^\nu(t; u_0) \in H^1$ satisfying (H0)-(H3). Instead we prefer to keep talking about the processes $u^\nu(t)$ as solutions of eq. (B), where the assumptions (7.0.1) hold.

Note that if $u^f(t; u_0)$ is a solution of equation (0.0.12), (0.0.13), (1.1.5), (1.1.6), equal to $u_0 \in H^1$ at $t = 0$, then it satisfies (H2) by the result of Problem 2.2.4, so it also meets (H0). Relation (H1) holds for u^f in view of the result of Problem 2.1.4. Using (H1) and (H2) we immediately see that u^f satisfies the balance relation (2.3.1), so (2.3.2) is valid for u^f by the same argument as in Section 2.3, and (H3) also holds. Accordingly the results of Sections 7.1-7.3 remain true for the solutions u^f.

In Section 10.1 we consider the Burgers equation (B) with forces η, different from (1.1.5), (1.1.6). We discuss the validity for them of relations (H0)–(H3) and of the results obtained below in this chapter.

The constants in estimates below may depend on $\vec{B} = (B_0, B_1, \dots)$, but not on $\nu \in (0, 1]$ and on T, σ as in (7.1.2) below.

7.1. Dissipation scale.

We recall that for a real-valued random process $\zeta^\omega(t)$ we denote by $\langle\langle\zeta\rangle\rangle = \langle\langle\zeta\rangle\rangle_{T,\sigma}$ its averaging in ensemble and local averaging in time:

$$\langle\langle\zeta\rangle\rangle = \frac{1}{\sigma}\int_T^{T+\sigma} \mathbb{E}\zeta^\omega(t)\,dt. \tag{7.1.1}$$

Here the parameters T and σ satisfy

$$T \geq 1, \quad \sigma \geq \sigma_0, \tag{7.1.2}$$

where σ_0, defined in (2.3.3), depends only on $B_0 > 0$ and B_4.

Elaborating a bit the definition (6.0.2) we arrive at the following

DEFINITION 7.1.1. Consider an H^1-smooth complex or real-valued function $u = u^\nu(x)$, $x \in S^1$, depending on a parameter $\nu \in (0,1]$,

$$u^\nu(x) = \sum_{k\in\mathbb{Z}^*} \hat{u}_k^\nu e^{2i\pi kx}. \tag{7.1.3}$$

(1) Its mathematical dissipation space scale $l_d(u)$ (if it exists) equals $\nu^{-\gamma}$ with the exponent $\gamma > 0$ defined as $\gamma = \inf\Gamma$, where Γ is the set formed by all real numbers $\gamma' > 0$ such that $\hat{u}_k^\nu$ as a function of k decays fast, if $|k| \geq \nu^{-\gamma'}$:

$$\forall N \in \mathbf{N} \;\; \exists C_{N,\gamma'} > 0 \quad \text{such that} \quad |\hat{u}_k^\nu| \leq C_{N,\gamma'}|k|^{-N} \quad \forall\, |k| \geq \nu^{-\gamma'}, \quad \forall \nu \in (0,1]. \tag{7.1.4}$$

If $\Gamma = \emptyset$, or if $\inf\Gamma = 0$, then the dissipation scale $l_d(u)$ is not defined.

(2) Consider a random field

$$u^{\nu,\omega}(t,x), \;\; 0 \leq t \leq T_1, \;\; x \in S^1, \;\; \text{where } T_1 \geq T+\sigma,$$

which defines a random process $u^{\nu,\omega}(t)$ in H^1.[1] Decompose $u^{\nu,\omega}(t,\cdot)$ in the Fourier series (7.1.3) with $\hat{u}_k^\nu = \hat{u}_k^{\nu,\omega}(t)$. Then $l_d(u) = \nu^{-\gamma}$, $\gamma > 0$, if (1) holds for the quantities $|\hat{u}_k^\nu|$ replaced by $\langle\langle|\hat{u}_k^\nu|^2\rangle\rangle^{\frac{1}{2}}$.

(3) The physical dissipation space scale $l_d(u)$ of a function or a random field u is $Cl_d(u)$, where $l_d(u)$ is as above and C is any ν-independent positive constant.

By a small abuse of notation, we denote by the same symbol the physical and mathematical space-scales.

Obviously, the set Γ in item (1) either is the closed interval $[\gamma,\infty)$, or the open interval (γ,∞). If a random field u^ν is time-independent, then in the item (2) of the definition the averaging $\langle\langle\cdot\rangle\rangle$ may be replaced by $\mathbb{E}$.

Note that since the dissipation scale is $l_d = \nu^{-\gamma}$ with $\gamma > 0$, then for a function $u^\nu(x) = u(x)$ which does not depend on ν the scale l_d is not defined, since it "should be $l_d = \nu^0 = 1$" (cf. the problem below with $a = 0$).

PROBLEM 7.1.2. Verify that if $v(x)$ is a smooth not identically zero function on S^1 with zero mean-value, if $a > 0$ and $u^\nu(x) = v(\lceil\nu^{-a}\rceil x)$ (see (0.0.16)), then $l_d(u) = \nu^a$.

Let $u(t) = u(t;u_0)$ be a solution of (B) with $u_0 \in H^1$. Since its Fourier coefficients $\hat{u}_k(t) = \hat{u}_k^{\nu,\omega}(t;u_0)$ satisfy (2.1.18), then in view of Corollary 2.1.5

$$\langle\langle|\hat{u}_k|^2\rangle\rangle \leq Ck^{-2} \quad \forall k \geq 1. \tag{7.1.5}$$

[1] We recall that in our book random processes are assumed to have continuous trajectories.

THEOREM 7.1.3. *For every $u_0 \in H^1$ the mathematical dissipation space scale of solution $u(t;u_0)$ equals ν^{-1}.*

PROOF. Theorem 2.3.4 implies that

$$\langle\langle|\hat{u}_k|^2\rangle\rangle \le |k|^{-2m}\langle\langle\|u\|_m^2\rangle\rangle \le C_m\nu^{1-2m}|k|^{-2m}$$

for every k and each $m \in \mathbf{N}$. So for $|k| \ge \nu^{-\gamma'}$ with $\gamma' > 1$ we get:

$$\langle\langle|\hat{u}_k|^2\rangle\rangle \le C_m|k|^{(2m-1)/\gamma'-2m} \le C_m|k|^{-2m(1-1/\gamma')},$$

for each $m \in \mathbf{N}$. Thus the set Γ in Definition 7.1.1 contains all $\gamma' > 1$.

On the contrary, if $\gamma' \in \Gamma$, then from (7.1.5) and (7.1.4), for every $m, N \in \mathbf{N}$ we have:

$$\langle\langle\|u\|_m^2\rangle\rangle = \sum |k|^{2m}\langle\langle|\hat{u}_k^\nu|^2\rangle\rangle \le \sum_{|k|\le\nu^{-\gamma'}} C|k|^{-2+2m} + \sum_{|k|\ge\nu^{-\gamma'}} |k|^{2m}C_{N,\gamma'}|k|^{-2N}.$$

Choosing $N = m+1$ we obtain $\langle\langle\|u\|_m^2\rangle\rangle \le C_m\nu^{-\gamma'(2m-1)} + C(m,\gamma')$. But by Theorem 2.3.4, $\langle\langle\|u\|_m^2\rangle\rangle \ge C_m^{-1}\nu^{-(2m-1)}$. Thus $\gamma' \ge 1$ and we have seen that $\{\gamma' > 1\} \subset \Gamma \subset \{\gamma' \ge 1\}$. So $\inf\Gamma = 1$ and the assertion is proved. □

So the (physical) dissipation range for solutions of (1.1.12) in the Fourier presentation is

$$I_{diss} = (C\nu^{-1}, \infty).$$

Energy and inertial ranges.

Elaborating the definition of the energy range $I_{energy} = [1, l_e)$ given in Chapter 6, we adopt the following rigorous modification of it for a random field $u^\nu(t,x)$ with $0 \le t \le T_1$, $T_1 \ge T + \sigma$, which defines a random process in H: the mathematical energy scale l_e of u, if it exists, is $l_e(u) = \nu^{-c_e}$, where the exponent $c_e \ge 0$ is defined by the following relation:

For any $\delta > 0$ there exists $C_e > 0$, depending on δ and u, such that

$$\sum_{|k|\le C_e\nu^{-c_e}} \langle\langle|\hat{u}_k|^2\rangle\rangle \ge (1-\delta)\langle\langle\|u\|^2\rangle\rangle, \tag{7.1.6}$$

for all $\nu \in (0,1]$. Moreover, if $c_e > 0$, then for any $c \in (0, c_e)$ and any choice of $0 < \delta < 1$, $C_e > 0$ the relation above with the exponent c_e replaced by c is violated for some value of ν.

Note that due to this definition the exponent $c_e \ge 0$, if it exists, is defined in a unique way. Similarly to the dissipation range l_d, the physical energy range of a random field u is the interval $[1, Cl_e(u)) = [1, C\nu^{-c_e})$, where C is any ν-independent positive constant.

PROPOSITION 7.1.4. *For any $u_0 \in H^1$, the mathematical energy scale l_e of solution $u(t;u_0)$ is $l_e = \nu^0 = 1$.*

PROOF. For $c_e = 0$ and $u = u(t;u_0)$ relation (7.1.6) holds with a suitable positive C_e since due to Theorem 2.3.15 $\langle\langle\|u\|^2\rangle\rangle \ge C^{-1}$, and by (7.1.5)

$$\sum_{|k|>C_e} \langle\langle|\hat{u}_k|^2\rangle\rangle \le C_1 \sum_{k>C_e} k^{-2} \le C_2C_e^{-1}.$$

So $\sum_{|k|\le C_e}\langle\langle|\hat{u}_k|^2\rangle\rangle \ge \langle\langle\|u\|^2\rangle\rangle - C_2C_e^{-1}$, which is bigger than $(1-\delta)\langle\langle\|u\|^2\rangle\rangle$ if C_e is large enough. □

Accordingly for burgulence the (physical) energy range in the Fourier presentation is

$$I_{energ} = [1, C),$$

and the (physical) inertial range in the Fourier presentation is

$$I_{inert} = [C, C'\nu^{-1}].$$

7.2. Structure function and intermittency.

Structure function.

DEFINITION 7.2.1. For a random field $u^\omega(t,x)$, $0 \le t \le T_1$, where $T_1 \ge T + \sigma$, which for every $p < \infty$ defines a random process $u^\omega(t)$ in L_p, we define its structure function $S_{p,l}(u)$ as follows:

$$S_{p,l} = S_{p,l}(u) = \langle\langle s_{p,l}(u^\omega(t,\cdot))\rangle\rangle, \quad |l| \le 1/2, \ p > 0, \tag{7.2.1}$$

where

$$s_{p,l}(v) = \int_{S^1} |v(x+l) - v(x)|^p dx$$

for any $v(x) \in L_p$.

This definition applies to solutions $u(t,x;u_0)$ of (B) with $u_0 \in H^1$. If in (1.1.6) $b_s \equiv b_{-s}$ and $u_0^\omega(x)$ is a space-homogeneous random field such that $u_0 \in H^1$ (e.g. if $u_0 = 0$), then by Theorem 1.5.6 solution $u(t,x;u_0)$ also is space-homogeneous. Then

$$S_{p,l} = \langle\langle |u(x+l) - u(x)|^p\rangle\rangle \quad \text{for any} \quad x, \tag{7.2.2}$$

in direct analogy with the definition (6.0.5) of the hydrodynamical longitudinal structure function.

The two theorems below give lower and upper estimates for the structure function of solutions for (B). They are asymptotically sharp as $l \to 0$, separately in the dissipation and inertial ranges. The constants entering the theorems depend only on the random force (1.1.6).[2]

We recall (7.0.1).

THEOREM 7.2.2. *For every $p > 0$, there exists $C_p > 0$ such that if $u = u^\omega(t;u_0)$ with any $u_0 \in H^1$, then $S_{p,l} = S_{p,l}(u)$ satisfies:*

$$\begin{cases} S_{p,l} \le C_p |l|^p, & \text{if } p \in (0,1) \\ S_{p,l} \le C_p |l|^p \nu^{1-p}, & \text{if } p \ge 1 \end{cases} \quad \text{for } |l| \le 1/2 \text{ and } \nu \in (0,1]\,. \tag{7.2.3}$$

Moreover, in the dissipation range the estimates above are sharp. Namely there exist constants $c_1', c_* > 0$, and for every $p > 0$ there exists $c_p > 0$ such that

$$\begin{cases} c_p |l|^p \le S_{p,l}, & \text{if } p \in (0,1) \\ c_p |l|^p \nu^{1-p} \le S_{p,l}, & \text{if } p \ge 1 \end{cases} \quad \text{for } |l| \le c_1'\nu \text{ and } \nu \in (0,c_*]\,. \tag{7.2.4}$$

THEOREM 7.2.3. *There exists a constant $c_2' \in (0,1/2]$, and for every $p > 0$, there exist $\tilde c_p$, $\tilde C_p > 0$ such that with the same choice of $c_1', c_* > 0$ as above, in the inertial range the function $S_{p,l}$ satisfies the following asymptotically sharp estimates:*

$$\begin{cases} \tilde c_p |l|^p \le S_{p,l} \le \tilde C_p |l|^p, & \text{if } p \in (0,1) \\ \tilde c_p |l| \le S_{p,l} \le \tilde C_p |l|, & \text{if } p \ge 1 \end{cases} \quad \text{for } |l| \in [c_1'\nu, c_2'] \text{ and } \nu \in (0,c_*]. \tag{7.2.5}$$

[2]Obviously, the constants C_p and c_p also depend on p.

Since $s_{p,l}(u) = \int |u(x+l) - u(x)|^p dx = \int |u(y) - u(y-l)|^p dy = s_{p,-l}(u)$ and for $l = 0$ the assertions are obvious, then proving both theorems we may assume that $l \in (0, 1/2]$.

PROOF OF (7.2.3) IN THEOREM 7.2.2. If $p \geq 1$, then

$$S_{p,l} \leq \langle\langle \int |u(x+l) - u(x)| dx \cdot \max_x |u(x+l) - u(x)|^{p-1} \rangle\rangle.$$

By Hölder's inequality we get:

$$\text{(7.2.6)} \quad S_{p,l} \leq \underbrace{\langle\langle \Big(\int |u(x+l) - u(x)| dx \Big)^p \rangle\rangle^{1/p}}_{=:I} \underbrace{\langle\langle \max_x |u(x+l) - u(x)|^p \rangle\rangle^{(p-1)/p}}_{=:J}.$$

On one hand, Lemma 2.1.1, applied to the function $x \mapsto u(x+t) - u(x)$, yields:

$$\int |u(x+l) - u(x)| dx = 2 \int (u(x+l) - u(x))^+ dx \leq 2 \sup_x (u_x)^+ \cdot l,$$

and therefore $I \leq 2l \langle\langle [\sup_x (u_x)^+]^p \rangle\rangle^{1/p} \leq c_p l$ in view of Corollary 2.1.5. On the other hand,

$$J \leq l^{p-1} \langle\langle |u|_{1,\infty}^p \rangle\rangle^{(p-1)/p}.$$

The Gagliardo–Nirenberg inequality (1.3.9) (with a sufficiently large m) and Hölder's inequality imply that

$$\langle\langle |u_{1,\infty}|^p \rangle\rangle^{\frac{p-1}{p}} \leq \left[c \langle\langle \|u\|_m^{\frac{2p}{2m-1}} |u|_{1,1}^a \rangle\rangle \right]^{\frac{p-1}{p}} \leq c \langle\langle \|u\|_m^2 \rangle\rangle^{\frac{p-1}{2m-1}} \langle\langle |u|_{1,1}^b \rangle\rangle^c,$$

for suitable constants $a, b, c > 0$. Now, using Corollary 2.1.5 and Theorem 2.3.4, we obtain that

$$J \leq C l^{p-1} \nu^{-(p-1)}.$$

Thus we have proved (7.2.3) if $p \geq 1$.

If $p \in (0, 1)$, then Hölder's inequality and the estimate (7.2.3) with $p = 1$ imply that

$$S_{p,l} \leq \langle\langle \int |u(x+l) - u(x)| dx \rangle\rangle^p = S_{1,l}^p \leq (c_1' l)^p.$$

This completes the proof of (7.2.3).

Assertion (7.2.4) is verified below after the proof of Theorem 7.2.3. □

PROOF OF THEOREM 7.2.3. 1) *Proof of the upper bounds.* Having obtained (7.2.3), we only need to consider the case $p > 1$. Proceeding in the same way as when deriving the upper bounds in (7.2.3), we get that $S_{p,l} \leq IJ$ (see (7.2.6)). As above, $I \leq c_p l$, and in view of (2.1.16)

$$J \leq \langle\langle |u|_{1,1}^p \rangle\rangle^{(p-1)/p} \leq C_p.$$

This yields that $S_{p,l} \leq \tilde{C}_p l$.

2) *Proof of the lower bounds.* Denote by Q_1 the event in (2.3.9) for $m = 1$:

$$Q_1 = \{(t, \omega) \in [T, T+\sigma] \times \Omega : \alpha \nu^{-\frac{1}{2}} \leq \|u^\omega(t)\|_1 \leq \alpha^{-1} \nu^{-\frac{1}{2}}\}.$$

Then $\rho(Q_1) \geq C(1)$. Let $M > 0$ and

$$Q_2 = \{(t, \omega) \in Q_1 : |u_x^{\omega+}(t)|_\infty + |u_x^\omega(t)|_1 + \nu^{\frac{3}{2}} \|u^\omega(t)\|_2 + \nu^{\frac{5}{2}} \|u^\omega(t)\|_3 \leq M\}.$$

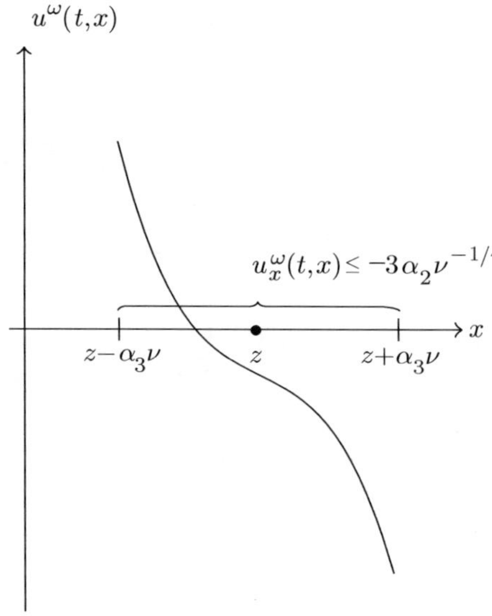

FIGURE 7.2.1. Behaviour of $x \mapsto u^\omega(t,x)$ near $x = z$ for $(t,\omega) \in Q_2$ and $0 < \nu \leq c_*$.

By Corollary 2.1.5, relation (2.3.7) with $m = 2, 3$ and Markov's inequality we get:

$$\rho(Q_2) \geq C(1) - C_1 M^{-1} \geq \tfrac{1}{2} C(1),$$

provided that M is large enough. Let $(t,\omega) \in Q_2$ and denote $v(x) := u^\omega(t,x)$. Now we prove that for any $p > 0$,

$$s_{p,l}(v) = \int |v(x+l) - v(x)|^p dx \geq C l^{\min(1,p)}, \qquad l \in [c_1'\nu, c_2'], \ (t,\omega) \in Q_2, \tag{7.2.7}$$

where $C = C(c_1', c_2', p) > 0$. To visualise the behaviour of the function v in the region which plays the key role in our proof, see Figure 7.2.1.

Firstly suppose that $p \geq 1$. Note that

$$\alpha^2 \nu^{-1} \leq \int |v_x|^2 dx \leq |v|_{1,\infty} |v|_{1,1} \leq M |v|_{1,\infty},$$

with α from the definition of the event Q_1. Thence

$$|v|_{1,\infty} \geq \alpha^2 M^{-1} \nu^{-1} =: \alpha_2 \nu^{-1}. \tag{7.2.8}$$

Since $|v_x^+|_\infty \leq M$, we have $|v_x^+|_\infty \leq \frac{1}{2}\alpha_2 \nu^{-1}$ for $\nu \leq c_*$, where

$$c_* = \tfrac{1}{2} \alpha_2 M^{-1}.$$

Since $v_x = v_x^+ - v_x^-$, then by the above bound on v_x^+ and (7.2.8), we get

$$|v_x^-|_\infty = |v_x|_\infty \geq \alpha_2 \nu^{-1} \quad \text{if } \nu \in]0, c_*].$$

Denote by $z = z(t,\omega)$ the quantity

$$z = \min\{x \in [0,1) : v_x^-(x) \geq \alpha_2 \nu^{-1}\}. \tag{7.2.9}$$

If $\nu \in]0, c_*]$, then z is a well-defined measurable function on Q_2.

We have

$$\text{(7.2.10)}\qquad \int |v(x+l)-v(x)|^p dx \geq \int_{z-\frac{l}{2}}^{z} \left| \int_x^{x+l} v_y^-(y)dy - \int_x^{x+l} v_y^+(y)dy \right|^p dx.$$

By the Gagliardo–Nirenberg inequality (1.3.9), $|v_{xx}|_\infty \leq C_2\|v\|_2^{\frac{1}{2}}\|v\|_3^{\frac{1}{2}}$ which implies that $|v_{xx}|_\infty \leq C_2M\nu^{-2}$ since $v = u^\omega(t)$ with $(t,\omega) \in Q_2$. Thus,

$$\text{(7.2.11)}\quad v_x^-(y) \geq \alpha_2\nu^{-1} - \alpha_3C_2M\nu^{-1} = \frac{3}{4}\alpha_2\nu^{-1} \qquad \text{for } y \in [z-\alpha_3\nu, z+\alpha_3\nu],$$

where $\alpha_3 = \alpha_2/4C_2M > 0$. Now suppose that $l \geq \alpha_3\nu$. Since $v_x^+ \leq M$, we have for $x \in [z-l/2, z]$:

$$\int_x^{x+l} v_x^-(y)dy \geq \int_z^{z+\alpha_3\nu/2} v_x^-(y)dy \geq \frac{3}{8}\alpha_2\alpha_3 \quad \text{and} \quad \int_x^{x+l} v_x^+(y)dy \leq Ml.$$

So by (7.2.10) we obtain that

$$\int |v(x+l)-v(x)|^p dx \geq \int_{z-\frac{l}{2}}^{z} \left|\frac{3}{8}\alpha_2\alpha_3 - Ml\right|^p dx \geq \frac{l}{2}\left(\frac{1}{4}\alpha_2\alpha_3\right)^p =: \varkappa$$

for all $(t,\omega) \in Q_2$, provided that $l \in [\alpha_3\nu,\ \alpha_2\alpha_3/8M]$ and $\nu \in (0, c_*]$. This proves (7.2.7) for $p \geq 1$ with

$$c_* = \frac{1}{2}\alpha_2M^{-1}, \quad c_1' = \alpha_3, \quad c_2' = \alpha_2\alpha_3/8M.$$

Replacing, if needed, c_* by a smaller number from $(0, \frac{1}{2}\alpha_2M^{-1}]$ we obtain that $[c_1'\nu, c_2'] \neq \emptyset$ if $\nu \leq c_*$.

Now suppose that $p \in (0,1)$. Let f be any positive function. Choosing in the Riesz-Thorin inequality (11.1.5) $p_0 = p, p_1 = 2$ and $\theta = (2-2p)/(2-p)$ we find that $p_\theta = 1$. So that for any non-negative function f we have $|f|_1^{2-p} \leq |f|_p^p|f|_2^{2-2p}$, or

$$\left(\int f\right)^{2-p} \leq \left(\int f^p\right)\left(\int f^2\right)^{1-p}.$$

Consequently,

$$\int |v(x+l)-v(x)|^p dx \geq \int \left((v(x+l)-v(x))^+\right)^p dx$$
$$\geq \left(\int \left((v(x+l)-v(x))^+\right)^2 dx\right)^{p-1} \left(\int \left((v(x+l)-v(x))^+\right) dx\right)^{2-p}.$$

Since $v_x^+ \leq M$, then $(v(x+l)-v(x))^+ \leq Ml$. Therefore, as $p-1<0$, the first factor in the r.h.s. above is bounded from below by $\left(M^2l^2\right)^{p-1}$.

As the space average of $v(\cdot+l)-v(\cdot)$ vanishes, then

$$\int (v(x+l)-v(x))^+ dx = \frac{1}{2}\int |v(x+l)-v(x)|dx,$$

and by (7.2.7) with $p=1$, we obtain that the second factor is bounded from below by Cl^{2-p}. This completes the proof of (7.2.7) for $p \in (0,1)$.

Since $\rho(Q_2) \geq \frac{1}{2}C(1)$, then (7.2.7) implies the lower bounds in (7.2.5). □

PROOF OF (7.2.4) IN THEOREM 7.2.2. Firstly, let $p \geq 1$. As in the proof of Theorem 7.2.3, for any $(t,\omega) \in Q_2$ let us define $z = z(t,\omega)$ by relation (7.2.9). Next we bound from below the integral of $|v(x+\ell)-v(x)|^p$ in dx over the interval $[z-\alpha_3\nu, z]$ (instead of the interval $[z-\ell/2, z]$, used earlier). This is easier than before since now the contribution of v_x^+ does not have to be taken into account. Indeed, since now $0 \leq l \leq c_1'\nu = \alpha_3\nu$, then in view of (7.2.11)

$$v_y(y) = -v_y^-(y) \leq -\frac{3}{4}\alpha_2\nu^{-1} \quad \text{for } y \in [z-\alpha_3\nu, z+l].$$

Therefore

$$v(x+l)-v(x) = \int_x^{x+l} v_x(y)dy \leq -\frac{3}{4}\alpha_2 l\nu^{-1}$$

if $z-\alpha_3\nu \leq x \leq z$, and so

$$s_{p,l}(v) \geq \int_{z-\alpha_3\nu}^{z} |v(x+l)-v(x)|^p dx \geq \alpha_3\nu\big(\frac{3}{4}\alpha_2\big)^p l^p\nu^{-p} =: c_p' l^p\nu^{1-p},$$

or

$$s_{p,l}(v) \geq c_p' l^p \nu^{1-\max(1,p)} \tag{7.2.12}$$

for any $(t,\omega) \in Q_2$ if $p \geq 1$.

Arguing exactly as in the proof of Theorem 7.2.3, using Hölder's inequality, the upper bound $v_x^+ \leq M$ and estimate (7.2.12) with $p=1$ we derive that (7.2.12) remains valid for $p<1$, when $(t,\omega) \in Q_2$.

Since (7.2.12) holds if $(t,\omega) \in Q_2$, then

$$S_{p,l}(u) \geq \rho(Q_2)c_p' l^p\nu^{1-p} \geq C(p)l^p\nu^{1-p},$$

for any $p>0$. □

The function $S_{p,l}$ is 1-periodic in l and $S_{p,0} = S_{p,1} = 0$. It is tempting to assume that (7.2.4) holds for $|l| \in [c_1', 1/2]$, i.e. that $c_2' = 1/2$. But in general this is not true:

PROBLEM 7.2.4. Use the result in Problem 1.3.7 to show that estimates (7.2.4) may not hold for all $|l| \in [c_1', 1/2]$.

Dissipation range in the x-presentation.

Let us denote

$$C_p^* = \max\big(C_p^{\pm 1}, c_p^{\pm 1}, (\tilde C_p)^{\pm 1}, (\tilde c_p)^{\pm 1}\big) \geq 1, \quad A = \ln(c_2')^{-1},\ B = \ln(c_1')^{-1},$$

where C_p etc. are the constants from Theorems 7.2.2, 7.2.3. Introducing the notation

$$x \overset{*}{\sim} y \quad \text{if} \quad |x-y| \leq \tfrac{1}{p}\ln C_p^*,$$

we derive from Theorems 7.2.2, 7.2.3 the following assertions:

PROPOSITION 7.2.5. *Let $0<\nu\leq c_*$, where c_* is the constant from (7.2.4), (7.2.5).*

i) If $p \geq 1$, then

$$\ln\big(S_{p,l}^{-1/p}\big) \overset{*}{\sim} \begin{cases} \frac{1}{p}\ln|l|^{-1}, & \text{if} \quad A \leq \ln|l|^{-1} \leq \ln\nu^{-1}+B, \\ \ln|l|^{-1} - \frac{p-1}{p}\ln\nu^{-1}, & \text{if} \quad \ln|l|^{-1} \geq \ln\nu^{-1}+B. \end{cases}$$

ii) If $0<p<1$, then $\ln\big(S_{p,l}^{-1/p}\big) \overset{*}{\sim} \ln|l|^{-1}$ for $\ln|l|^{-1} \geq A$.

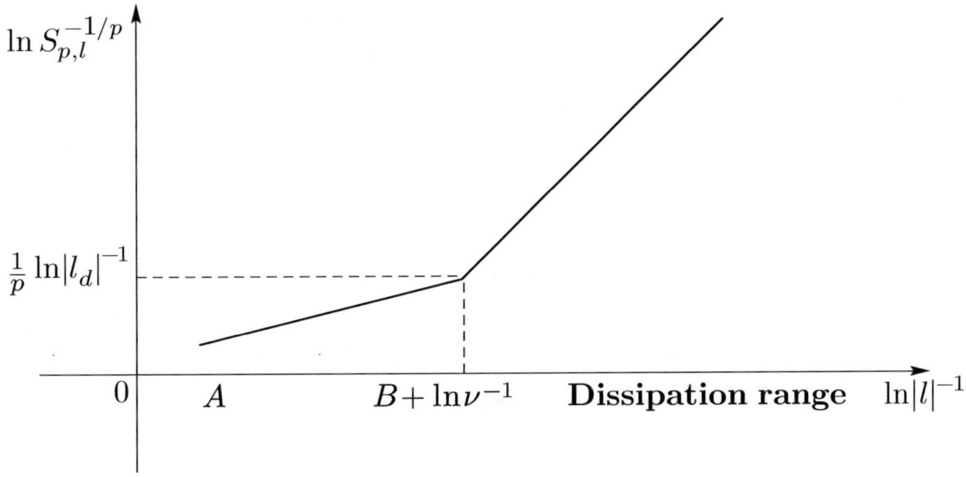

FIGURE 7.2.2. Behaviour of the structure function $S_{p,l}$, $p > 1$.

For $p > 1$ see Figure 7.2.2. We observe that the slope of the graph of the function $\ln|l|^{-1} \mapsto \ln\left(S_{p,l}^{-1/p}\right)$ changes from $1/p$ to 1 at the point $\ln\nu^{-1} + B = \ln l_d^{-1}$.

From the first assertion of Proposition 7.2.5 we see that for burgulence the dissipation range in the x-presentation I_{diss}^x is well-defined as an interval $(0, l_d^x)$, characterised by the following relation:

(7.2.13)

for $r \in I_{diss}^x$ the averaged increments $S_{p,r}^{1/p}$ are approximately proportional to $|r|$ for all p, while for $|r| \gg l_d^x$ they are not.

As usual in this context, the dissipation scale in x-presentation l_d^x is defined up to multiplication by a ν-independent constant, is assumed to be of the form ν^{c_d}, and the assertion above means that for r of the form $r = \nu^c$, $c < c_d$, the functions $S_{p,r}^{1/p}$ do not all behave linearly in r.

Intermittency.

The vorticity $\zeta = u_x$ of a solution u for (1.1.13) with small ν is highly intermittent as a function of x: in view of (2.3.13), for typical (t, ω), its L_1-norm in x is of order 1, while its L_∞-norm is of order ν^{-1}. If we measure space-increments of u using not the vorticity ζ, but the structure function $S_{p,l}(u)$, then for l in the inertial range in x we will see strong intermittency of u as a random variable. Indeed, in view of Theorem 7.2.3, the flatness F (see (6.0.11)), which for solutions of the Burgers equation is defined as

$$F(l) = S_{4,l}/S_{2,l}^2,$$

satisfies

$$c_3'|l|^{-1} \le F(l) \le c_4'|l|^{-1} \quad \text{for} \quad |l| \in [c_1'\nu, c_2'], \tag{7.2.14}$$

if ν is sufficiently small. So for small l from the inertial range $F(l)$ is a large quantity of order $|l|^{-1}$.

In the space-homogeneous case (see (7.2.2)) $S_{p,l}$ is the p-th absolute moment of the r.v.

$$(t,\omega)\mapsto u^\omega(t,x+l)-u^\omega(t,x). \tag{7.2.15}$$

Since for a Gaussian r.v. $\zeta_\sigma \sim N(0,\sigma^2)$ its p-th absolute moment scales as σ^p (indeed, $\mathbb{E}|\zeta_\sigma|^p = \sigma^p\mathbb{E}|\zeta_1|^p$), then the flatness of a Gaussian r.v. is an absolute constant (in fact, it equals 3). So we see from (7.2.14) that for small l from the inertial range the r.v.'s (7.2.15) are "strongly non-Gaussian", and accordingly the random field $u^\omega(t,x)$ as a function of the random parameter (t,ω) also is. For a stationary solution u_{stat} of a space-homogeneous equation (B) the quantity $S_{p,l}$ equals to the absolute moment of the r.v. $\omega\mapsto u^\omega_{stat}(t,x+l)-u^\omega_{stat}(t,x)$ (for any t and x). Thus by the same argument the stationary solutions are non-Gaussian as functions of ω, and for small ν their space-increments above, corresponding to l which are 'short, but not too short' are far from Gaussian r.v.'s.

7.3. Energy spectrum.

The definition of the energy spectrum, given below, depends on a real parameter $M>1$. We recall (7.1.1).

DEFINITION 7.3.1. For a random process $u(t)=u^\omega(t)\in H$, $0\le t\le T_1$, $T_1\ge T+\sigma$, we define the averaged *energy of a wave number* $k\ge 1$ as

$$E_k = E_k(u) = \langle\langle e_k(u)\rangle\rangle, \tag{7.3.1}$$

where for a function $H\ni v(x)=\sum_k \hat v_k e^{2i\pi k}$ we set

$$e_k(v) = \frac{1}{2k(M-M^{-1})}\sum_{M^{-1}k\le|n|\le Mk}\tfrac12|\hat v_n|^2. \tag{7.3.2}$$

The function $k\mapsto E_k(u)$ is called the *energy spectrum* of u.

Note that if $u(t)$ is a stationary process, then $E_k(u)=\mathbb{E}e_k(u(t))$ for any t, and that this definition of the energy spectrum agrees with (6.0.6).

If $u(t;u_0)$ is a solution of (B) with $u_0\in H^1$, then by Theorem 7.1.3 for k much larger than the dissipation scale $l_d=C\nu^{-1}$, the energy spectrum $E_k(u)$ decreases faster than any negative power of k, while for $k\ll\nu^{-1}$ it decreases at a slower rate. In the theorem below we specify the behaviour of $E_k(u)$ for $k\lesssim\nu^{-1}$. The constants there, including M, depend only on the random force (1.1.6), and M should be sufficiently large. So in (7.3.2) we use the averaging over large layers around the zero-dimensional spheres $S^0(k)=\{-k,k\}$ (cf. (6.0.6)), and due to the theorem the energy spectrum, defined in this way, possesses physically correct properties. We do not know if the theorem's assertion remains true (with amended multiplicative constants) if in (7.3.2) we take for M just any constant, larger than one.

THEOREM 7.3.2. *Consider the same constants $c_*, c_1'>0$ as in Theorem 7.2.3. Then there exist $c_5', c_6'>0$ and $M>1$ such that for any solution $u=u(t;u_0)$ of (B) with $u_0\in H^1$, for $0<\nu<c_*$ and for k from the inertial range, i.e. for*

$$1\le k\le (c_1'\nu)^{-1},$$

we have

$$c_5'k^{-2}\le E_k(u)\le c_6'k^{-2}. \tag{7.3.3}$$

PROOF. In this proof, we denote by $C, C', C_k, \dots$ various quantities, independent from M (and from ν).

The upper bound in (7.3.3) is an immediate consequence of Oleinik's estimate which implies that $\langle\langle|\hat{u}_n|^2\rangle\rangle \leq Cn^{-2}$ (see (7.1.5)).

Now we prove the lower bound. Since $\langle\langle|\hat{u}_n|^2\rangle\rangle \leq Cn^{-2}$, then

$$\sum_{|n|\leq M^{-1}k} |n|^2\langle\langle|\hat{u}_n|^2\rangle\rangle \leq CM^{-1}k, \tag{7.3.4}$$

and

$$\sum_{|n|\geq Mk} \langle\langle|\hat{u}_n|^2\rangle\rangle \leq C'M^{-1}k^{-1}. \tag{7.3.5}$$

Put $\mathcal{S}_k = \sum_{|n|\leq Mk} |n|^2\langle\langle|\hat{u}_n|^2\rangle\rangle$. Since $|\alpha| \geq |\sin\alpha|$, we get

$$\begin{aligned} \mathcal{S}_k &\geq \frac{k^2}{\pi^2}\sum_{|n|\leq Mk} \sin^2(n\pi k^{-1})\langle\langle|\hat{u}_n|^2\rangle\rangle \\ &= \frac{k^2}{\pi^2}\left(\sum_{n\in\mathbb{Z}^*} \sin^2(n\pi k^{-1})\langle\langle|\hat{u}_n|^2\rangle\rangle - \sum_{|n|>Mk} \sin^2(n\pi k^{-1})\langle\langle|\hat{u}_n|^2\rangle\rangle\right). \end{aligned} \tag{7.3.6}$$

By Parseval's identity,

$$\|u(t,\cdot+y) - u(t,\cdot)\|^2 = 4\sum_{n\in\mathbb{Z}^*} \sin^2(n\pi y)|\hat{u}_n(t)|^2.$$

As the l.h.s. above equals $s_{2,y}(u(t))$, then (7.3.5) and (7.3.6) imply that

$$\begin{aligned} \mathcal{S}_k &\geq \frac{k^2}{\pi^2}\left(\frac{1}{4}\langle\langle\|u(\cdot+k^{-1}) - u(\cdot)\|^2\rangle\rangle - \sum_{|n|>Mk} \langle\langle|\hat{u}_n|^2\rangle\rangle\right) \\ &\geq k^2C_1S_{2,k^{-1}}(u) - C'M^{-1}k. \end{aligned}$$

So by (7.2.5) with $p = 2$ and $l = k^{-1}$, we get

$$\mathcal{S}_k \geq k^2C_1\tilde{c}_2k^{-1} - C'M^{-1}k = k(C'' - C'M^{-1}) \tag{7.3.7}$$

if $k \geq 1/c_2'$. Noting that

$$E_k \geq \frac{1}{2k^3M^3}\sum_{M^{-1}k\leq|n|\leq Mk} |n|^2\langle\langle|\hat{u}_n|^2\rangle\rangle$$

and using (7.3.4), (7.3.7), we obtain for a suitable $C_3 > 0$:

$$\begin{aligned} E_k &\geq \frac{1}{2k^3M^3}\left(\mathcal{S}_k - \sum_{|n|\leq M^{-1}k} |n|^2\langle\langle|\hat{u}_n|^2\rangle\rangle\right) \\ &\geq \frac{C'' - C'M^{-1} - CM^{-1}}{2M^3k^2} > M^{-3}C_3k^{-2}, \end{aligned}$$

for $M \gg 1$. This proves the left inequality in (7.3.3) if $k \geq 1/c_2'$. But if $1 \leq k < 1/c_2'$, then choosing $M \geq 1/c_2'$ and using Theorem 2.3.15 jointly with (7.1.5) we get that

$$E_k(u) \geq \frac{1}{4kM} \sum_{|n|\leq Mk} \langle\langle |\hat{u}_n|^2 \rangle\rangle \geq \frac{1}{4kM}\Big(\langle\langle |u|_2^2\rangle\rangle - \frac{C}{Mk}\Big) \geq \frac{1}{8kM}\langle\langle |u|_2^2\rangle\rangle \geq \frac{C'}{kM},$$

if M is large enough. So (7.3.3) holds for all $k \geq 1$, if we suitably adjust the constant c_5'. □

Note that to prove the power law (7.3.3) for $1 \leq k \leq (c_1'\nu)^{-1}$ we used nothing but the bound (7.1.5) (which follows from (2.1.16)) and estimate (7.2.5) with $p = 2$ and $|l| \leq c_1'\nu$ for the structure function. In particular we have not used that $l_d(u) = \nu^{-1}$.

7.4. The limiting in time behaviour.

We recall (7.0.1).

The energy spectrum.

The energy spectrum of the stationary measure μ^ν is the function

$$\mathbf{N} \ni k \mapsto \langle e_k, \mu^\nu \rangle,$$

where $e_k(u)$ is defined in (7.3.2).

THEOREM 7.4.1. *There exist positive constants $c_1 - c_3$ depending only on $\vec{B}$, such that*

$$c_1 k^{-2} \leq \langle e_k, \mu^\nu\rangle \leq c_2 k^{-2} \qquad \text{if} \quad 1 \leq k \leq (c_3\nu)^{-1}. \tag{7.4.1}$$

The proof is similar to that of Theorem 3.4.1 and we omit it.

Now let us consider the instant energy spectrum of a solution u at a moment t, i.e. the function

$$\mathbf{N} \ni k \mapsto \mathbb{E}(e_k(u(t))).$$

It converges to the spectrum of the measure μ^ν, uniformly in ν:

THEOREM 7.4.2. *There exist $c_4, C > 0$, depending on $\vec{B}$, such that for any solution $u(t) = u(t; u_0)$ with $u_0 \in H^1$ and for $t \geq 3$ we have*

$$|\mathbb{E}(e_k(u(t)) - \langle e_k, \mu^\nu\rangle| \leq C(\ln t)^{-1/24} \quad \text{if} \quad 1 \leq k \leq (c_4\nu)^{-1},$$

uniformly in $\nu \in (0, 1]$.

PROOF. Consider the complex linear functional on L_1 which sends $u(x)$ to $\hat{u}_n = \int u(x)e^{-2\pi i n x}dx$. Its norm is one. So if for $K \geq 1$ we set $\phi_n^K = \min(K^2, |\hat{u}_n|^2)$, then

$$\|\phi_n^K\|_{L(L_1)} \leq K^2 + 2K < (K+1)^2$$

(see (1.7.2)). Therefore by (4.2.5)

$$\Big|\mathbb{E}\phi_n^K(u(t)) - \langle\phi_n^K, \mu^\nu\rangle\Big| \leq (K+1)^2 \varkappa(\ln t)^{-1/8}, \tag{7.4.2}$$

if $t \geq 3$. On the other hand, for any $t \geq 1$ in view of Corollary 2.1.5 we have

$$\begin{aligned}\mathbb{E}|\phi_n^K(u(t)) - |\hat{u}_n(t)|^2| &\leq \mathbb{E}|\hat{u}_n(t)|^2 \chi_{\{|\hat{u}_n(t)|\geq K\}} \leq K^{-1}\mathbb{E}|\hat{u}_n(t)|^3 \\ &\leq K^{-1}\mathbb{E}|u(t)|_1^3 \leq CK^{-1}.\end{aligned}$$

Similarly, using (3.4.1) instead of Corollary 2.1.5 we find that also

$$\Big|\langle\phi_n^K, \mu^\nu\rangle - \langle|\hat{u}_n|^2, \mu^\nu\rangle\Big| \leq CK^{-1}.$$

From here and (7.4.2),

$$\left|\mathbb{E}|\hat{u}_n(t)|^2 - \langle|\hat{u}_n|^2, \mu^\nu\rangle\right| \le C_1 K^2 (\ln t)^{-1/8} + 2CK^{-1}, \quad t \ge 3,$$

for all $n \in \mathbb{Z}^*$ and $K \ge 1$. Choosing $K = C_2(\ln t)^{1/24}$ we get that

$$\left|\mathbb{E}|\hat{u}_n(t)|^2 - \langle|\hat{u}_n|^2, \mu^\nu\rangle\right| \le C_3(\ln t)^{-1/24}.$$

Averaging in $n \in [-Mk, -M^{-1}k] \cup [M^{-1}k, Mk]$ we obtain the theorem's assertion. □

From the last two theorems we get:

COROLLARY 7.4.3. *If* $1 \le k \le (c_4\nu)^{-1}$ *and* $t \ge C' e^{k^{48}}$, *then* $\mathbb{E}e_k(u(t))$ *obeys the estimates in* (7.4.1) *with modified* ν*-independent constants* c_1, c_2.

The structure function.

Similarly to the above we define the structure function of the stationary measure μ^ν and the instant structure function of a solution u at time t as $\langle s_{p,l}, \mu^\nu\rangle$ and $\mathbb{E}s_{p,l}(u(t))$, respectively. Obviously for $p \ge 1$, $|s_{p,l}(u)| \le C_p |u|_p^p$, while for $0 < p \le 1$,

$$|s_{p,l}(u)| \le \int (|u(x+l)|^p + |u(x)|^p)dx \le 2|u|_1^p.$$

Moreover, the functional $s_{p,l}$ is continuous on $L_{\max(p,1)}$:

LEMMA 7.4.4. *For any* l,
1) if $p \ge 1$, *then the functional* $s_{p,l}$ *is continuous on* L_p.
2) If $0 < p < 1$, *then* $s_{p,l}$ *is continuous on* L_1.

PROOF. 1) The functional is continuous as a composition of the continuous linear operator in L_p which sends $u(x)$ to $u(x+l) - u(x)$, and the continuous functional $u(x) \mapsto |u|_p^p$.

2) As above, the functional is the composition of a continuous operator in L_1 and the functional

$$f_p : L_1 \to \mathbb{R}, \qquad u(x) \mapsto \int |u(x)|^p dx, \quad 0 < p < 1.$$

It remains to show that f_p is continuous. To do this consider in L_1 any convergent sequence $u_n \to u$. We have to verify that $f_p(u_n) \to f_p(u)$ as $n \to \infty$. Consider any sequence $n_j \to \infty$. It contains a subsequence $n'_j \to \infty$ such that $u_{n'_j}(x) \to u(x)$ almost surely. But the system of functions $|u_{n'_j}(x)|^p$ is uniformly integrable since $\int \left(|u_n(x)|^p\right)^{1/p} dx = |u_n|_1 \le C$ for all n. So by Theorem 11.1.8, $f_p(u_{n'_j}) \to f_p(u)$. This proves that $f_p(u_n) \to f_p(u)$, see Problem 11.2.1. □

Using the continuity of the functional $s_{p,l}$ on $L_{\max(p,1)}$ and the bound on it, and arguing as proving Theorems 7.4.1, 7.4.2 and using Theorem 4.2.5 we get:

THEOREM 7.4.5. *For any* $p > 0$ *the function* $\langle s_{p,l}, \mu^\nu\rangle$ *satisfies estimates in* (7.2.3), (7.2.4) *for* $0 < l \le c'_1\nu$, *and estimates* (7.2.5) *for* $c'_1\nu \le l \le c'_2$. *Moreover, for any solution* $u(t) = u(t; u_0)$ *with* $u_0 \in H^1$ *we have*

$$\mathbb{E}s_{p,l}(u(t)) \to \langle s_{p,l}, \mu^\nu\rangle \quad \text{as} \quad t \to \infty, \ \forall p, l, \tag{7.4.3}$$

where the rate of convergence depends on p, *but not on* l *and* ν.

We recall that if $b_s \equiv b_{-s}$, then the stationary measure μ^ν is space-homogeneous (see Corollary 3.3.5). In this case

$$\langle s_{p,l}, \mu^\nu \rangle = \int_\Omega \mu^\nu(du) \int_{S^1} dx \big| u(x+l) - u(x) \big|^p = \int_\Omega \mu^\nu(du) \big| u(x+l) - u(x) \big|^p$$

for any $x \in S^1$, where to get the second equality we use Fubini's theorem and the x-independence of the integral in the r.h.s.

7.5. Appendix: High–probability versions of the results above.

In Theorems 7.1.3, 7.2.2, 7.2.3 and 7.3.2 we proved that if u is a solution of (B), then the corresponding laws of burgulence hold for the quantities $|\hat{u}_s|^2$, $s_{p,l}(u)$ and $e_k(u)$, averaged in ensemble and locally averaged in time. Below we show that if these quantities are locally averaged in time, without the averaging in ensemble, then the resulting random variables still satisfy the laws of burgulence with high probability. For this to hold, the pre-factors for the degrees of ν in the laws should be sufficiently large/small, and the time of local averaging – sufficiently large.

As before we assume (7.0.1), and the constants in estimates do not depend on ν or u_0.

Let us fix some numbers

$$T' \geq 1, \quad \bar{\sigma} \geq \sigma' = \max(1, \sigma_0), \quad A \geq \alpha(1)^{-1},$$

where σ_0 is as in Theorem 2.3.1 and $\alpha(1)$ is the constant from Corollary 2.3.8. We define the intervals

$$J^k = J^k(T', \bar{\sigma}) = [T' + k\bar{\sigma},\ T' + (k+1)\bar{\sigma}], \quad k \in \mathbf{N}_0.$$

Next for a solution $u^\omega(t; u_0)$, $u_0 \in H^1$, we consider the sets

$$\mathcal{Q}^k = \Big\{ (t, \omega) \in J^k \times \Omega : \ \nu^{1/2} \left\| u^\omega(t) \right\|_1 \geq A^{-1} \Big\}, \ k \in \mathbf{N}_0,$$

and we define the r.v.'s f^k by the relation

$$f^k(\omega) = f^k(\omega;\ T', \bar{\sigma}) = l \,\{ t \in J^k :\ (t, \omega) \in \mathcal{Q}^k \},$$

where $l = l_{\bar{\sigma}}$ is the normalised Lebesgue measure on the interval J^k given by $dl = \bar{\sigma}^{-1} dt$. As in (2.3.4) we set $\rho = \rho_{T', \bar{\sigma}} = dl \times \mathbf{P}$. Then, in view of Fubini's theorem and Corollary 2.3.8, there exists a constant $C(1) \in (0, 1]$ such that

$$\mathbb{E} f^k = \rho(\mathcal{Q}_k) \geq C(1).$$

Denote

$$q_k = \mathbf{P} \left\{ f^k \geq \tfrac{1}{2} \mathbb{E} f^k \right\}.$$

Since $0 \leq f^k \leq 1$, then we have

$$\mathbb{E} f^k = \mathbb{E} f^k \chi_{\{f^k \geq \frac{1}{2} \mathbb{E} f^k\}} + \mathbb{E} f^k \chi_{\{f^k < \frac{1}{2} \mathbb{E} f^k\}} \leq q_k + \tfrac{1}{2} \mathbb{E} f^k.$$

So $q_k \geq \frac{1}{2} \mathbb{E} f^k$, and

$$\mathbf{P} \left\{ f^k \geq \tfrac{1}{2} C(1) \right\} \geq q_k \geq \tfrac{1}{2} \mathbb{E} f^k \geq \tfrac{1}{2} C(1) \qquad \forall k \in \mathbf{N}_0, \tag{7.5.1}$$

for all $u_0 \in H^1$, all $\nu > 0$ and for any $T', \bar{\sigma}$ as above.

LEMMA 7.5.1. *Let $\bar\sigma \ge 2\sigma'$ and $T' \ge 1$. Then for each $m \ge 1$ the event*

$$Q^m = \{\omega:\ f^k < \tfrac14 C(1) \text{ for } k = 0,\dots,m-1\}$$

satisfies

$$\mathbf{P}(Q^m) \le q^m, \tag{7.5.2}$$

where $q = 1 - \frac12 C(1) \in [0,1)$.

PROOF. We proceed by induction. For $m = 1$ the assertion follows from (7.5.1). To implement the induction step, we take the probability space Ω for eq. (B) of the form $\Omega_1 \times \Omega_2$ and define the Wiener processes $\beta_s(t)$ in (1.1.12) by relation (1.6.11) with $t_1 = T' + m\bar\sigma$ and $t_2 = T' + (m+1)\bar\sigma$. Then

$$Q^{m+1} = \{(\omega_1,\omega_2):\ \omega_1 \in Q^m,\ f^m(\omega_1,\omega_2) < \tfrac14 C(1)\}.$$

For any fixed $\omega_1 \in Q^m$, the section

$$Q^{m+1}(\omega_1) = \{\omega_2:\ (\omega_1,\omega_2) \in Q^{m+1}\}$$

is defined by the relation $f^m(\omega_1,\omega_2) < \frac14 C(1)$. Denoting by $|X|$ the Lebesgue measure of a set $X \subset \mathbb{R}$, we have:

$$\mathbf{P}^{\omega_2}(Q^{m+1}(\omega_1)) = \mathbf{P}^{\omega_2}\left[\frac{1}{\bar\sigma}\big|\{t \in [0,\bar\sigma]:\ \nu^{1/2}\,\|u^{\omega_2}(t;u')\|_1 \ge A^{-1}\}\big| < \tfrac14 C(1)\right],$$

where $u' = u^{\omega_1}(t_1;u_0)$. But for any ω_1 the r.h.s. is bounded from above by

$$\mathbf{P}^{\omega_2}\left[\frac{1}{\bar\sigma}\big|\{t \in [\tfrac12\bar\sigma,\bar\sigma]:\ \nu^{1/2}\,\|u^{\omega_2}(t;u')\|_1 \ge A^{-1}\}\big| < \tfrac14 C(1)\right]$$

$$= \mathbf{P}^{\omega_2}\left[\tfrac12\, l_{\frac12\bar\sigma}\{t \in [\tfrac12\bar\sigma,\bar\sigma]:\ \nu^{1/2}\,\|u^{\omega_2}(t;u')\|_1 \ge A^{-1}\} < \tfrac14 C(1)\right] \le 1 - \tfrac12 C(1) = q,$$

in view of (7.5.1) with $k = 0$, applied to the case when $\bar\sigma$ and T both equal $\bar\sigma/2$.

Therefore

$$\mathbf{P}(Q^{m+1}) \le \mathbf{P}^{\omega_1}(Q^m)q,$$

and since as previously mentioned (7.5.2) holds for $m = 1$, the general statement is proved by induction. □

Now for $q < 1$ as in the lemma above and any $\varepsilon > 0$ we choose an $m \in \mathbf{N}$ large enough so that $q^m \le \varepsilon/2$. Then let us take $\bar\sigma_\varepsilon = m\bar\sigma$ and consider the intervals $J^k_\varepsilon = J^k(T',\bar\sigma_\varepsilon)$ and the corresponding r.v.'s $f^k = f^k_\varepsilon$, defined in terms of the measure l_ε given by $dl_\varepsilon = \frac{1}{m\bar\sigma_\varepsilon}dt$. Let us set $\rho_\varepsilon = l_\varepsilon \times \mathbf{P}$.

If ω does not belong to the set Q^m defined in the lemma above, there exists an integer $j \in [0, m-1]$ such that $f^j(\omega;T',\bar\sigma) \ge \frac14 C(1)$. Then

$$f^0_\varepsilon(\omega) \ge l_\varepsilon\{t \in J^j:\ \nu^{1/2}\,\|u(t)\|_1 \ge A^{-1}\} \ge \frac{1}{4m}C(1), \quad \forall\,\omega \notin Q^m.$$

Using (7.5.2), where $q^m \le \varepsilon/2$, we get

LEMMA 7.5.2. *For any $\varepsilon > 0$ there exist $R_\varepsilon > 1$ and $\bar\sigma_\varepsilon \ge 2\sigma'$ with the property that for every $T' \ge 1$ the sections*

$$Q^1_\varepsilon(\omega) = \{t \in J^0_\varepsilon:\ (t,\omega) \in \mathcal{Q}^1_\varepsilon\}$$

of the set

$$\mathcal{Q}^1_\varepsilon := \mathcal{Q}^k\,|_{k=1,\bar\sigma=\bar\sigma_\varepsilon} = \left\{(t,\omega) \in J^0_\varepsilon \times \Omega =: \mathcal{Q}_\varepsilon:\ \nu^{1/2}\,\|u^\omega(t)\|_1 \ge A^{-1}\right\}$$

satisfy the estimate

$$\mathbf{P}\left[\ l_\varepsilon(Q^1_\varepsilon(\omega)) \geq R_\varepsilon^{-1}\right] \geq 1 - \tfrac{1}{2}\varepsilon.$$

Now let us consider the functional

(7.5.3)

$$\Phi(u) = \sum_{r=1}^{\infty} (K_r^0)^{-1}(|u_x^+|_\infty + |u_x|_1)^r + \sum_{r=1}^{\infty}\sum_{m=1}^{\infty} (K_r^m)^{-1}(\nu^{m-1/2}\,\|u\|_m)^r \leq \infty,$$

where K_r^m are positive constants, fixed below. By Problem 2.1.6 the mapping from $\mathcal{Q}_\varepsilon = J_\varepsilon^0 \times \Omega$ to $H^{5/3}$ given by $(t,\omega) \mapsto u^\omega(t;u_0)$ is measurable for every $u_0 \in H^1$. As the functions $u \mapsto \|u^\omega(t;u_0)\|_m$, $m \geq 1$, all are measurable on the space $H^{5/3}$,[3] as well as the function $u \mapsto |u_x^+|_\infty + |u_x|_1$, then the function

$$(t,\omega) \mapsto \Phi(u^\omega(t;u_0)) \leq \infty$$

also is. Choosing the constants $K_r^m = K_r^m(\vec{B},\varepsilon)$ large enough, we derive from (2.1.15), (2.1.16) and (2.3.6) that

(7.5.4)
$$\langle\langle\ \Phi(u^\omega(t;u_0))\ \rangle\rangle = \langle \Phi(u^\omega(t;u_0)), \rho_\varepsilon\rangle \leq 1,$$

uniformly in ν and in $u_0 \in H^1$.

Next for a fixed constant $B_\varepsilon > 0$ we define the sets

$$\mathcal{Q}^2_\varepsilon = \{(t,\omega) \in \mathcal{Q}_\varepsilon :\ \Phi(u^\omega(t)) \leq B_\varepsilon\}, \quad \mathcal{S}_\varepsilon = \mathcal{Q}_\varepsilon \backslash \mathcal{Q}^2_\varepsilon,$$

and accordingly the sections

$$Q^2_\varepsilon(\omega) = \left\{t \in J^0_\varepsilon :\ (t,\omega) \in \mathcal{Q}^2_\varepsilon\right\}, \quad S_\varepsilon(\omega) = \left\{t \in J^0_\varepsilon :\ (t,\omega) \in \mathcal{S}_\varepsilon\right\}.$$

By (7.5.4), $\rho_\varepsilon(\mathcal{S}_\varepsilon) \leq B_\varepsilon^{-1}$ (we recall that $\rho_\varepsilon = l_\varepsilon \times \mathbf{P}$ is a probability measure). Denoting $f_\varepsilon = l_\varepsilon(S_\varepsilon(\omega))$, we have

$$\mathbb{E} f_\varepsilon = \rho_\varepsilon(\mathcal{S}_\varepsilon) \leq B_\varepsilon^{-1}.$$

So (we recall that R_ε is the constant in Lemma 7.5.2)

$$\mathbf{P}(f_\varepsilon \geq \tfrac{1}{2}R_\varepsilon^{-1}) \leq 2R_\varepsilon B_\varepsilon^{-1}.$$

Choosing $B_\varepsilon \gg 1$ allows to bound the r.h.s. from above by $\varepsilon/2$. Consequently,

$$\mathbf{P}\Big(\ \{\omega :\ l_\varepsilon(Q^2_\varepsilon(\omega)) \geq 1 - \tfrac{1}{2}R_\varepsilon^{-1}\}\ \Big) \geq 1 - \varepsilon/2.$$

With such a choice of B_ε, in view of Lemma 7.5.2, we have

LEMMA 7.5.3. *The event*

$$Q^{12}_\varepsilon = \{\omega :\ l_\varepsilon(Q^1_\varepsilon(\omega) \cap Q^2_\varepsilon(\omega)) \geq \tfrac{1}{2}R_\varepsilon^{-1}\}$$

satisfies $\mathbf{P}(Q^{12}_\varepsilon) \geq 1 - \varepsilon$.

Obviously, for every $(t,\omega) \in \mathcal{Q}^1_\varepsilon \cap \mathcal{Q}^2_\varepsilon$ we have

(7.5.5)

$$\nu^{1/2}\,\|u(t)\|_1 \geq A^{-1},\ |u_x^+(t)|_\infty + |u_x^+(t)|_1 \leq K_1^0 B_\varepsilon,\ \nu^{m-1/2}\,\|u(t)\|_m \leq K_1^m B_\varepsilon, \quad \forall m \in \mathbf{N}.$$

[3]Since they are point-wise limits as $N \to \infty$ of the obviously measurable functions $\|\Pi_N u^\omega\|_m$, $N \in \mathbf{N}$.

For any real process $f^\omega(t)$ we will denote by $\overline{f}^\omega$ it local averaging in time:

$$\overline{f}^\omega = \frac{1}{\bar\sigma^\varepsilon}\int_{J^0_\varepsilon} f^\omega(t)dt = \frac{1}{\bar\sigma^\varepsilon}\int_{T'}^{T'+\bar\sigma^\varepsilon} f^\omega(t)dt. \tag{7.5.6}$$

Then $\mathbb{E}\overline{f} = \langle\langle f\rangle\rangle$. In particular, in view of (7.5.4), $\mathbb{E}\overline{\Phi(u^\omega)} \le 1$.

So the event $Q^3_\varepsilon = \left\{\overline{\Phi(u^\omega)} \le \varepsilon^{-1}\right\}$ satisfies

$$\mathbf{P}(Q^3_\varepsilon) \ge 1-\varepsilon.$$

For each $\omega \in Q^3_\varepsilon$ we have

$$\overline{(|u_x^+|_\infty + |u_x|_1)^r} \le K^0_r\varepsilon^{-1}, \forall r\in\mathbf{N}, \quad \overline{(\nu^{m-1/2}\left\|u\right\|_m)^r} \le K^m_r\varepsilon^{-1},\ \forall r,m\in\mathbf{N}, \tag{7.5.7}$$

where $u = u^\omega(t;u_0)$. Finally we define the event

$$Q^4_\varepsilon = Q^{12}_\varepsilon \cap Q^3_\varepsilon.$$

Then $\mathbf{P}(Q^4_\varepsilon) \ge 1-2\varepsilon$.

The result below shows that the assertion of Theorem 7.2.3 remains true with high probability if we replace the averaging $\langle\langle\cdot\rangle\rangle$ by the local in time averaging given by (7.5.6). We recall that the function $s_{p,l}(u)$ was introduced in Definition 7.2.1.

THEOREM 7.5.4. *For any $\varepsilon>0$ there exists $\bar\sigma=\bar\sigma_\varepsilon$, and constants $\tilde{\underline{c}}_p, \tilde{\underline{C}}_p, \underline{c}'_1, \underline{c}'_2, \underline{c}_*$ >0 such that in the inertial range the function $\overline{s_{p,l}(u)}$ satisfies the asymptotically sharp estimates:*

$$\begin{cases} \tilde{\underline{c}}_p|l|^p \le \overline{s_{p,l}(u)} \le \tilde{\underline{C}}_p|l|^p, & \text{if } p\in(0,1) \\ \tilde{\underline{c}}_p|l| \le \overline{s_{p,l}(u)} \le \tilde{\underline{C}}_p|l|, & \text{if } p\ge 1 \end{cases} \quad \text{for } |l|\in[\underline{c}'_1\nu, \underline{c}'_2] \ \text{ and } \nu\in(0,\underline{c}_*], \tag{7.5.8}$$

with probability at least $1-\varepsilon$, for any $T'\ge 1$. The high probability event, where the relations above hold, depends on ν and u_0, but it may be chosen independent from p and l.

PROOF. The proof follows closely the argument in Section 7.2. For the p,l–independent event of high probability in the assertion of the theorem (where estimates (7.5.8) are valid) we take the set Q^4_ε, which satisfies $\mathbf{P}(Q^4_\varepsilon)\ge 1-2\varepsilon$.

1) To prove the upper bounds in (7.5.8), for any fixed $\omega\in Q^4_\varepsilon \subset Q^3_\varepsilon$ we repeat the proof of estimates (7.2.3), as well as of the upper bounds in (7.2.5), replacing there the averaging $\langle\langle\cdot\rangle\rangle$ by the local in time averaging given by (7.5.6). In view of (7.5.7), for such an ω various time-averaged functional norms of the curves $t\mapsto u^\omega(t;u_0)$ meet (up to multiplicative constants) the same upper estimates as the $\langle\langle\cdot\rangle\rangle$-averaged norms of the process $u^\omega(t;u_0)$. So literally repeating the corresponding calculations in Section 7.2, we derive the required upper bounds.

2) To establish the lower bounds, we again follow the derivation of those in Theorem 7.2.3. Instead of the set Q_2 used there, we consider the set

$$\mathcal{Q}'_2 = (J^0_\varepsilon\times Q^4_\varepsilon)\cap\mathcal{Q}^1_\varepsilon\cap\mathcal{Q}^2_\varepsilon.$$

By (7.5.5), for every $(t,\omega)\in\mathcal{Q}'_2$ the function $v(x)=u^\omega(t,x)$ meets the estimates defining the set Q_2, with a suitable $M=M_\varepsilon>0$ and $\alpha>0$. So the argument in the proof of Theorem 7.2.3 implies that for every $(t,\omega)\in\mathcal{Q}'_2$ the function $v(x)$ satisfies (7.2.7) with some $C=C_\varepsilon>0$. Moreover, for any other point $(t',\omega)\in\mathcal{Q}'_2$ with the same ω as for the function $v(x)=u^\omega(t,x)$ and with a t' different from t, the function

$u^\omega(t', x)$ satisfies the same estimates. By definition of the set $Q_\varepsilon^{12} \supset Q_\varepsilon^4$, for any $\omega \in Q_\varepsilon^4$, the l_ε-measure of all such points t' (i.e., those satisfying $(t', \omega) \in \mathcal{Q}_\varepsilon^1 \cap \mathcal{Q}_\varepsilon^2$) is at least $\frac{1}{2} R_\varepsilon^{-1}$. So

$$\overline{\int_{S^1} |v(x+l) - v(x)|^p \, dx} \geq \frac{1}{2} R_\varepsilon^{-1} C_\varepsilon |l|^{\min(1,p)}$$

for each $\omega \in Q_\varepsilon^4$, and the desired lower bound follows. □

Theorems 7.1.3, 7.2.2 and 7.3.2 admit obvious similar "high probability" reformulations, which we do not state. Their proofs also follow those in Sections 7.1-7.3, using estimates (7.5.5) and (7.5.7).

In particular, an analogue of Theorem 7.3.2 tells that for any $\varepsilon > 0$ there exists $\bar{\sigma}_\varepsilon \geq 2\bar{\sigma}'$ and constants $M(\varepsilon), c_1'(\varepsilon), c_5'(\varepsilon), c_6'(\varepsilon) > 0$ such that for any $T' \geq 1$ the relations

$$c_5'(\varepsilon) k^{-2} \leq \overline{e_k(u)} \leq c_6'(\varepsilon) k^{-2} \qquad \forall 1 \leq k \leq (c_1'(\varepsilon)\nu)^{-1}, \tag{7.5.9}$$

are valid with probability at least $1 - \varepsilon$. Cf. below Example 10.3.3.

CHAPTER 8

The Inviscid limit and Inviscid Burgulence

In this chapter we study the limiting behaviour of solutions for eq. (B) as $\nu \to 0$. The corresponding dynamics for the free inviscid Burgers equation $(\mathrm{B})_{\eta=0}$ is given by the celebrated Lax–Oleinik formula [**116, 129**]. In the papers [**140, 141**] rigorous study of the inviscid stochastic Burgers equation was initiated, and in [**56**] the Lax-Oleinik formula was used to get a description of the limiting inviscid dynamics for eq. (B) using stochastic Lagrangians. The approach in [**56**] initiated many further developments, mainly due to Iturriaga, Khanin, Zhang and their collaborators [**70, 78, 83**]. Also see the paper of Saussereau and Stoica [**135**], where similar techniques are applied to equations (1.1.13), where the random force ξ is a fractional Brownian motion as a function of time.

Unlike in those works we do not evoke the Lax–Oleinik formula and its consequences, but instead use Oleinik's estimates from Chapter 2 to prove that the inviscid limit of solutions $u^\nu(t,x)$ as $\nu \to 0$ exists in the L_p-sense for any finite $p \geq 1$, and show that this convergence allows to pass to the limit as $\nu \to 0$ in some results from Chapters 4 and 7. Our interpretation of the Oleinik estimate and the way in which we use it to prove the inviscid limit are similar to the approach in [**95, 96**] and in Appendix to [**18**].

Other methods were used to study the stochastic inviscid problem, among others, by Gyöngy and Rovira [**73**], Kim [**86**], Dirr and Souganidis [**49**], Feng and Nualart [**62**], Debussche and Vovelle [**47, 48**] (using kinetic solutions introduced in the deterministic case by Lions, Perthame and Tadmor [**120**]) and Bauzet, Vallet and Wittbold [**9**]. A recent paper by Chen and Pang [**37**] contains a comprehensive literature survey of the works on the inviscid stochastic limit.

Unlike in other parts of the book, in this chapter we sometimes use spaces of functions of x with non-zero mean. We distinguish them by bars. So $\bar{L}_p$ stands for the L_p-space of functions on S^1 with any mean value; the norm in $\bar{L}_p$ still is denoted $|\cdot|_p$. Similarly for $m \in \mathbf{N}_0$, $\bar{H}^m$ stands for the Sobolev space $\bar{H}^m = \{u \in \bar{H} : u^{(m)} \in \bar{H}\}$, equipped with the non-homogeneous norm $\|u\|^2_{\bar{H}^m} = \|u\|^2 + \|u^{(m)}\|^2$. Naturally, $\bar{X}^m_T = C(0,T;\bar{H}^m)$, etc.

8.1. Cauchy problem for an auxiliary deterministic equation.

We begin by considering an auxiliary initial-value problem, formally obtained from (1.1.13) by integrating in dx and denoting $\int^x u(t,y)dy = \phi(t,x)$, $\int^x \xi(t,y)dy = \zeta(t,x)$:

$$
\left\{
\begin{array}{l}
\phi_t + \phi_x^2/2 - \nu\phi_{xx} = \zeta_t, \\
\phi(0,\cdot) = \phi_0(\cdot), \\
t \in [0,T],\ x \in S^1,\ \nu \in (0,1].
\end{array}
\right.
\tag{8.1.1}
$$

Here

$$\phi_0 \in \bar{H}^m, \quad \zeta \in \dot{\bar{X}}_T^{m_*}, \tag{8.1.2}$$

with $\mathbf{N} \ni m \geq 3$.

We immediately see that if $\zeta(t,x)$ is a smooth function and $\phi(t,x)$ is a smooth solution of the equation in (8.1.1), then $u = \phi_x(t,x)$ solves (B) with $\eta = \partial \zeta_t / \partial x$. If ζ is only $\bar{X}_T^{m_*}$–smooth, then we define solutions of (8.1.1) similarly to those of the problem (1.1.13):

DEFINITION 8.1.1. We say that $\phi \in \bar{X}_T^m$ solves the problem (8.1.1) if for all $t \in [0,T]$ we have the following equality:

$$\phi(t) - \phi_0 + \int_0^t \big((\phi_x^2(s)/2 - \nu\phi_{xx}(s)\big)ds = \zeta(t). \tag{8.1.3}$$

It is straightforward that if $\phi \in \bar{X}_T^m$ solves (8.1.1) with ϕ_0 and ζ as in (8.1.2), then $u = \phi_x$ solves the initial-value problem (1.1.13), which we repeat here for convenience:

$$u_t(t,x) + u(t,x)u_x(t,x) - \nu u_{xx}(t,x) = \partial_t \xi(t,x), \quad u(0,x) = u_0(x), \tag{8.1.4}$$

with $u_0 = (\phi_0)_x \in H^{m-1}$ and $\xi = \zeta_x \in \dot{X}_T^{m_*-1}$. Conversely, if $u(t,x) \in X_T^\mu$ is a solution of (8.1.4) with $\xi \in \dot{X}_T^{\mu_*}$ and $u_0 \in H^\mu$, where $\mu_* > \mu \geq 2$,[1] $\mu \in \mathbf{N}$, then the function $\tilde{\phi}(t,x) = \int_0^x u(t,y)dy$ satisfies

$$\tilde{\phi}(t,x) - \tilde{\phi}_0(x) + \int_0^t (\tfrac{1}{2}\tilde{\phi}_x^2(s,x) - \nu\tilde{\phi}_{xx}(s,x))ds = \zeta(t,x) + \Theta(t),$$

where $\tilde{\phi}_0(x) = \int_0^x u_0(y)\,dy$, $\zeta(t,x) = \int_0^x \xi(t,y)\,dy$, and

$$\Theta(t) = \int_0^t (\tfrac{1}{2}u(s,0)^2 - \nu u_x(s,0))\,ds. \tag{8.1.5}$$

Then $\phi(t,x) = \tilde{\phi}(t,x) - \Theta(t) \in \bar{X}_T^{\mu+1}$ solves problem (8.1.1); so the latter has a solution. If $\phi^1(t,x)$ is another solution of (8.1.1), then ϕ_x and ϕ_x^1 both solve the same problem (1.1.13). From here $\phi_x = \phi_x^1$ and $\phi(t,x) = \phi^1(t,x) + \tilde{\Theta}(t)$ for a suitable function $\tilde{\Theta}$. Since ϕ and ϕ^1 both satisfy Definition 8.1.1, we immediately get that $\tilde{\Theta} = 0$. Since the problem (8.1.4) with $u_0 \in H^\mu$ and $\xi \in \dot{X}_T^{\mu_*}$, where $\mu_* > \mu \geq 2$, has a unique solution by Theorem 1.3.6, then we have proved

PROPOSITION 8.1.2. *1) For any ϕ_0, ζ as in (8.1.2), where $m_* > m \geq 3$, $m \in \mathbf{N}$, the problem (8.1.1) has a unique solution $\phi \in \bar{X}_T^m$. The function $u = \phi_x \in X_T^{m-1}$ solves the problem (8.1.4) with $u_0 = (\phi_0)_x \in H^{m-1}$ and $\xi = \zeta_x \in \dot{X}_T^{m_*-1}$.*

2) Conversely, if $u(t,x)$ solves the problem (8.1.4) with $u_0 \in H^{m-1}$ and $\xi \in \dot{X}_T^{m_*-1}$, then $\phi(t,x) := \int_0^x u(t,y)\,dy - \Theta(t)$, where Θ is defined in (8.1.5), solves (8.1.1) with

$$\phi_0(x) = \int_0^x u_0(y)\,dy, \qquad \zeta(t,x) = \int_0^x \xi(t,y)\,dy. \tag{8.1.6}$$

[1]The restriction $\mu \geq 2$ is needed since for $u(t,x) \in X_T^1$ the term $\nu u_x(s,0)$ in the expression for Θ below is not defined.

8.2. Inviscid limit for the deterministic equation.

Let us consider the deterministic problem (8.1.4), where

$$u_0 \in H^2, \quad \xi \in \dot{X}_T^4, \quad 0 < \nu \leq 1.$$

We denote its solution u^ν and define $\phi^\nu(t,x) = \int_0^x u^\nu(t,y)\,dy - \Theta(t)$ as in Proposition 8.1.2. Then ϕ^ν solves (8.1.1) with ϕ_0 and ζ as in (8.1.6). Let the constant $B = B_T(u_0, \xi)$ be as in Theorem 2.1.7, i.e.

$$B = \max(|u_{0x}^+(0,\cdot)|_\infty + \|\xi\|_{X_T^4},\ 4\|\xi\|_{X_T^4} + \|\xi\|_{X_T^4}^{1/2}). \tag{8.2.1}$$

Then by that theorem

$$\begin{aligned} &|(\phi_{xx}^\nu)_+(t,\cdot)|_\infty = |(u_x^\nu)_+(t,\cdot)|_\infty \leq B, \quad |\phi_{xx}^\nu(t,\cdot)|_1 = |u_x^\nu(t,\cdot)|_1 \leq 2B, \\ &|\phi_x^\nu(t,\cdot)|_\infty = |u^\nu(t,\cdot)|_\infty \leq B, \end{aligned} \tag{8.2.2}$$

for all $0 \leq t \leq T$ and $\nu \in (0,1]$. Our goal in this section is to prove the following inviscid limit theorem:

THEOREM 8.2.1. *Let $1 \geq \nu_2 > \nu_1 > 0$. Then:*
i) for any $p \in [1,\infty)$ the solutions u^{ν_2} and u^{ν_1} satisfy

$$|u^{\nu_2}(t) - u^{\nu_1}(t)|_p \leq C_p B^{1-\alpha_p} \bar{\nu}^{\alpha_p} e^{Bt\alpha_p}, \quad t \in [0,T], \tag{8.2.3}$$

where $\bar{\nu} = \nu_2 - \nu_1 > 0$, $\alpha_p = \min(1/4, 1/3p)$ and $C_p > 0$ is a suitable constant.
ii) There exists a function $u^e(t,x) = u^e(t,x;u_0,\xi) \in \bigcap_{p\in[1,\ \infty)} C(0,T;L_p)$ such that

$$u^\nu \underset{\nu\to 0}{\longrightarrow} u^e \quad \textit{in} \quad C(0,T;L_p) \quad \textit{for every} \quad 1 \leq p < \infty. \tag{8.2.4}$$

iii) Relation (8.2.3) remains true for the limiting function u^e:

$$|u^\nu(t) - u^e(t)|_p \leq C_p B^{1-\alpha_p} \nu^{\alpha_p} e^{Bt\alpha_p}, \qquad t \in [0,T],\ 0 < \nu \leq 1, \tag{8.2.5}$$

and

$$\|u^e\|_{C(0,T;L_p)} \leq B. \tag{8.2.6}$$

PROOF. i) Firstly we show that the difference $b(t,x) = \phi^{\nu_1}(t,x) - \phi^{\nu_2}(t,x)$ meets the inequality

$$|b(t)|_2 \leq (8B)^{1/2} e^{Bt/2} \bar{\nu}^{1/2}, \quad 0 \leq t \leq T. \tag{8.2.7}$$

Indeed, b vanishes at $t = 0$ and satisfies the equation

$$b_t + (\phi_x^{\nu_1} + \phi_x^{\nu_2}) b_x / 2 = \nu_1 \phi_{xx}^{\nu_1} - \nu_2 \phi_{xx}^{\nu_2}.$$

Let us denote by $E(t)$ the quantity $|b(t,\cdot)|_2^2$. Then multiplying the relation above by $2b(t)$ in H and integrating by parts we get:

$$\begin{aligned}\frac{dE(t)}{dt} &= \frac{1}{2}\int_{S^1} (\phi^{\nu_1}_{xx}(t,x)+\phi^{\nu_2}_{xx}(t,x))b^2(t,x)\,dx\\ &\quad -2\int_{S^1} (\nu_1\phi^{\nu_1}_x(t,x)-\nu_2\phi^{\nu_2}_x(t,x))(\phi^{\nu_1}_x(t,x)-\phi^{\nu_2}_x(t,x))\,dx\\ &\le \frac{1}{2}\int_{S^1} (\phi^{\nu_1}_{xx}(t,x)+\phi^{\nu_2}_{xx}(t,x))b^2(t,x)\,dx\\ &\quad +2\bar\nu\int_{S^1} \phi^{\nu_2}_x(t,x)(\phi^{\nu_1}_x(t,x)-\phi^{\nu_2}_x(t,x))\,dx.\\ &\le \frac{1}{2}\Big(|(\phi^{\nu_1}_{xx})^+(t)|_\infty+|(\phi^{\nu_2}_{xx})^+(t)|_\infty\Big)E(t)+2\bar\nu(|\phi^{\nu_1}_x(t)|_\infty+|\phi^{\nu_2}_x(t)|_\infty)^2\\ &\le BE(t)+8B^2\bar\nu.\end{aligned}$$

Since $b(0)=0$, then by Gronwall's inequality $E(t)\le 8B\bar\nu(e^{Bt}-1)\le 8Be^{Bt}\bar\nu$. This proves (8.2.7).

Next, applying the Gagliardo–Nirenberg inequality (1.3.9) to $\phi^{\nu_1}(t,\cdot)-\phi^{\nu_2}(t,\cdot)$, recalling that $(\partial/\partial x)(\phi^{\nu_1}(t,x)-\phi^{\nu_2}(t,x))=u^{\nu_1}(t,x)-u^{\nu_2}(t,x)$ and using estimates (8.2.2), (8.2.7) we obtain that

$$|u^{\nu_1}(t)-u^{\nu_2}(t)|_{4/3}\le C|\phi^{\nu_1}(t)-\phi^{\nu_2}(t)|_2^{1/2}|u^{\nu_1}_x(t)-u^{\nu_2}_x(t)|_1^{1/2}\le C_1B^{1/4}e^{Bt/4}\bar\nu^{1/4}B^{1/2}.$$

Using this inequality, (8.2.2) and the Riesz–Thorin inequality (11.1.5) we get that for any $4/3\le p<\infty$ we have

$$|u^{\nu_1}(t)-u^{\nu_2}(t)|_p\le|u^{\nu_1}(t)-u^{\nu_2}(t)|_\infty^{1-4/3p}|u^{\nu_1}(t)-u^{\nu_2}(t)|_{4/3}^{4/3p}\le C_pB^{1-1/3p}\bar\nu^{1/3p}e^{Bt/3p}.$$

Since $|u^{\nu_1}(t)-u^{\nu_2}(t)|_{p'}\le|u^{\nu_1}(t)-u^{\nu_2}(t)|_{4/3}$ for $1\le p'\le 4/3$, then the last estimate implies (8.2.3).

ii) By (8.2.3) the mapping $(0,1]\ni\nu\mapsto u^\nu\in C(0,T;L_p)$ is *Cauchy–continuous* as $\nu\to 0$ [2] for each $p<\infty$, so the limit (8.2.4) exists for every $1\le p<\infty$. Clearly the limiting curve $u^e(t)$ does not depend on p.

iii) Passing in (8.2.3) to the limit as $\nu_1\to 0$ using (8.2.4) we recover (8.2.5), while the inequality (8.2.6) follows from (8.2.2) and (8.2.4). □

PROBLEM 8.2.2. If $u(t,x)$ satisfies eq. (0.0.12), then $\phi(t,x)=\int_0^x u(t,y)dy-\Theta^f(t)$ with a suitable function Θ^f is a solution of equation $\phi_t+f(\phi_x)-\nu\phi_{xx}=\zeta(t,x)$ with $\zeta=\int_0^x\xi(t,y)dy$. Use this fact to derive an analogue of estimate (8.2.3) for $u^{\nu_2}(t)-u^{\nu_1}(t)$, where $u^{\nu_j}(t)$ solves (0.0.12) with $\nu=\nu_j$ and f satisfies (0.0.13). **Hint:** Use relation (1.3.32).

By (8.2.2), $|u^\nu(t)|_\infty\le B$ for each $t\in[0,T]$ and every ν. Since $L_\infty=(L_1)^*$, then by the Banach-Alaoglu theorem (e.g. see [**91**, Section IV.3.4] and [**52**, Section V.4]) for any t there exists a sequence $\nu_j\to 0$ and a function $v\in L_\infty$ such that $u^{\nu_j}(t)\to v$ in the $*$-weak topology of L_∞ and $|v|_\infty\le B$. In view of (8.2.4) we must have $v=u^e(t)$, so $|u^e(t)|_\infty\le B$ for every t. Next, due to (8.2.2), for any $t\in[0,T]$ and any ν the total variation of the function $x\mapsto u^\nu(t,x)$ is bounded by $2B$. Since $|u_\nu(t,\cdot)|_\infty\le B$, then by Helly's selection theorem (see [**91**, Section VI.6.5]) there

[2] That is, for any sequence $\nu_j\to 0$ the sequence $\{u^{\nu_j}\}\subset C(0,T;L_p)$ possesses the Cauchy property.

exists a sequence $\nu_j \to 0$ such that, as $\nu_j \to 0$, the function $x \mapsto u^{\nu_j}(t,x)$ converge a.s. to a function, whose total variation is bounded by $2B$. The latter must equal $u^e(t,x)$ a.s., so we get

COROLLARY 8.2.3. *For any $t \in [0,T]$, $|u^e(t)|_\infty \le B$ and after a modification on a set of zero measure, the total variation of $u^e(t,\cdot)$ is bounded by $2B$.*

Similarly to (1.3.34) and (1.3.35), and based on Theorem 8.2.1 and Corollary 8.2.3, for a $T > 0$ we denote by $\mathcal{M}^0$ and $\mathcal{M}^0_t$, $0 \le t \le T$, the operators

$$\mathcal{M}^0 : H^2 \times \dot{X}^4_T \to \bigcap_{1\le p<\infty} C(0,T;L_p), \quad (u_0,\xi) \mapsto u^e,$$

$$\mathcal{M}^0_t : H^2 \times \dot{X}^4_T \to L_\infty, \quad (u_0,\xi) \mapsto u^e(t).$$

They are well-defined and

$$\|\mathcal{M}^0(u_0,\xi)\|_{C(0,T;L_p)} \le B \;\; \forall p < \infty, \qquad |\mathcal{M}^0_t(u_0,\xi)|_\infty \le B \;\; \forall 0 \le t \le T, \tag{8.2.8}$$

where $B = B_T(u_0,\xi)$ is defined in (8.2.1). By Theorem 1.3.11, for each given value of $\nu > 0$ the mapping $\mathcal{M}^\nu : H^2 \times \dot{X}^4_T \to C(0,T;L_p)$ is continuous for every p. So in view of convergence (8.2.4) (which is uniform on bounded subsets of $H^2 \times \dot{X}^4_T$) each mapping

$$\mathcal{M}^0 : H^2 \times \dot{X}^4_T \to C(0,T;L_p), \quad 1 \le p < \infty, \tag{8.2.9}$$

also is continuous, as well as the mappings

$$\mathcal{M}^0_t : H^2 \times \dot{X}^4_T \to L_p, \quad 1 \le p < \infty,\ 0 \le t \le T.$$

PROBLEM 8.2.4. Let u_0 and ξ be as in Theorem 8.2.1. Then for any $m_1 < m$ a solution u^ν of (8.1.4) continuously depends on $\nu \in (0,1]$ as a curve in $C(0,T;H^{m_1})$. **Hint:** use (8.2.3) and Theorem 1.3.6.

PROBLEM 8.2.5. Denote by L^*_∞ the space L_∞ given the weak-$*$ topology (see [**52, 152**]), and prove that for each $0 \le t \le T$ the map

$$\mathcal{M}^0_t : H^2 \times \dot{X}^4_T \to L^*_\infty \tag{8.2.10}$$

is a point-wise limit of the maps $\mathcal{M}^\nu_t$ as $\nu \to 0$. So this mapping is weak-$*$ measurable. That is, for any $\phi(x) \in L_1$ the function

$$H^2 \times \dot{X}^4_T \ni (u,\xi) \mapsto \int \mathcal{M}^0_t(u,\xi)(x)\phi(x)\,dx$$

is measurable.

The function $u^e(t,x)$ is the inviscid limit for solutions $u^\nu(t,x)$. Obviously it equals u_0 at $t = 0$, and it is easy to derive from (8.2.4) that this is a solution of the initial-value problem for the inviscid Burgers equation

$$u_t(t,x) + u(t,x)u_x(t,x) = \partial_t\xi(t,x), \quad u(0,x) = u_0(x), \tag{8.2.11}$$

in the sense of generalised functions. That is, $u^e(0) = u_0$ and for any smooth function $w(t,x)$ with compact support in the open cylinder $(0,T) \times S^1 = Q$ we have

$$(u^e, -w_t)_Q + (\tfrac{1}{2}(u^e)^2, -w_x)_Q = (\xi, -w_t)_Q,$$

where $(\cdot,\cdot)_Q$ stands for the pairing between $C^\infty_0(Q)$ and $\mathcal{D}'(Q)$. The initial-value problem (8.2.11) has many generalised solutions. The limit (8.2.4) selects among them the unique solution u^e, called an *entropy solution* for the problem (8.2.11).

For the case of continuous functions $\xi_t(t,x)$ (e.g. when $\xi=0$) the entropy solutions of (8.2.11) and of similar equations have been intensively studied since the seminal works of Lax, Oleinik and Kruzhkov [**95, 116, 129**], e.g. see [**45, 59**]. Entropy solutions for equations with irregular forces ξ_t have got less attention (see the beginning of the chapter for the corresponding references). The statement that an initial-value problem (8.2.11) with any $u_0\in H^2$, $\xi\in\dot X^4_T$ has a unique entropy solution for each $T>0$ is in sharp contrast with the fact that for $\xi=0$ and a smooth $u_0(x)$, a classical solution of this problem exists for all $t\ge 0$ only if $u_0=0$:

PROBLEM 8.2.6. Prove that if $\xi=0$ and $u_0(x)$ is a smooth function with zero mean, then a C^1-smooth solution of the problem (8.2.11) exists for all $t\ge0$ only if $u_0=0$.
Hint: Construct a C^1-solution of the problem, using the method of characteristics. Note that in the case of the inviscid Burgers equation $u_t+uu_x=0$ the method takes the especially simple form:

$$u(t,y^x(t))=\text{Const}=u_0(x)$$

along any ray $y^x(t)=x+tu_0(x), t\ge 0$. Certainly an analogous statement of non-existence of smooth solutions for (8.2.11) for all $t\ge0$ holds for a large choice of right-hand sides and initial conditions, and also for solutions of various inviscid equations $(0.0.12)_{\nu=0}$.

8.3. Inviscid limit for the stochastic equation.

Now let us return to the stochastic problem (1.1.13), assuming in Sections 8.3-8.4 that

$$B_4<\infty.$$

As usual, the initial data u_0 is a r.v., independent from the process ξ^ω. At first we will assume that $u_0\in H^2$; next this restriction will be relaxed.

By Theorem 1.4.2 a strong solution of (1.1.13) is $u^{\nu,\omega}(t,x)=\mathcal{M}^\nu(u_0^\omega,\xi^\omega)$, and by Theorem 8.2.1

$$u^{\nu,\omega}\underset{\nu\to0}{\longrightarrow}u^{e,\omega}=\mathcal{M}^0(u_0^\omega,\xi^\omega)\quad\text{in}\quad C(0,T;L_p)\quad\forall\omega, \tag{8.3.1}$$

for every $1\le p<\infty$. Since the mapping (8.2.9) is measurable, then the entropy solution $u^{e,\omega}=\mathcal{M}^0(u_0^\omega,\xi^\omega)$ as a function of ω is a r.v. in any space $C(0,T;L_p)$, $1\le p<\infty$, satisfying the first estimate in (8.2.8). It is easy to see that in view of the result of Problem 8.2.5, for any $t\ge0$ and $u_0\in H^2$ the map $\omega\to u^{e,\omega}(t;u_0)\in L_\infty$ is weak-$*$ measurable and meets the second estimate in (8.2.8).

The entropy solutions $u^{e,\omega}$ inherit Oleinik's estimates (2.1.16) in a weaker form:

PROPOSITION 8.3.1. *Let $u_0\in H^2$ and $u^{e,\omega}=\mathcal{M}^0(u_0^\omega,\xi^\omega)$ be an entropy solution of* (8.2.11). *Then for any $0<\theta\le1$ and $1\le p,\alpha<\infty$,*

$$\mathbb{E}|u^e(t)|_p^\alpha\le C(\alpha,B_4)\theta^{-\alpha}\quad\forall t\ge\theta. \tag{8.3.2}$$

PROOF. By Corollary 2.1.5 estimate (8.3.2) holds for u^e replaced by u^ν for any $\nu>0$, and by (8.2.4) $|u^\nu(t)|_p^\alpha\underset{\nu\to0}{\longrightarrow}|u^e(t)|_p^\alpha$ for each ω, so the validity of (8.3.2) follows by Fatou's lemma. $\square$

The result below estimates the rate of convergence of the law of u^ν to that of u^e if u_0 is non-random:

PROPOSITION 8.3.2. *Let $u_0 \in H^2$ be a non-random vector and $u^\nu(t) = u^\nu(t; u_0)$. Then for any $t \geq 0$ and any $1 \geq \nu_2 > \nu_1 \geq 0$ we have*

$$\|\mathcal{D}u^{\nu_2}(t) - \mathcal{D}u^{\nu_1}(t)\|^*_{L(L_1)} \leq C\bar{\nu}^{1/4} e^{\frac{1}{2}\max(1,t)(1+\|u_0\|_2)+4\max(1,t)^3 B_4}, \tag{8.3.3}$$

where $\bar{\nu} = \nu_2 - \nu_1 > 0$ and we denote $u^0(t) := u^e(t)$.

PROOF. Let $f \in \bar{B}^1_{L(L_1)}$. Then using (8.2.3) and (8.2.5) with $T = t$ and $p = 1$ and noting that

$$B_t(u_0, \xi) \leq \tilde{B}_t := \|u_0\|_2 + 4\|\xi\|_{X^4_t} + \|\xi\|^{1/2}_{X^4_t} \leq \|u_0\|_2 + \tfrac{9}{2}\|\xi\|_{X^4_t} + \tfrac{1}{2},$$

we get:

$$d(t) := \Big|\mathbb{E}\big(f(u^{\nu_2})(t) - f(u^{\nu_1})(t)\big)\Big| \leq \mathbb{E}|u^{\nu_2}(t) - u^{\nu_1}(t)| \leq C\bar{\nu}^{1/4}\mathbb{E}(\tilde{B}_t^{3/4} e^{\tilde{B}_t t/4}) =: d_1(t).$$

Assume firstly that $t \geq 1$. Then

$$d_1(t) \leq C_\gamma \bar{\nu}^{1/4}\mathbb{E}e^{\gamma \tilde{B}_t t}, \qquad \forall \gamma > \tfrac{1}{4}.$$

Choosing $\gamma = \frac{4}{9}$ we find that

$$d_1(t) \leq C_1 \bar{\nu}^{1/4} e^{t(1+\|u_0\|_2)/2}\mathbb{E}e^{2t\|\xi\|_{X^4_t}}.$$

By (1.2.19) with $\alpha = (4tB_4)^{-1}$, $\mathbb{E}\exp\left(\frac{1}{4tB_4}\|\xi\|^2_{X^4_t}\right) \leq C_2$. As $2tx \leq \frac{x^2}{4tB_4} + 4t^3B_4$ for any x, then using this inequality with $x = \|\xi\|_{X^4_t}$ in the estimate above we get that

$$d(t) \leq d_1(t) \leq C_1C_2\bar{\nu}^{1/4}e^{t(1+\|u_0\|_2)/2}e^{4t^3B_4} =: d_2(t).$$

Since this holds for any $f \in \bar{B}^1_{L(L_1)}$, then (8.3.3) follows for $t \geq 1$.

If $0 \leq t \leq 1$, then $d(t) \leq d_1(t) \leq d_1(1) \leq d_2(1)$. This proves (8.3.3) for $0 \leq t \leq 1$. □

The limiting process in L_1.

Recalling that $B_4 < \infty$, let us consider the mappings $\mathcal{M}^0$ and $\mathcal{M}^0_t$ (see (8.2.9)). In view of (3.2.7) and the convergence (8.2.4), for each $\xi \in \dot{X}^4_T$ the map

$$\mathcal{M}^0(\cdot, \xi) : H^2 \to C(0, T; L_1)$$

is 1–Lipschitz with respect to the L_1-norms in the range and the domain. Considering the mappings

$$\mathcal{M}^0(\Phi_n \star \cdot, \cdot) : L_1 \times \dot{X}^4_T \to C(0, T; L_1), \quad n \geq 1, \tag{8.3.4}$$

where $\Phi_n\star$ is the convolution with Fejér's kernel, and arguing as in Section 4.1 we find that the operators $\mathcal{M}^0$ and $\mathcal{M}^0_t$ extend by continuity to Borel mappings

$$\mathcal{M}^e : L_1 \times \dot{X}^4_T \to C(0, T; L_1), \qquad \mathcal{M}^e_t : L_1 \times \dot{X}^4_T \to L_1\,, \tag{8.3.5}$$

1–Lipschitz in the first argument, where the first is the limit of the mappings in (8.3.4) as $n \to \infty$ and the second is defined in the obvious way. In view of the convergence (8.2.4), the entropy solutions u^e with $u_0 \in H^2$ satisfy estimate (3.2.11) since this relation holds for viscous solutions u^ν. Approximating any $u_0 \in L_1$ by smooth functions and using that the maps $\mathcal{M}^e_t(\cdot, \xi)$ are 1-Lipschitz, we get from here that

$$|\mathcal{M}^e_t(u_0, \xi)|_1 \leq |u_0|_1 + 4\|\xi\|_{X^4_t} + \|\xi\|^{1/2}_{X^4_t} \quad \forall\, (u_0, \xi) \in L_1 \times \dot{X}^4_t.$$

Since the mappings $\mathcal{M}^T$ in (4.1.2) are 1–Lipschitz in u_0 with respect to the L_1–norm, then equivalently the maps $\mathcal{M}^e$ and $\mathcal{M}_t^e$ in (8.3.5) may be obtained as limits of the mappings in (4.1.2) as $\nu \to 0$. Indeed, for any $u_0 \in L_1$ and $\xi \in \dot{X}_T^4$ taking a sequence $\{u_j\}$ in H^2, converging to u_0 in L_1, we have:

$$\begin{aligned}\mathcal{M}^e(u_0,\xi) &= \lim_{j\to\infty} \mathcal{M}^e(u_j,\xi) = \lim_{j\to\infty}\lim_{\nu\to 0} \mathcal{M}^\nu(u_j,\xi) = \lim_{\nu\to 0}\lim_{j\to\infty} \mathcal{M}^\nu(u_j,\xi)\\ &= \lim_{\nu\to 0} \mathcal{M}^\nu(u_0,\xi),\end{aligned}$$

where the convergences hold in $C(0,T;L_1)$. That is,

$$\mathcal{M}^e(u_0,\xi) = \lim_{\nu\to 0} \mathcal{M}^\nu(u_0,\xi) \quad \text{in } C(0,T;L_1), \quad \forall\, u_0 \in L_1,\ \xi \in \dot{X}_T^4, \tag{8.3.6}$$

where $\mathcal{M}^\nu = \mathcal{M}^{\nu,T}$ is the operator in (4.1.2), and similarly for the maps $\mathcal{M}_t^e$.

The maps (8.3.5) allow to define L_1-entropy solutions of the problem (8.1.4) with $\nu = 0$, $u_0 \in L_1$ and $B_4 < \infty$ by the relation

$$u^{e,\omega}(t;u_0) = \mathcal{M}^e(u_0;\xi^\omega).$$

Since $\mathcal{M}^e$ is 1-Lipschitz in u_0, then

$$|u^{e,\omega}(t;u_0) - u^{e,\omega}(t;u_1)|_1 \le |u_0 - u_1|_1, \quad \forall\, u_0, u_1 \in L_1,\ \forall \omega,\ \forall t \ge 0. \tag{8.3.7}$$

PROPOSITION 8.3.3. *Estimates* (8.3.2), *where* $1 \le p, \alpha < \infty$, *remain true for entropy solutions* $u^e(t;u_0)$ *with* $u_0 \in L_1$.

PROOF. Approximating $u_0 \in L_1$ by H^2-smooth initial data u_{0n} and using (8.3.2) for solutions $u^e(t;u_{0n})$ we derive the estimates from (8.3.7) and Theorem 11.1.6. □

Now we extend by continuity the transition probability Σ, defined in Section 1.6, to a transition probability Σ^e, corresponding to the entropy solutions. To do this we note that in view of (8.3.6), for any $f \in C_b(L_1)$ the function

$$\mathbb{R}_+ \times L_1 \ni (t,u_0) \mapsto \langle f, \mathcal{D}u^e(t;u_0)\rangle$$

is a point-wise limit as $\nu \to 0$ of the measurable functions (4.1.4). Accordingly it is measurable, so

$$\Sigma_t^e(u_0) = \mathcal{D}u^e(t,u_0) \in \mathcal{P}(L_1)$$

is a transition probability, see Theorem 11.3.2. In view of (8.3.7),

$$\text{the mapping } L_1 \to \mathcal{P}(L_1),\ u \mapsto \Sigma_t^e(u) \in \mathcal{P}(L_1), \text{ is 1-Lipschitz for } t \ge 0 \tag{8.3.8}$$

(the space $\mathcal{P}(L_1)$ is given the Lipschitz-dual distance).

Similarly to Section 4.1 the L_1-entropy solutions define a Markov process in L_1 (see Appendix 11.6). The corresponding Markov semigroup in $\mathcal{P}(L_1)$ is formed by the continuous operators S_t^{e*}, where

$$S_t^{e*}(\mu) = \mathcal{M}_t^e \circ (\mu \times \pi^4), \qquad S_t^{e*}(\mu)(Q) = \int \Sigma_t^e(u)(Q)\,\mu(du).$$

8.4. Inviscid Burgulence.

Till the end of this section we assume that

$$u_0 \in H^2 \text{ is a non-random vector and } B_0 > 0,\ B_4 < \infty.$$

Firstly we will discuss the structure function of solutions u^ν with small ν and of the entropy solutions u^e. We recall (see Section 7.2) that the structure function $S_{p,l}$ of a solution u^ν of (1.1.13) is defined by the relation

$$S_{p,l} = S_{p,l}(u^\nu) = \langle\langle s_{p,l}(u^{\nu,\omega})(t,\cdot)\rangle\rangle, \qquad s_{p,l}(u(\cdot)) = \int |u(x+l) - u(x)|^p dx,$$

where $\langle\langle\cdot\rangle\rangle$ stands for the expectation (2.3.5), corresponding to the probability space $(Q,\mathcal{T},\rho) = ([T,T+\sigma]\times\Omega, \mathcal{L}\times\mathcal{F}, \frac{dt}{\sigma}\times \mathbf{P})$, where $T \geq 1$ and $\sigma \geq \sigma_0 > 0$ (see (2.3.4)). On this space consider the random variables

$$\mathcal{S}^\nu_{p,l}(t,\omega) = s_{p,l}(u^{\nu,\omega}(t;u_0)), \qquad 0 < \nu \leq 1,\ p > 0,\ 0 < |l| \leq 1/2,$$

where u^ν is a solution of (1.1.13).

LEMMA 8.4.1. *For any $p > 0$ and any l, the family $\{\mathcal{S}^\nu_{p,l}, 0 < \nu \leq 1\}$ of r.v.'s on Q is uniformly integrable.*

PROOF. Let firstly $p \geq 1$. Then obviously $\mathcal{S}^\nu_{p,l}(t,\omega) \leq 2^p |u^{\nu,\omega}(t)|_p^p$. So for any $p_1 > p$ we have

$$\mathcal{S}^\nu_{p,l}(t,\omega)^{p_1/p} \leq 2^{p_1} |u^{\nu,\omega}(t)|_p^{p_1},$$

and by Corollary 2.1.5 with $\theta = 1$,

$$\langle\langle (\mathcal{S}^\nu_{p,l})^{p_1/p}\rangle\rangle = \mathbb{E}^Q (\mathcal{S}^\nu_{p,l})^{p_1/p} \leq C(p_1, B_4) < \infty,$$

for all $\nu > 0$. If $0 < p < 1$, then $\mathcal{S}^\nu_{p,l}(t,\omega) < 2|u^{\nu,\omega}(t)|_1$, and arguing as above we obtain that $\mathbb{E}^Q (\mathcal{S}^\nu_{p,l})^2 \leq C(u_0) < \infty$, for all ν. This proves the lemma. ☐

Due to Lemma 7.4.4 and convergence (8.3.1),

$$s_{p,l}(u^{\nu,\omega}(t)) \underset{\nu\to 0}{\longrightarrow} s_{p,l}(u^{e,\omega}(t)) \qquad \forall\, \omega, t, p, l.$$

By this result, Lemma 8.4.1 and Theorem 11.1.8, $S_{p,l}(u^\nu) \underset{\nu\to 0}{\longrightarrow} S_{p,l}(u^e)$ for all p and l. Passing in the assertion of Theorem 7.2.3 to the limit as $\nu \to 0$, using the just obtained convergence, we get the first main result of this section:

THEOREM 8.4.2. *There exists $c_2' > 0$ and for every $p > 0$ there exist $\tilde c_p, \tilde C_p > 0$ such that the structure function $S_{p,l}(u^e)$ of the entropy solution u^e satisfies the estimates*

$$\begin{cases} \tilde c_p |l|^p \leq S_{p,l}(u^e) \leq \tilde C_p |l|^p, & \text{if } p \in (0,1), \\ \tilde c_p |l| \leq S_{p,l}(u^e) \leq \tilde C_p |l|, & \text{if } p \geq 1, \end{cases} \tag{8.4.1}$$

for every $|l| \leq c_2'$.

Now we consider the energy spectrum $E_k(u^e) = \langle\langle e_k(u^e)\rangle\rangle$ of an entropy solution u^e (see (7.3.1)). Since the functions $u \mapsto \hat u_n$ are continuous linear functionals on L_1, then the functionals e_k also are continuous on L_1, and in view of (8.3.1)

$$e_k(u^{\nu,\omega}(t)) \underset{\nu\to 0}{\longrightarrow} e_k(u^{e,\omega}(t)) \qquad \forall\, t, \omega.$$

As $|\hat u_n| \leq |u|_1$, then by (8.2.2) and Theorem 1.2.3 the family of random variables $\{e_k(u^{\nu,\omega}(t)), 0 < \nu \leq 1\}$ on the space Q is uniformly integrable. So Theorems 7.3.2 and 11.1.8 imply the second main result:

THEOREM 8.4.3. *There exist positive constants c and C such that*

$$ck^{-2} \le E_k(u^e) \le Ck^{-2} \qquad \text{for } k \ge 1.$$

Thus, for the limiting entropy solutions of the inviscid Burgers equation (8.2.11) we have established the algebraic behaviour of the structure function $S_{p,l} \sim |l|^{\min(p,1)}$ and the spectral power law $E_k \sim k^{-2}$. Moreover, the two results hold for all space-scales. This allows to say that the entropy solutions $u^{e,\omega}(t,x)$ describe *inviscid burgulence.*

8.5. Mixing for the entropy solutions.

From now on, till the end of this chapter, we assume that

$$B_5 < \infty.$$

Inviscid limit for the stationary measure for eq. (1.1.12).

Let μ^ν be the stationary measure for eq. (1.1.12) as in Theorem 3.3.4. Then we have

THEOREM 8.5.1. *The limit $\mu^\nu \rightharpoonup \mu^e$ as $\nu \to 0$ exists in $\mathcal{P}(L_1)$ and*

$$\|\mu^\nu - \mu^e\|^*_{L(L_1)} \le C\Big(\ln\big[\ln c\,\nu^{-1}\big]\Big)^{-1/8} \quad \text{if} \quad 0 < \nu \le 1/2, \tag{8.5.1}$$

for some $c, C \ge 1$.

Moreover, the measure μ^e is supported by the linear space $\bigcap_{p<\infty} L_p$, i.e. $\mu^e(\cap L_p) = 1$, and there exists $C_\alpha(B_5) > 0$ such that

$$|\langle \mu^e, |u|_p^\alpha\rangle| \le C_\alpha < \infty \quad \forall\, 1 \le p, \alpha < \infty. \tag{8.5.2}$$

PROOF. Let us denote by $u^\nu(t)$ a solution of (1.1.13) with $u_0 = 0$. Then by Theorem 4.2.4, $\|\mathcal{D}u^\nu(t) - \mu^\nu\|^*_{L(L_1)} \le \kappa(\ln t)^{-1/8}$, if $t \ge 3$. So for any $1/2 \ge \nu_2 > \nu_1 > 0$, using Proposition 8.3.2 we have

$$\begin{aligned}\|\mu^{\nu_2} - \mu^{\nu_1}\|^*_{L(L_1)} &\le 2\kappa(\ln t)^{-1/8} + \|\mathcal{D}u^{\nu_2}(t) - \mathcal{D}u^{\nu_1}(t)\|^*_{L(L_1)}\\ &\le 2\kappa(\ln t)^{-1/8} + C_1\bar\nu^{1/4}e^{t/2+4t^3B_4}\\ &\le 2\kappa(\ln t)^{-1/8} + C_1\bar\nu^{1/4}e^{5t^3B_4^+}, \quad \forall t \ge 3,\end{aligned} \tag{8.5.3}$$

where $\bar\nu = \nu_2 - \nu_1$ and $B_4^+ = \max(B_4, 1)$. Consider $t = t(\bar\nu)$ such that $t^3 = \ln(\bar\nu^{-1})/40B_4^+$. Then $\exp(5t(\bar\nu)^3B_4^+) = \bar\nu^{-1/8}$, and $t(\bar\nu) \ge 1$ if $\nu_2 \ll 1$ (since $\bar\nu \le \nu_2$). Choosing in (8.5.3) $t = t(\bar\nu)$ we obtain that

$$\begin{aligned}&\|\mu^{\nu_2} - \mu^{\nu_1}\|^*_{L(L_1)}\\ &\quad \le C_2\Big(\ln\big[\ln\bar\nu^{-1}/40B_4^+\big]\Big)^{-1/8} + C_1\bar\nu^{1/8} \le C\Big(\ln\big[\ln\bar\nu^{-1}/40B_4^+\big]\Big)^{-1/8},\end{aligned}$$

if $\nu_2 \ll 1$. So the mapping $(0,1/2] \ni \nu \mapsto \mu^\nu \in \mathcal{P}(L_1)$ is Cauchy–continuous at zero, the limit $\mu^e := \lim_{\nu\to 0}\mu^\nu$ exists, and passing in the inequality above to the limit as $\nu_1 \to 0$ we get (8.5.1).

As measures μ^ν meet estimates (3.4.3), then (8.5.2) follows from convergence (8.5.1) and Corollary 11.1.7. So $\mu^e(L_p) = 1$ for each $p < \infty$, and $\mu^e(\cap_{1\le p<\infty} L_p) = \mu^e(\cap_{p\in\mathbf{N}} L_p) = 1$. ∎

Since all moments for the L_p-norms for the measures μ^ν and μ^e are bounded by (3.4.3) and (8.5.2), then repeating the proof of Theorem 4.2.5 we get that the just established estimate (8.5.1) implies the L_p-convergence of the measures: for any $1 \le p < \infty$,

$$\|\mu^\nu - \mu^e\|^*_{L(L_p)} \le C_p(\ln[\ln c\nu^{-1}])^{-c_p} \quad \text{if} \quad 0 < \nu \le 1/2, \tag{8.5.4}$$

where C_p and $c_p > 0$ depend on B_5.

Mixing for the entropy solutions.

THEOREM 8.5.2. *For $t \ge 3$ and $u_0 \in L_1$,*

$$\|\Sigma^e_t(u_0) - \mu^e\|^*_{L(L_1)} \le 2\kappa(\ln t)^{-1/8}, \tag{8.5.5}$$

where $\kappa(B_5)$ is the same positive constant as in Theorem 4.2.4.

PROOF. Let first $u_0 \in H^2$. Denote $u^e(t) = u^e(t; u_0)$ and $u^\nu(t) = u^\nu(t; u_0)$. Then $\Sigma^e_t(u_0) = \mathcal{D}(u^e(t))$ and for any $0 < \nu \le 1/2$ and $t \ge 1$ we have

$$\begin{aligned}&\|\mathcal{D}(u^e(t)) - \mu^e\|^*_{L(L_1)} \\ &\quad \le \|\mathcal{D}(u^e(t)) - \mathcal{D}(u^\nu(t))\|^*_{L(L_1)} + \|\mathcal{D}(u^\nu(t)) - \mu^\nu\|^*_{L(L_1)} + \|\mu^\nu - \mu^e\|^*_{L(L_1)}.\end{aligned}$$

By Proposition 8.3.2, Theorem 4.2.4 and Theorem 8.5.1 the r.h.s. is bounded by

$$C_1\nu^{1/4}e^{\frac{1}{2}t(1+\|u_0\|_2)+4t^3B_4} + \kappa(\ln t)^{-1/8} + C\Big(\ln\big[\ln c\nu^{-1}\big]\Big)^{-1/8}.$$

For any $t \ge 1$, choosing $\nu = \nu(t)$ so small that the first and the third terms above are smaller than $\kappa(\ln t)^{-1/8}/2$, we get (8.5.5) if $u_0 \in H^2$. The general case follows from this result and (8.3.8) if we approximate $u_0 \in L_1$ by H^2-functions. □

Similarly to the above (see (8.5.4)), in view of Proposition 8.3.3 and estimate (8.5.2) the L_1-mixing (8.5.5) implies the L_p-mixing for any $1 \le p < \infty$:

$$\|\Sigma^e_t(u_0) - \mu^e\|^*_{L(L_p)} \le C'_p(\ln t)^{-c'_p}, \quad t \ge 3, \tag{8.5.6}$$

for any $u_0 \in L_1$, where C'_p and $c'_p > 0$ depend on B_5.

The measure μ^e is invariant for the semigroup $\{S^{e*}_t\}$. Indeed, since by (8.5.5) $S^{e*}_t\delta_0 \rightharpoonup \mu^e$, then applying to this convergence the continuous operator $S^{e*}_\tau, \tau \ge 0$, we get that $S^{e*}_{t+\tau}\delta_0 \rightharpoonup S^{e*}_\tau\mu^e$. So $S^{e*}_\tau\mu^e = \mu^e$ for any $\tau \ge 0$.

Writing $S^{e*}_t\mu$ as $\int \Sigma^e_t(u)\mu(du)$ and applying (8.5.5) we find that

$$\|S^{e*}_t\mu - \mu^e\|^*_{L(L_1)} \le 2\kappa(\ln t)^{-1/8} \quad \forall \mu \in \mathcal{P}(L_1).$$

We see that the Markov process defined by the entropy solutions in L_1 is logarithmically mixing.

If $b_s \equiv b_{-s}$, then in view of (8.5.1) and Corollary 3.3.5 the stationary measure μ^e is space-homogeneous.

As we wrote at the beginning of this chapter, another way to study the L_1-entropy solutions of the inviscid stochastic Burgers equation goes back to the paper [**56**], and the measure μ^e is exactly the stationary measure for the inviscid dynamics constructed in that work using the Lax–Oleinik formula. The approach of [**56**] allows to prove that the rate of mixing in the corresponding Markov process in L_1 is exponential. I.e. that in fact the l.h.s. of (8.5.5) is bounded by $Ce^{-\kappa t}$, for suitable $C, \kappa > 0$. See [**31**, **79**].

Time-asymptotic for the inviscid energy spectrum and structure function.

Let us write $e_k(u)$ (see (7.3.2)) as

$$e_k(u) = C_{k,M}^{-1} \sum_{\{n:\ M^{-1}k \le |n| \le Mk\}} \tfrac{1}{2}|\hat{u}_n|^2, \qquad C_{k,M} = 2(M - M^{-1})k.$$

The constants below may depend on M, but not on k.

THEOREM 8.5.3. *Assume that $B_0 > 0$, $B_m < \infty$ for all m, and let $c_1 - c_3$ be the constants from Theorem 7.4.1. Then*

$$\tfrac{1}{2}c_1 k^{-2} \le \langle e_k, \mu^e \rangle \le 4c_2 k^{-2} \qquad \textit{if} \quad k \ge 1. \tag{8.5.7}$$

PROOF. Let us denote $\mu^0 = \mu^e$ and for $R \ge 1$ set $e_k^R(u) = \min(e_k(u), R)$. Since

$$e_k(u) \le \tfrac{1}{2} C_{k,M}^{-1} \|u\|^2 \le Ck^{-1}\|u\|^2,$$

then using (3.4.1) if $\nu > 0$ and (8.5.2) if $\nu = 0$, for any $\nu \ge 0$ and any $R \ge 1$ we have

(8.5.8)

$$\langle e_k - e_k^R, \mu^\nu \rangle = \langle e_k \chi_{e_k > R}, \mu^\nu \rangle \le R^{-1}\langle (e_k)^2, \mu^\nu \rangle \le k^{-2}R^{-1}\langle \|u\|^4, \mu^\nu \rangle \le C_1 k^{-2} R^{-1}.$$

Taking any $\nu > 0$ and choosing R sufficiently large, we derive from (8.5.8) and (7.4.1) that

$$\tfrac{3}{4}c_1 k^{-2} \le \langle e_k^R, \mu^\nu \rangle \le 2c_2 k^{-2} \qquad \text{if} \quad 1 \le k \le (c_3\nu)^{-1}. \tag{8.5.9}$$

Since $e_k^R \in C_b(L_1)$, then we find from (8.5.9) and Theorem 8.5.1 that $\langle e_k^R, \mu^e \rangle$ also satisfies the estimates in (8.5.9) if $k \ge 1$. Next using (8.5.8) with $\nu = 0$ and a large enough R we obtain that (8.5.7) holds for all $k \ge 1$. □

Now let us take any non-random $u_0 \in L_1$ and consider the entropy solution $u^e(t) = u^e(t; u_0)$ of (8.2.11). We wish to examine for large t its instantaneous energy spectrum $\mathbb{E}e_k(u^e(t))$.

THEOREM 8.5.4. *The instantaneous energy spectrum $\mathbb{E}e_k(u^e(t))$ converges for each k, as $t \to \infty$, to the energy spectrum of the stationary measure $\langle e_k, \mu^e \rangle$ (satisfying estimates* (8.5.7)*). The rate of the convergence depends on B_5, but not on k.*

PROOF. Since the function $e_k^R(u) = \min(e_k(u), R)$, $R > 0$, is bounded and continuous on L_1, then by Theorem 8.5.2

$$\mathbb{E}e_k^R(u^e(t)) \to \langle e_k^R, \mu^e \rangle \quad \text{as} \quad t \to \infty,$$

for each R. Similarly to the derivation of (8.5.8) with $\nu = 0$ we obtain from Proposition 8.3.3 that

$$\left|\mathbb{E}(e_k - e_k^R)(u^e(t))\right| \le Ck^{-2}R^{-1}, \quad \forall t \ge 1,\ \forall R > 0.$$

This estimate jointly with $(8.5.8)_{\nu=0}$ and the convergence above imply the assertion. □

We may examine the structure function of the stationary measure μ^e in the same way as we proceeded above for the energy spectrum. Namely, since by Theorem 7.4.5 for $\nu > 0$ the structure function $\langle s_{p,l}, \mu^\nu \rangle$ satisfies estimates (7.2.3)-(7.2.5), then using the L_p-convergence (8.5.4) jointly with estimates (3.4.3) (and Lemma 7.4.4) we get that the structure function of the measure μ^e satisfies (8.4.1) with modified constants. Next, the convergence (8.5.6) and the L_p-estimates from Proposition 8.3.3 imply similar bounds for the instant structure function of entropy solutions:

THEOREM 8.5.5. *There exists* $c_2' > 0$ *and for each* $p > 0$ *there are constants* $\tilde{c}_p, \tilde{C}_p > 0$ *such that the structure function of* μ^e *satisfies estimates* (8.4.1) *with* $S_{p,l}(u^e)$ *replaced by* $\langle s_{p,l}, \mu^e \rangle$*, for any* $|l| \le c_2'$*. For every* $u_0 \in L_1$ *and* l *as above the instantaneous structure function* $\mathbb{E}s_{p,l}(u^e(t; u_0))$ *converges to* $\langle s_{p,l}, \mu^e \rangle$*, as* $t \to \infty$*. The rate of the convergence depends on* p *and* B_5*, but not on* l*.*

CHAPTER 9

Notes and comments, II

The presentation of the K41 theory in Chapter 6 follows [**114**] and [**68**], discussion of 2d turbulence is based on [**103**], and that of burgulence – on [**1**, **10**].

The rigorous definition of the dissipative space scale l_d adopted in Chapter 7 is taken from [**107**] and is similar to a definition in [**6**]; Problem 7.1.2 shows that it agrees with the intuitive definition of the scale. The approach to estimate l_d using upper and lower bounds for Sobolev norms of solutions was suggested in [**107**] for complex Ginzburg–Landau equations and next was used in [**17**, **18**, **26**, **28**] for solutions of the Burgers equation. The derivation of two-sided estimates for the structure function $S_{p,l}(u^\nu)$ in the inertial range from the asymptotically sharp estimates on Sobolev norms of solutions (Theorem 7.2.3 and the first part of Theorem 7.2.2) and obtaining from the latter asymptotics for the energy spectrum $E_k(u^\nu)$ (Theorem 7.3.2) were suggested in the works [**26**, **28**], where the study of the structure function $S_{p,l}$ was inspired by the physical argument in [**1**, Section 7]. The second part of Theorem 7.2.2 and Proposition 7.2.5 were proved for this book. As we point out at the beginning of Chapter 7 (see there assumptions (H0)–(H3)), the proofs of the asymptotics for $S_{p,l}$ and E_k when ν goes to zero while $|l|^{-1}$ and k remain in the inertial ranges use the specifics of the Burgers equation essentially only via Oleinik's maximum principle and the asymptotically sharp estimates on Sobolev norms of solutions. We are not aware of any (non-trivial) nonlinear PDE, except the Burgers equation and its modifications like (0.0.12), (0.0.13) which admits such asymptotically sharp estimates, still see below Section 10.4.3 for non-sharp lower and upper bounds for averaged Sobolev norms of solutions for complex Ginzburg–Landau equations, and see [**13**, **14**] for a non-local equation on the whole space for which results similar to that in this book are obtained. The time-asymptotic results in Section 7.4 and the high-probability results in Section 7.5 were obtained for this book and are new.

The results of Chapters 1-3 and Sections 7.1-7.3 were the material for an M2 course which SK has twice taught at the University Paris Diderot. The lecture notes for this course [**82**] were at the origin of our book.

The results of the first two sections of Chapter 8 are well known and the references are discussed in the introduction to the chapter. What makes our presentation a bit specific is that we do not evoke the Lax–Oleinik formula to construct directly the entropy solutions of the inviscid stochastic Burgers equation, but instead use the inviscid limit strategy and derive properties of the entropy solutions from those of solutions for the viscous equation. In that we follow the papers [**95**, **96**] and their presentation in an appendix to [**18**]. Construction of the L_1 stochastic entropy solutions in Sections 8.3 and the results of the three propositions there may be new. Results of Sections 8.4, 8.5 were obtained by us for this book and are new. In the same time, the logarithmic mixing for the L_1-entropy solutions, established in

Theorem 8.5.2 as an inviscid limit of the assertion of Theorem 4.2.4, is not optimal and the Lax–Oleinik approach to the entropy dynamics, introduced and developed in [**56**], allows to show that for $\nu = 0$ the mixing is exponential [**31, 79**]. Still the result on the L_p-mixing (8.5.6) which follows from Theorem 8.5.2 seems to be new.

Part 3

Additional Material

CHAPTER 10

Miscellanea

In this chapter we are concerned with other settings related to the Burgers equation, where we also observe turbulence and mixing phenomena similar to those described above. Our goal is to present and discuss results without the proofs. The only exception is Section 10.2, whose content is closely related with the stochastic Burgers equation, our main topic. The results in that section are obtained specifically for this work, and the proofs are less detailed than in other parts of the book.

10.1. Other types of forces: Estimates, Burgulence and mixing.

10.1.1. Free Burgers equation. The first setting that we consider is the unforced or free Burgers equation:

$$\begin{aligned} &u_t(t,x) + u(t,x)u_x(t,x) - \nu u_{xx}(t,x) = 0, \\ &t \geq 0, \quad x \in S^1 = \mathbb{R}/\mathbb{Z}, \end{aligned} \tag{10.1.1}$$

where we assume $u(0,\cdot) = u_0 \in H^2$ and discard the trivial case $u_0 = 0$. We recall that a solution of (10.1.1), equal to u_0 at $t = 0$, exists and is unique by Theorem 1.8.6, and is smooth for $t > 0$ by Corollary 1.8.10. The behaviour of solutions to (10.1.1) for small (but positive) values of ν was studied in [**18**] and continued in [**29**]; see also the references cited in the latter paper. We also recall (see Chapter 8) that the limits of the solutions as $\nu \to 0$ exist and that vast literature is dedicated to its study; we do not touch here these results.

Since (10.1.1) is a particular case of the stochastic equation (1.1.12) with $b_s \equiv 0$, then many results from Chapters 2-8 hold for (10.1.1) without any modification, in particular Theorem 2.2.1 and the upper estimates in Chapter 7 (where the expectation sign should be dropped).

However, the estimates of Section 2.3 and the lower bounds in Chapter 7 only hold provided that the forcing does not vanish (cf. Theorem 2.3.1, where we assumed that $B_0 > 0$). It is easy to see that we have no hope to get time-uniform lower bounds since the system is purely dissipative, in the sense that

$$d\|u(t)\|^2/dt = -2\nu\|u(t)\|_1^2 \leq -2\nu\|u(t)\|^2, \tag{10.1.2}$$

so the norm $\|u(t)\|$ exponentially decays to zero with time. Moreover, Problem 2.1.3 implies that

$$|u(t)|_{1,1} \leq t^{-1}, \quad t > 0,$$

so the convergence to zero is uniform in $\nu > 0$ (and in u_0). On the other hand, by Theorem 2.2.1 with $B_0 = 0$ we have

$$\|u(t)\|_{2m} \leq C(m)\nu^{-(4m-1)/2}, \quad m \in \mathbf{N}, \quad t \geq 1.$$

Combining the two upper bounds and using (1.3.11), we obtain the time-decay estimate

$$\begin{aligned}\|u(t)\|_m &\le C(m)\ |u(t)|_{1,1}^{2m/(4m-1)}\|u(t)\|_{2m}^{(2m-1)/(4m-1)}\\ &\le C(m)\ t^{-2m/(4m-1)}\nu^{-(2m-1)/2},\quad m\in\mathbf{N},\quad t\ge 1.\end{aligned}\tag{10.1.3}$$

However, there are asymptotically sharp lower bounds both for Sobolev norms and for small-scale quantities, which hold on a fixed time interval of order one [**29**, Section 4]. To state them, we begin by defining the quantities

$$D=\max(|u_0|_1^{-1},\ |u_0|_{1,\infty})\ge 1;\quad T_1=\tilde{C}^{-1}D^{-4};\quad T_2=2D,$$

where $\tilde{C}>1$ is an absolute constant. Replacing the averaging $\langle\langle\cdot\rangle\rangle$ by the time–averaging along the interval $[T_1,T_2]$ we can repeat for solutions of (10.1.1) (up to multiplicative constants) the statements of Theorem 2.3.4 and almost copy their proofs to get the following result:

THEOREM 10.1.1. *For every $m\ge 1$, there exists $C_m(D)>1$ such that for each $u_0\in H^2$ the solution $u=u(t;u_0)$ of* (10.1.1) *satisfies*

$$C_m^{-1}\nu^{-(2m-1)}\le\frac{1}{T_2-T_1}\int_{T_1}^{T_2}\|u(s)\|_m^2ds\le C_m\nu^{-(2m-1)}.\tag{10.1.4}$$

In the same way, we can also almost copy the statements of Theorems 7.2.3 and 7.3.2 to describe the small–scale properties of solutions for eq. (10.1.1). A solution with a fixed initial data u_0 of order one for $t\in[T_1,T_2]$ satisfies natural deterministic analogues of the conditions (H0)–(H3), and the proofs of the results on its small-scale features closely follow the arguments in Sections 7.1–7.3. In difference with the latter, now instead of depending on the forcing, the multiplicative constants depend on the initial condition through the single parameter D.

In the language of hydrodynamics, the behaviour of $u(t,x)$ for $t\in[T_1,T_2]$ may be called *decaying burgulence*. Actually, such behaviour can also be observed for time intervals arbitrarily far from 0, but we need to take into account the energy decay (10.1.3) and to rescale the estimates appropriately.

If we assume that the initial condition u_0 is C^∞-smooth, then the estimates (10.1.4) remain true with the averaging on $[T_1,T_2]$ replaced by averaging on an interval $[0,T_3]$. However, the multiplicative constants and T_3 depend on u_0 in a more involved way, see [**18**]. For the particular case of the time-averaged squared H^1 norm, see Problem 2.3.2. For the general case, see Problem 2.3.5.

10.1.2. Kick force. Now we consider the Burgers equation with a *kick force* η, which acts, for some fixed time-period $\theta>0$, by adding independent identically distributed impulses at discrete moments of time $k\theta$, $k\in\mathbf{N}$, to solutions of the free equation (10.1.1). Such forces are popular in the analytic study of randomly perturbed equations (e.g. see [**10**, Section 8] and [**103**, Section 2.3]). Moreover, they serve as discrete-time analogues for the white in time random forces, so kick-forces with short time-step θ appear in computer modelling of equations with randomness since then the time inevitably is discrete.

To construct such a force, we begin by considering independent identically distributed real-valued random variables

$$\xi_{sk},\ s\in\mathbb{Z}^*,\ k\in\mathbf{N},\qquad \mathbb{E}\xi_{sk}^2\equiv 1,$$

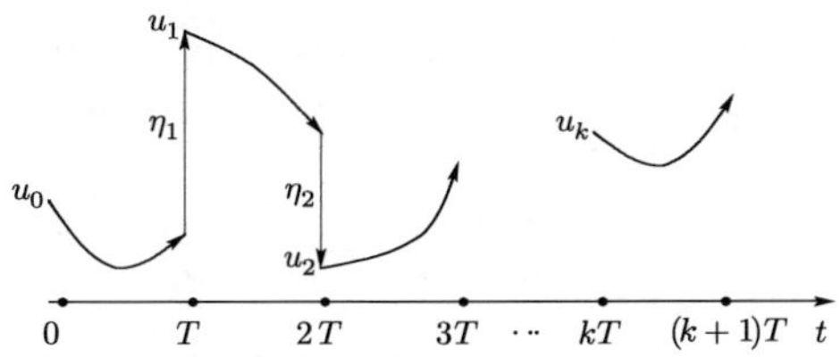

FIGURE 10.1.1. Evolution defined by the kicked Burgers equation

defined on some probability space. We assume that for each s and k

$$|\xi_{sk}| \le C_* \text{ for all } \omega, \text{ and } \mathcal{D}(\xi_{sk}) = p(x)dx, \quad \text{where } p \text{ is a Lipschitz function on } [-C_*, C_*],\ p(0) \neq 0, \tag{10.1.5}$$

for some $C_* > 0$. The boundedness condition $|\xi_{sk}| \le C_*$ may be replaced by assuming that the r.v.'s ξ_{sk} have finite second exponential moments, but then the proof of the mixing property (10.1.12) becomes significantly more complicated, see in [**103**].

Next, we take a sequence of real numbers b_s, $s \in \mathbb{Z}^*$, such that $B_1 < \infty$, where as before

$$B_m = \sum_{s\in\mathbb{Z}^*} |2\pi s|^{2m} b_s^2 \le \infty,$$

and consider r.v.'s η_k, $k \in \mathbf{N}$, given by

$$\eta_k(x) = \sum_{s\in\mathbb{Z}^*} b_s\, \xi_{sk}\, e_s(x). \tag{10.1.6}$$

Here e_s is the standard trigonometric basis (1.1.2). Note that η_k's are i.i.d. random variables in H^1. Now we define the random forcing η as a generalised function in $\mathcal{D}'(\mathbb{R}^+, H^1)$:

$$\eta^\theta(t,x) = \sum_{k=1}^{+\infty} \delta_{t=k\theta}\ \eta_k(x). \tag{10.1.7}$$

where $\delta_{t=k\theta}$ denotes the Dirac measure at a time moment $k\theta$ (see [**118**] concerning the space $\mathcal{D}'(\mathbb{R}^+, H^1)$). In the context of formula (10.1.7) the r.v.'s η_k are called *the kicks*. Then we consider the equation

$$u_t(t,x) + u(t,x)u_x(t,x) - \nu u_{xx}(t,x) = \eta^\theta(t,x), \quad t \ge 0, \quad x \in S^1 = \mathbb{R}/\mathbb{Z}, \tag{10.1.8}$$

with a prescribed $u(0,\cdot) = u_0 \in H^1$. We write $\eta^\theta(t,x)$ as $\partial\xi^\theta(t,x)/\partial t$, where

$$\xi^\theta(t,x) = \int_0^t \eta^\theta(s,x)ds, \tag{10.1.9}$$

and understand solutions of (10.1.8), (10.1.7) in the sense of (1.1.10). The integral $\int_0^t \delta_{s=k\theta}ds$ is the Heaviside function, vanishing for $t < k\theta$ and equal 1 for $t > k\theta$. We normalise it to be right continuous, so at $k\theta$ it equals 1. Then $\xi^\theta(t)$ is a right-continuous curve in H^1.

With this definition of the forcing, for integers $k \ge 1$ at the moments $k\theta$ the solution $u(x)$ instantly increases by the kicks $\eta_k^\theta(x) \in H^1$, and between these moments, u solves the unforced Burgers equation (10.1.1). See Figure 10.1.1. By

Theorem 1.8.6 this uniquely defines a pathwise solution of (10.1.8) in H^1 for all moments of time except the points $k\theta$, $k \in \mathbf{N}$, where we again impose that the curve $u(t)$ is right-continuous like $\xi^\theta(t)$ (so the difference $u(t) - \xi^\theta(t)$ is continuous for $t \geq 0$). Denoting by $\mathcal{S}_t$ the flow-maps of the free equation (10.1.1), i.e. $\mathcal{S}_t(u_0) = u(t; u_0, 0)$, where, as before, $u(t; u_0, \eta^\theta)$ stands for a solution of (10.1.8), we see that

$$u(k\theta; u_0, \eta^\theta) = \mathcal{S}_\theta\big(u((k-1)\theta; u_0, \eta^\theta)\big) + \eta_k, \qquad k \geq 1, \tag{10.1.10}$$

$$u(k\theta + l; u_0, \eta^\theta) = \mathcal{S}_l(u(k\theta; u_0, \eta^\theta)) \quad \text{for} \quad 0 \leq l < \theta. \tag{10.1.11}$$

Thus, solutions for (10.1.10), evaluated at times which are integer multiples of θ, define in the space H^1 a Markov chain $\{u(k\theta; u_0, \eta^\theta),\ k \in \mathbf{N}_0\}$, while the solutions $\{u(t; u_0, \eta^\theta), t \geq 0\}$ define in H^1 a non-homogeneous Markov process.

The random kick-forces (10.1.7) are natural discrete-time approximations for white in time forces (1.1.5), (1.1.6). When the time-step θ goes to zero, it is expected that solutions of a kick-forced equation converge in distribution to solutions of that equation, perturbed by a white force. See below Section 10.2 for a proof of this convergence in the case of the Burgers equation, and see in [**103**] Sections 2.4.3 and 4.3.2, where the result is proved for the 2d Navier-Stokes system.

Below in this section we assume that

$$B_M < \infty$$

for some $M \geq 4$, where the numbers B_m are defined as in (1.1.7). This ensures that the kicks $\eta_k^\theta(x)$ and solutions $u(t,x)$, $t > 0$, are H^M–smooth functions of x, and that solutions of (10.1.8), evaluated at integer times, define a Markov chain in the space H^M.

The statement of Theorem 2.3.4 with m replaced by M holds true for solutions of the kick forced equation (10.1.8). In this case the constant C_M depends on B_M, while σ in the definition of $\langle\langle\cdot\rangle\rangle$ should be no smaller than $\max(2\theta, \sigma_0)$, where $\sigma_0 > 0$ depends on B_4 and on $B_0 > 0$.

PROBLEM 10.1.2. Prove for solutions of eq. (10.1.8), (10.1.7) Oleinik's estimates (2.1.1), (2.1.2) and (2.2.2). The function ξ in (2.1.1), (2.1.2) is defined in (10.1.9) and its norm $\|\xi\|_{X_T^4}$ should be replaced by $\sup_{0 \leq t \leq T} \|\xi(t)\|_4$.
Hint: The first two estimates immediately follow from Theorem 2.1.2. To establish the third, differentiate the squared norm $\mathbb{E}\|u(t)\|_M^2$: between the kicks it evolves as for the free equation and therefore the relation (2.2.4) holds with 0 instead of B_M, while at any moment $k\theta$, $k \in \mathbf{N}$, we have

$$\mathbb{E}\|u(k\theta)\|_M^2 = \mathbb{E}\|u(k\theta - 0)\|_M^2 + \mathbb{E}\|\eta_k^\theta\|_M^2.$$

PROBLEM 10.1.3. Prove for solutions of eq. (10.1.8) an analogue of the energy balance (2.3.1) and derive the left inequality in (2.3.6), if $\sigma \geq \max(2\theta, \sigma_0(B_4, B_0))$.
Hint: Proceed as above by observing that $\mathbb{E}|u(t)|^2$ increases at kick moments and dissipates between the kicks like in the unforced equation.

Unlike in the rest of the book, solutions $u^\omega(t; u_0)$ for eq. (10.1.8), (10.1.7) are not continuous on $\mathbb{R}_+$, but only on intervals $[k\theta, (k+1)\theta)$ and continuously extend to the intervals' closures. So regarding them as random processes on $\mathbb{R}_+$ is a minor violation for the definition used in our book (see in Appendix 11.5). Modulo this small discrepancy, the family of processes $u^\omega(t; u_0)$ meets the conditions (H0)-(H3) in Chapter 7. Indeed, (H0) is obvious, while (H1)-(H3) follow from the two problems above. In agreement with the thesis stated at the beginning of Chapter 7, the small-scale properties of solutions for (B), established in Sections 7.1-7.3, remain true for solutions of (10.1.8), (10.1.7).

The long-time behaviour of solutions for (10.1.10) is governed by the mixing property. Literally repeating the Bogolyubov–Krylov argument, presented in Section 3.1, we get that the family of the Bogolyubov–Krylov ansatzes $\{m_j, j \in \mathbf{N}\}$, where $m_r = \frac{1}{r}\sum_{s=1}^{r} \mathcal{D}u(s\theta; 0, \eta^\theta)$, is tight in H^{M-1}, and any its limiting point μ^θ is a measure in $\mathcal{P}(H^{M-1})$, stationary for the Markov chain (10.1.10). Moreover, repeating closely the proof of Theorem 3.3.4 we see that the system (10.1.10) is mixing, uniformly in $1 \geq \theta \geq \theta_0$, for any $\theta_0 > 0$. That is,

$$\|\mathcal{D}u(t; u_0, \eta\theta) - \mu^\theta\|^*_{L(H^{M-1})} \to 0 \quad \text{as } \theta\mathbf{N} \ni t \to \infty, \tag{10.1.12}$$

uniformly in $1 \geq \theta \geq \theta_0 > 0$ and $u_0 \in H^{M-1}$.

10.1.3. Red noise. Now we consider equation

$$u_t(t,x) + u(t,x)u_x(t,x) - \nu u_{xx}(t,x) = \eta_{red}(t,x), \tag{10.1.13}$$

where the force η_{red} is a *coloured noise.* That is, it has the form

$$\eta_{red}(t,x) = \sum_{|s|\leq N, s\neq 0} b_s \eta^s(t) e_s(x), \qquad B_m = \sum |2\pi s|^{2m} b_s^2 < \infty \ \forall m, \tag{10.1.14}$$

where N is either a finite number or $N = \infty$, $\{b_s\}$ are non-zero real numbers, $\{e_s\}$ is the standard trigonometric basis (1.1.2) and $\{\eta^s\}$ are i.i.d. bounded real-valued random processes such that

$$\mathcal{D}(\eta^s(\cdot)) = \mathcal{D}(\tilde{\eta}(\cdot)) \qquad \forall s, \tag{10.1.15}$$

where the process $\tilde{\eta}(t)$ is constructed below. Then for every $m \in \mathbf{N}$ the force $\eta_{red}(t)$ is bounded in H^m uniformly in t and ω, so equation (10.1.13) is well posed in that space by Theorem 1.8.6.

Let $\{h_l, l \geq 0\} \cup \{h_{jl}, j, l \geq 0\}$, be the Haar system of functions on $\mathbb{R}_+$, where h_l is the characteristic function of the interval $[l, l+1]$, and for $j \geq 0$, $l \geq 0$

$$h_{jl}(t) = \begin{cases} 0 & \text{for } t < l2^{-j} \text{ or } t \geq (l+1)2^{-j}, \\ 2^{j/2} & \text{for } l2^{-j} \leq t < \left(l + \frac{1}{2}\right)2^{-j}, \\ -2^{j/2} & \text{for } \left(l+\frac{1}{2}\right)2^{-j} \leq t < (l+1)2^{-j}. \end{cases}$$

(see Figure 10.1.2). The system of functions $\mathcal{H} = \{h_0\} \cup \{h_{jl}, j \geq 0, 2^j - 1 \geq l \geq 0\}$ forms an orthonormal basis in $L^2(0,1)$. Indeed, this system obviously is orthonormal, so it remains to show that if $f \in L^2(0,1)$ is orthogonal to all functions in $\mathcal{H}$, then $f = 0$. To do this denote $F(t) = \int_0^t f(u)du$. It is a continuous function on $[0,1]$, vanishing at zero. Since $\langle f, h_0\rangle = 0$, then also $F(1) = 0$. Next, as $\langle f, h_{00}\rangle = 0$, then

$$0 = \int_0^{1/2} f(t)dt - \int_{1/2}^1 f(t)dt = -F(0) + 2F(1/2) - F(1),$$

so $F(1/2) = 0$. Then the orthogonality of f to h_{10} and h_{11} yields that $F(1/4) = F(3/4) = 0$. Continuing by induction we find that $F(k/2^n) = 0$ for every $k/2^n \in [0,1]$ with $k, n \in \mathbf{N}_0$. As F is continuous, then $F \equiv 0$, so $f = 0$ and $\mathcal{H}$ is a basis.

From the result proved above, it follows that the system $\{h_l, l \geq 0\} \cup \{h_{jl}, j, l \geq 0\}$ forms an orthonormal basis of $L^2(\mathbb{R}_+)$.

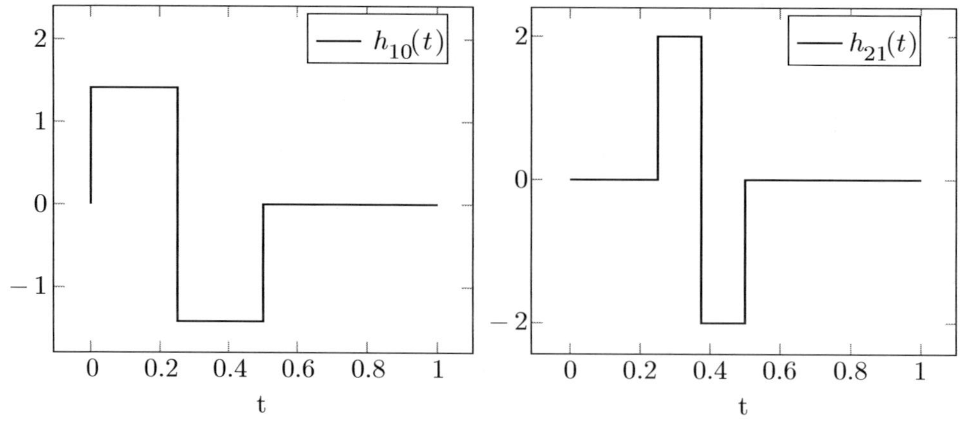

FIGURE 10.1.2. Examples of Haar basis elements supported on $[0, 1]$.

Now we define the random process $\tilde{\eta}(t)$ (see (10.1.15)) as

$$\tilde{\eta}(t) = c_{-1} \sum_{l=0}^{\infty} \xi_{-1l} h_l(t) + \sum_{j=0}^{\infty} c_j \sum_{l=0}^{\infty} \xi_{jl} h_{jl}(t), \tag{10.1.16}$$

where $\{c_j\}$ is a sequence of non-zero numbers, and $\{\xi_{jl}\}$ is a family of i.i.d. real-valued random variables as in (10.1.5). Obviously the process $\tilde{\eta}$ is statistically periodic:

$$\mathcal{D}(\tilde{\eta}(\cdot)) = \mathcal{D}(\tilde{\eta}(k + \cdot)) \in \mathcal{P}(X_\infty^m), \qquad \forall k \in \mathbf{N}. \tag{10.1.17}$$

Besides,

$$\text{random variables } \eta^\omega|_{[k,k+1)},\ k \in \mathbf{N}_0, \text{ are independent.} \tag{10.1.18}$$

Random processes of this form (and their variations) are called *coloured noises* and are widely used in engineering sciences; see [**58**]. If $c_j \equiv 1$ and $\{\xi_{jl}\}$ are independent Gaussian variables with law $N(0, 1)$, then $\tilde{\eta}(t)$ is a white noise, see [**113**, Chapter 2.2]. We consider smoother processes, satisfying (10.1.5) and the condition

$$\sum_{j \ge -1} c_j 2^{j/2} \max(1, j)^q < \infty, \quad \text{for some } q > 1. \tag{10.1.19}$$

Since for all $m \in \mathbf{N}$, $B_m < \infty$, then (10.1.19) guarantees that the force $\eta(t)$ is bounded in H^m, uniformly in t and ω, for each m. This and other processes (10.1.16) such that $c_j \to 0$ as $j \to \infty$ are called *red noises*.

For $N \in \mathbf{N}$ let us consider the process

$$\beta_N^\omega(T) = \frac{1}{\sqrt{N}} \int_0^{NT} \tilde{\eta}^\omega(t) dt = c_{-1} \frac{1}{\sqrt{N}} \sum_{l=0}^{\lfloor NT \rfloor} \xi_{-1l}^\omega + O\big(\tfrac{NT - \lfloor NT \rfloor}{\sqrt{N}}\big) \tag{10.1.20}$$

(we have used that all functions h_{jl} have zero integrals, while $\int h_l = 1$ for each l). Its trajectories are Lipschitz functions of T. So if $\mathbb{E}\xi_{sk} \equiv 0$, then by Donsker's theorem (see [**16**, Theorem 37.8]) the process $\beta_N(T)/\sigma$, $\sigma^2 = c_{-1}^2 \mathbb{E}\xi_{-10}^2$, converges in distribution to the Wiener process. That is, on large time-scales $\int \tilde{\eta}$ behaves as a Wiener process.

A solution $u^\omega(t;u_0)$ of equation (10.1.13), (10.1.14)-(10.1.16), equal at $t=0$ to $u_0 \in H^m$, $m \in \mathbf{N}$, is a random process in H^m. As in Section 10.1.2, due to (10.1.17) and (10.1.18) the process $\{u(k;u_0),\ k\in \mathbf{N}_0\}$, is a Markov chain in H^m. But in difference with the kick-forced equation, now the solutions $u(t;u_0), t\geq 0$, do not define in H^m a non-homogeneous Markov process.

For the same reason as the kick-forced equation (10.1.8), (10.1.7), equation (10.1.13), (10.1.14)–(10.1.16) satisfies conditions (H0)–(H2) from Chapter 7, where in Oleinik's estimates (2.1.1), (2.1.2), involved in (H1), $\xi = \int \eta dt$, and in (H2) the relation (2.2.2) holds uniformly in ω (so it is valid even without averaging).

However eq. (10.1.13), unlike the kicked equation, does not meet the balance of energy relation (2.3.1) (with an u_0–independent constant B_0), so we cannot repeat the argument in Section 2.3 to verify for eq. (10.1.13) the condition (H3). Accordingly, we cannot establish for it the small-scale results as in Sections 7.1-7.3. However let us consider eq. (10.1.13) with η_{red} replaced by a high–frequency red noise $\eta_N(t) = N^{-1/2}\eta_{red}(Nt)$, where $N \gg 1$. Then its solution $u_N^\omega(t;u_0)$, equal u_0 at $t=0$, satisfies

$$u_t + uu_x - \nu u_{xx} = \frac{\partial}{\partial t}\int_0^t \eta_N(s)ds = \frac{\partial}{\partial t}\beta_N^\omega(t) \tag{10.1.21}$$

(see (10.1.20)). Due to what was said above,

$$\mathcal{D}(\beta_N(\cdot)) \rightharpoonup \mathcal{D}(\xi(\cdot)) \quad \text{in} \quad X_\infty^m \quad \forall m,$$

where ξ is the Wiener process (0.0.2) with $b_s := b_s c_{-1}\big(\mathbb{E}\xi_{-10}^2\big)^{1/2}$ and the rate of convergence depends on ν. Accordingly, in view of Theorem 1.3.11,

$$\mathcal{D}(u_N(\cdot;u_0)) \rightharpoonup \mathcal{D}(u(\cdot;u_0)), \tag{10.1.22}$$

where $u(t;u_0)$ solves the problem (1.1.13) with ξ as above (cf. item i) of Theorem 10.2.1). So the results obtained in Parts I–II apply to solutions $u_N(t;u_0)$ with N large in terms of ν. We leave it as an (a bit advanced) exercise for a reader to quantify the condition "with large N". For example, what may a version of Theorem 7.3.2 for solutions of eq. (10.1.21) with $N \gg 1$ look like?

Naturally, what was just said about solutions of equation (10.1.13) with η_{red} replaced by the high–frequency force $\eta_N(t) = N^{-1/2}\eta_{red}(Nt)$ applies not only to red noises $\eta = \eta_{red}$ as in (10.1.14)-(10.1.16), but also to other random forces $\eta(t,x)$ defining mixing processes $\eta(t)$ in spaces H^m, see [**55**, Section 7.7] (processes $\eta(t) \in H^m$, corresponding to the red forces (10.1.14)-(10.1.16), are mixing in the sense of [**55**] since in view of (10.1.18) they have finite correlation time). Similarly, the results apply to the kick-forced equations (10.1.8) considered in the previous section (the kick-forces ξ^θ are mixing random processes in a suitable generalised sense). An analogy of convergence (10.1.22) for solutions of (10.1.8) and some related facts are discussed in the next section.

10.2. High–frequency kick forces.

In this section we discuss the Burgers equation, perturbed by the normalised high frequency kick force:

(10.2.1)

$$u_t(t,x) + u(t,x)u_x(t,x) - \nu u_{xx}(t,x) = \tilde\eta^\theta(t,x) := \sqrt{\theta}\eta^\theta(t,x), \quad 0<\theta\leq 1/2.$$

We assume that the numbers B_m, characterising the kick-force η^θ (see (10.1.7)), all are finite[1] and that the r.v.'s ξ_{sk} in (10.1.5) satisfy

$$\mathbb{E}\,\xi_{sk} \equiv 0 \tag{10.2.2}$$

(we recall the normalisation $\mathbb{E}\,\xi_{sk}^2 \equiv 1$). We denote solutions of (10.2.1) as $u(t;u_0,\tilde{\xi}^\theta)$ $=: u^\theta(t;u_0)$, where $\tilde{\xi}^\theta = \int_0^t \tilde{\eta}^\theta$ (as usual, the integral is normalised to be right-continuous), and abbreviate $u^\theta(k\theta;u_0) = u_k$, $k \in \mathbf{N}_0$. For any fixed $\theta > 0$ the results in Section 10.1.2 apply to the equation and its solutions u^θ.

Our goal is to show that, in a number of senses, when $\theta \to 0$, solutions of (10.2.1) become close in distribution to solutions of the stochastic Burgers equation (1.1.12). We recall that the random forces $\tilde{\eta}^\theta(t,x)$ with small θ naturally appear in the computer modelling as a time–discrete substitute for white in time random forces.

The viscosity ν, the numbers B_m and the constant C_* in (10.1.5) will stay fixed, and we will not indicate how the constants depend on them. On the contrary, the dependence of the latter on θ, u_0 and on the time interval $[0,T]$, $T \geq 1$, is crucial and will be always indicated.

Let us take any $u_0 \in H^m$, $m \in \mathbf{N}$.

1°. *Basic θ-uniform estimates.* Since $\tilde{\xi}^\theta(t) = \sqrt{\theta}\sum_{k\theta\leq t}\eta_k$, then for $T \in \theta\mathbf{N}_0$ we have $\mathbb{E}\|\tilde{\xi}^\theta(T)\|_m^2 = B_m T$, and by Doob's moment inequality (applied to the process $\Pi_N\tilde{\xi}^\theta$ and followed by an application of Fatou's lemma as $N \to \infty$), see [**80**, **139**],

$$\mathbb{E}\sup_{0\leq t\leq T}\|\tilde{\xi}^\theta(t)\|_m^2 \leq 4B_m T, \qquad T \in \theta\mathbf{N}_0. \tag{10.2.3}$$

Since $\|\mathcal{S}_\theta u\| \leq e^{-\nu\theta}\|u\|$ (see (10.1.2)) and η_k is independent from $\mathcal{S}(u_{k-1})$, then in view of (10.1.10) and (10.2.2)

$$\mathbb{E}\|u_k\|^2 \leq e^{-2\nu\theta}\mathbb{E}\|u_{k-1}\|^2 + \theta B_0, \quad k \geq 1,$$

as $\mathbb{E}\|\eta_k\|^2 = B_0$. So

$$\mathbb{E}\|u_k\|^2 \leq e^{-2k\nu\theta}\|u_0\|^2 + \theta B_0\frac{1-e^{-2k\nu\theta}}{1-e^{-2\nu\theta}} \leq e^{-2k\nu\theta}\|u_0\|^2 + \nu^{-1}B_0. \tag{10.2.4}$$

Arguing similarly and using (10.2.4) we get that

$$\mathbb{E}\|u_k\|^4 = \mathbb{E}\|\mathcal{S}_\theta u_{k-1} + \sqrt{\theta}\,\eta_k\|^4 \leq e^{-4\nu\theta}\mathbb{E}\|u_{k-1}\|^4 + \theta C(\|u_0\|)$$

(we recall the agreement concerning the constants). From here it follows that $\mathbb{E}\|u_k\|^4 \leq C_4(\|u_0\|)$, for all k. Arguing by induction we find that

$$\mathbb{E}\|u_k\|^{2r} \leq C_{2r}(\|u_0\|), \qquad k \geq 0.$$

for any $r \in \mathbf{N}$.

Now let us evoke estimate (1.4.18) and use it with $B_m = 0$ and with the random initial data u_0, distributed as u_k^ω (it is valid for the exact solutions $u(t)$ as well as

[1] Instead it suffices assume that $B_M < \infty$, where $M \geq M_*$ for a sufficiently large M_*. Then the assertions of Theorem 10.2.1 below would hold for $1 \leq m \leq m_M$, where m_M grows to infinity with M.

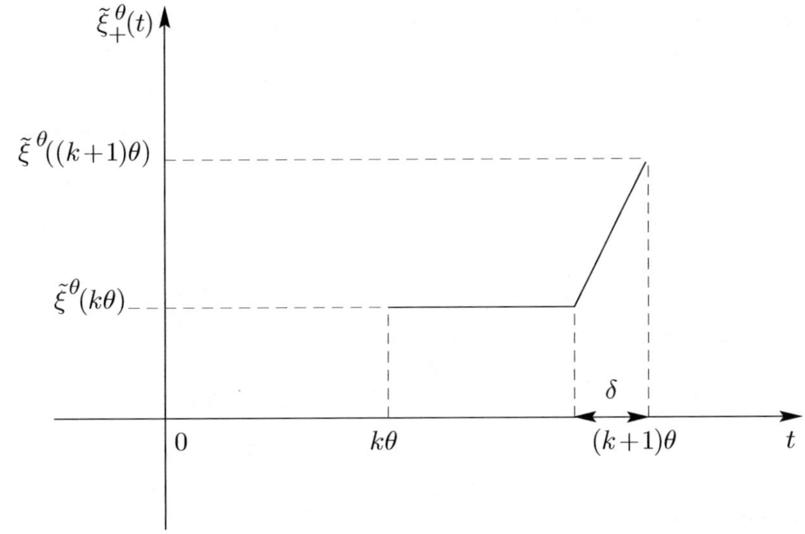

FIGURE 10.2.1. The curve $t \mapsto \tilde\xi_+^\theta(t)$.

for their Galerkin approximations $u^N(t)$ in view of Proposition 1.3.9). The estimate gives us that for any $0 \le \theta < \tau$

$$\begin{aligned}\mathbb{E}\|u^\theta(k\theta+\tau;u_0)\|_m^2 &\le e^{-\nu\tau}\mathbb{E}\|u_k\|_m^2 + \tau C_m \sup_{t\ge 0}\mathbb{E}\|u^\theta(t;u_0)\|^M \\ &= e^{-\nu\tau}\mathbb{E}\|u_k\|_m^2 + \tau C_m \sup_{k\ge 0}\mathbb{E}\|u_k\|^M.\end{aligned} \tag{10.2.5}$$

Using here the just obtained bound for the moments $\mathbb{E}\|u_k\|^{2r}$, and since

$$\mathbb{E}\|u_{k+1}\|_m^2 = \mathbb{E}\|u^\theta\big((k+1)\theta - 0;u_0\big)\|_m^2 + \theta B_m,$$

we get that

$$\mathbb{E}\|u_{k+1}\|_m^2 \le e^{-\nu\theta}\mathbb{E}\|u_k\|_m^2 + \theta C_m(\|u_0\|), \quad k \ge 0.$$

So $\mathbb{E}\|u_k\|_m^2 \le C'_m(\|u_0\|_m)$ for $k \in \mathbf{N}$. Using again (10.2.5) with $0 \le \tau < \theta$ we finally get that

$$\mathbb{E}\|u^\theta(t;u_0)\|_m^2 \le C_m(\|u_0\|_m), \quad \forall t \ge 0. \tag{10.2.6}$$

2°. *Smoothing the force $\tilde\xi^\theta$ and the solution u^θ.* For $0 < \delta \le \theta$ consider the continuous curve $\tilde\xi_+^\theta(t) = \tilde\xi_+^\theta(t;\delta)$, which on any interval $[k\theta,(k+1)\theta]$, $k \in \mathbf{N}$, equals $\tilde\xi^\theta(t)$ if $t \in [k\theta,(k+1)\theta-\delta]$ or if $t = (k+1)\theta$, and is linear on the intervals $[(k+1)\theta-\delta,\ (k+1)\theta]$. See Figure 10.2.1.

Let us abbreviate

$$\xi_1 = \tilde\xi^\theta;\ \xi_2 = \tilde\xi_+^\theta;\ \xi_3 = \tilde\xi^\theta - \tilde\xi_+^\theta.$$

Then $\xi_3(t)$ vanishes on intervals $[k\theta,(k+1)\theta-\delta]$ and for each ω and each $t \ge 0$ its H^{m+1}-norm is bounded by $C_m\sqrt{\theta}$. So for $1 \le p \le \infty$,

$$\|\xi_3\|_{L_p(0,T;H^{m+1})} \le C_m T^{1/p}\delta^{1/p}\theta^{1/2-1/p} \qquad \forall \omega \tag{10.2.7}$$

(we recall that $T \ge 1$).

The curves $X^\theta(t) := \tilde\xi^\theta_+(t;\theta)$ are important for us below. They are equal to $\tilde\xi^\theta$ at points $t = k\theta$, $k \in \mathbf{N}_0$, and are linear between them. In view of (10.2.7) with $\delta = \theta$,

$$\|X^\theta - \tilde\xi^\theta\|_{L_p(0,T;H^{m+1})} \le C_m T^{1/p}\theta^{1/2} \qquad \forall\omega,\ \forall 1\le p\le\infty. \tag{10.2.8}$$

Let us consider $u^\theta_+(t;u_0) := u(t;u_0,\tilde\xi^\theta_+)$, denote $u_1 := u^\theta$, $u_2 := u^\theta_+$ and write the difference $u^\theta - u^\theta_+ = u_1 - u_2$ as $u^\theta - u^\theta_+ = \xi_3 + w_3$. Then w_3 satisfies

$$\partial_t w_3 + \frac{1}{2}\frac{\partial}{\partial x} w_3(u_1+u_2) - \nu w_{3xx} = \nu\xi_{3xx} - \frac{1}{2}\frac{\partial}{\partial x}\xi_3(u_1+u_2) =: h(t,x), \qquad w_3(0) = 0.$$

We note that for all $m \in \mathbf{N}$,

$$\|u_1+u_2\|_{L_\infty(0,T;H^m)} \le C_m(\theta,T,\|u_0\|_m), \quad \|h\|_{L_2(0,T;H^{m-1})} \le \delta^{1/2}C_m(\theta,T,\|u_0\|_m).$$

Indeed, the first inequality holds since the norm of u_2 in $L_\infty(0,T;H^m)$ is bounded by Theorem 1.3.6, while the norm of u_1 obviously is bounded in view of (10.1.10), (10.1.11). The second inequality follows from the first and from (10.2.7). So in view of Proposition 1.3.14, $\|w_3\|_{X^m_T} \le \sqrt{\delta}\, C_m(\theta,T,\|u_0\|_m)$, for any $m \in \mathbf{N}$. Choosing δ small in terms of θ, T, m and $\|u_0\|_m$ and using $(10.2.7)|_{p=\infty}$ we achieve that u^θ_+ is $\sqrt{\theta}$-close to u^θ in $L_\infty(0,T;H^m)$. Again by $(10.2.7)|_{p=\infty}$, $\tilde\xi^\theta_+$ is $\sqrt{\theta}$-close to $\tilde\xi^\theta$ in $L_\infty(0,T;H^{m+1})$, so in view of $(10.2.8)\ |_{p=\infty}$ X^θ also is $\sqrt{\theta}$-close to $\tilde\xi^\theta$ in $L_\infty(0,T;H^{m+1})$:
(10.2.9)

$$\|u^\theta - u^\theta_+\|_{L_\infty(0,T;H^m)},\ \|\tilde\xi^\theta - \tilde\xi^\theta_+\|_{L_\infty(0,T;H^{m+1})},\ \|X^\theta - \tilde\xi^\theta_+\|_{\dot X^{m+1}_T} \le C_m\sqrt{\theta}, \quad \forall\omega,$$

where C_m does not depend on u_0 and T, and we use that for any r, on the space $\dot X^r_T$ the norm $\|\cdot\|_{L_\infty(0,T;H^r)}$ coincides with $\|\cdot\|_{\dot X^r_T}$. From now on the constant $\delta = \delta(\theta,T,\|u_0\|_m,m)$ is fixed.

Solutions $u(t;u_0,\tilde\xi^\theta_+(t;\delta))$ with $\delta \ll 1$ are good time-continuous approximations for the discontinuous solutions u^θ. This will allow us to apply to eq. (10.2.1) results of Section 1.3, instead of developing for their study a parallel theory of discontinuous solutions for eq. (B), using Skorokhod's spaces, cf. [**103**, Section 2.2].

3°. *Donsker's theorem.* Below for $j = 1,2\ldots$ and $\gamma > 0$ we denote by $\kappa^j_\gamma(\ldots)$ positive functions of γ, monotonically going to zero with γ and depending on the parameters in the brackets (dependence on $\|u_0\|_m$ tautologically assumes the dependence on m). By $\xi^\omega(t,x)$ we denote the Wiener process (1.1.6) with the same b_s's as in (10.1.6), and abbreviate $u(t;u_0,\xi^\omega)$ to $u(t;u_0)$.

Due to (10.2.2) and (10.1.5), Donsker's theorem applies to the process $X^\theta(t)$ and implies that

$$\|\mathcal{D}X^\theta - \mathcal{D}\xi\|^*_{L(\dot X^{m+1}_T)} \le \kappa^1_\theta(T,m). \tag{10.2.10}$$

Indeed, the classical theorem, e.g. see in [**16**], implies that for any $N \in \mathbf{N}$ the result holds with X^θ and ξ replaced with $\Pi_N X^\theta$ and $\Pi_N\xi$ (see (1.3.23)) and with $\kappa^1_\theta(T,m)$ replaced with some $\tilde\kappa^1_\theta(T,m,N)$. Then (10.2.10) follows since all numbers B_m are finite, so

$$\|\mathcal{D}\xi - \mathcal{D}\Pi_N\xi\|^*_{L(\dot X^{m+1}_T)} + \|\mathcal{D}X^\theta - \mathcal{D}\Pi_N X^\theta\|^*_{L(\dot X^{m+1}_T)} \to 0 \quad \text{as} \quad N\to\infty,$$

for each T.

Now we will use (10.2.10) to estimate the distance between the laws of $u(t;u_0)$ and $u^\theta(t;u_0)$. Since due to (10.2.9)

$$\|\mathcal{D}u_+^\theta(t;u_0)-\mathcal{D}u^\theta(t;u_0)\|^*_{L(H^m)}\le C_m\sqrt{\theta},\quad \forall\, 0\le t\le T, \tag{10.2.11}$$

then it suffices to estimate the distance between the laws of $u(t;u_0)$ and $u_+^\theta(t;u_0)$. To do this we note that firstly, by (10.2.9) and (10.2.10)

$$\begin{aligned}\|\mathcal{D}\tilde\xi_+^\theta-\mathcal{D}\xi\|^*_{L(\dot X_T^{m+1})}&\le \|\mathcal{D}\tilde\xi_+^\theta-\mathcal{D}X^\theta\|^*_{L(\dot X_T^{m+1})}+\|\mathcal{D}X^\theta-\mathcal{D}\xi\|^*_{L(\dot X_T^{m+1})}\\ &\le C_m\sqrt{\theta}+\kappa_\theta^1(T,m).\end{aligned}$$

Secondly, by (10.2.3), (10.2.9) and Theorem 1.2.3,

$$\mathbb{E}\|\tilde\xi_+^\theta\|^2_{\dot X_T^{m+1}},\ \mathbb{E}\|\xi\|^2_{\dot X_T^{m+1}}\le C_{m+1}(T). \tag{10.2.12}$$

By the above and Proposition 11.4.2,

$$\|\mathcal{D}\tilde\xi_+^\theta-\mathcal{D}\xi\|_{Kant,X_T^{m+1}}\le\kappa_\theta^2(m,T).$$

Thus by the Kantorovich-Rubinstein Theorem 11.4.3 for $0<\theta\le 1/2$ there exist random variables $\tilde\eta_+^\theta$ and η^θ, valued in $\dot X_T^{m+1}$, such that $\mathcal{D}\tilde\eta_+^\theta=\mathcal{D}\tilde\xi_+^\theta$, $\mathcal{D}\eta^\theta=\mathcal{D}\xi$ and

$$\mathbb{E}\|\tilde\eta_+^\theta-\eta^\theta\|_{\dot X_T^{m+1}}\le\kappa_\theta^2(m,T). \tag{10.2.13}$$

Denote $m_+^\theta=\mathcal{D}u_+^\theta(\cdot;u_0)\in\mathcal{P}(X_T^m)$ and $m_\xi=\mathcal{D}u(\cdot;u_0)\in\mathcal{P}(X_T^m)$. Then

$$m_+^\theta=\mathcal{D}\left(F\circ\tilde\xi_+^\theta\right)\in\mathcal{P}(X_T^m),\quad m_\xi=\mathcal{D}\left(F\circ\xi\right)\in\mathcal{P}(X_T^m),$$

where $F:\dot X_T^{m+1}\to X_T^m$, $F(\xi)=\mathcal{M}(u_0,\xi)$, and for any function $\phi\in\bar B^1_{L(X_T^m)}$ we have

$$\langle\phi,m_+^\theta\rangle-\langle\phi,m_\xi\rangle=\mathbb{E}\big(\phi\circ F(\tilde\eta_+^\theta)-\phi\circ F(\eta^\theta)\big).$$

For $R>0$ consider the event $\Omega_R=\Omega_R(\theta)=\{\|\tilde\eta_+^\theta\|_{\dot X_T^{m+1}}+\|\eta^\theta\|_{\dot X_T^{m+1}}\ge R\}$. Since $\mathcal{D}\tilde\eta_+^\theta=\mathcal{D}\tilde\xi_+^\theta$ and $\mathcal{D}\eta^\theta=\mathcal{D}\xi$, then due to Markov's inequality, (10.2.12) and Theorem 1.2.3, $\mathbf{P}(\Omega_R)\le C_m(T)R^{-1}$. Therefore,

$$\begin{aligned}&\langle\phi,m_+^\theta\rangle-\langle\phi,m_\xi\rangle\\ &\quad=\mathbb{E}\,\chi_{\Omega_R}\big(\phi\circ F(\tilde\eta_+^\theta)-\phi\circ F(\eta^\theta)\big)+\mathbb{E}\,\chi_{\Omega\setminus\Omega_R}\big(\phi\circ F(\tilde\eta_+^\theta)-\phi\circ F(\eta^\theta)\big)\\ &\quad\le C(T,\|u_0\|_m,R)\mathbb{E}\big(\|\tilde\eta_+^\theta-\eta^\theta\|_{\dot X_T^{m+1}}\big)+2R^{-1}C_m(T),\end{aligned}$$

where we used that the mapping $\mathcal{M}:H^m\times\dot X_T^{m+1}\to X_T^m$ is locally Lipschitz by Theorem 1.3.11. Choosing first R sufficiently big and next θ small and using (10.2.13) we bound the r.h.s. by $\kappa_\theta^3(T,\|u_0\|_m)$. Since this estimate holds for any $\phi\in\bar B^1_{L(X_T^m)}$, then the distance $\|m_+^\theta-m_\xi\|^*_{L(X_T^m)}$ is bounded by the same quantity. From here and (10.2.11) we get that

$$\|\mathcal{D}u(t;u_0)-\mathcal{D}u^\theta(t;u_0)\|^*_{L(H^m)}\le\kappa_\theta^4(T,\|u_0\|_m)\qquad\forall\,0\le t\le T. \tag{10.2.14}$$

4°. *The uniform in θ mixing.* In view of Theorem 3.3.4,

$$\|\mathcal{D}u(T;u_0)-\mu\|^*_{L(H^m)}\le\kappa_{T^{-1}}^5(m), \tag{10.2.15}$$

where μ is the stationary measure for eq. (1.1.12). As $\|\dots\|^*_{L(L_1)} \le \|\dots\|^*_{L(H^m)}$, then we derive from this estimate and (10.2.14) that

(10.2.16)
$$\|\mathcal{D}u^\theta(T;u_0) - \mathcal{D}u^\theta(T;u_1)\|^*_{L(L_1)} \le 2\kappa^5_{T^{-1}}(m) + 2\kappa^4_\theta(T, \max(\|u_0\|_m, \|u_1\|_m)).$$

Let μ^θ be the stationary measure in (10.1.12). Since it is a limiting point for the law of solution $u^\theta(t;0)$, evaluated at times $t = k\theta$, then in view of (10.2.6),

$$\int \|u\|^2_m \mu^\theta(du) \le C_m. \tag{10.2.17}$$

Let $u^\theta_{stat}(t)$ be a solution of (10.2.1) such that

$$\mathcal{D}u^\theta_{stat}(k\theta) \equiv \mu^\theta, \qquad k \in \mathbf{N}. \tag{10.2.18}$$

Then for any function $\phi \in \bar{B}^1_{L(L_1)}$ and any $T \in \theta\mathbf{N}$ we have

$$\begin{aligned}\langle \phi, \mathcal{D}u^\theta(T;u_0)\rangle - \langle \phi, \mu^\theta\rangle &= \mathbb{E}(\phi \circ u^\theta(T;u_0) - \phi \circ u^\theta_{stat}(T))\\ &= \int_{H^m} \mathbb{E}(\phi \circ u^\theta(T;u_0) - \phi \circ u^\theta(T;u))\mu^\theta(du).\end{aligned}$$

Splitting H^m into the ball $\bar{B}^R_{H^m}$ and its complement, we use (10.2.16) to bound the integral over the ball by $2\kappa^5_{T^{-1}}(m) + 2\kappa^4_\theta(T, \max(\|u_0\|_m, R))$, and (10.2.17) to estimate the integral over the complement to $\bar{B}^R_{H^m}$ by $2\mu^\theta(H^m \setminus \bar{B}^R_{H^m}) \le 2R^{-2}C_m$. Taking the supremum over all ϕ as above we arrive at the estimate

$$\|\mathcal{D}u^\theta(T;u_0) - \mu^\theta\|^*_{L(L_1)} \le 2\kappa^5_{T^{-1}}(m) + 2\kappa^4_\theta(T, \max(\|u_0\|_m, R)) + 2R^{-2}C_m, \quad T \in \theta\mathbf{N},$$

for any $R > 0$. Choosing $R = R(T)$ large enough so that $2R^{-2}C_m \le \kappa^5_{T^{-1}}(m)$ we get that

$$\|\mathcal{D}u^\theta(T;u_0) - \mu^\theta\|^*_{L(L_1)} \le 3\kappa^5_{T^{-1}}(1) + \kappa^6_\theta(T, \|u_0\|_m), \quad T \in \theta\mathbf{N}.$$

Noting that the assertion of Proposition 4.1.4 remains true for solutions of (10.2.1) with the same easy proof, we have that $\|\mathcal{D}u^\theta(t;u_0) - \mu^\theta\|^*_{L(L_1)} \le 3\kappa^5_{T^{-1}}(m) + \kappa^6_\theta(T, \|u_0\|_m)$, for all $\theta\mathbf{N} \ni t \ge T$. Then for any $\varepsilon > 0$,

$$\|\mathcal{D}u^\theta(t;u_0) - \mu^\theta\|^*_{L(L_1)} \le \varepsilon \quad \text{if } \theta\mathbf{N} \ni t \ge T^1_\varepsilon,\ \theta \le \theta_\varepsilon,$$

for a suitable $\theta_\varepsilon > 0$ and $T^1_\varepsilon \ge 1$. But for $1 \ge \theta \ge \theta_\varepsilon$, in view of (10.1.12), the distance $\|\mathcal{D}u^\theta(t;u_0) - \mu^\theta\|^*_{L(L_1)}$ also is bounded by ε if $\theta\mathbf{N} \ni t \ge T^2_\varepsilon$ for a suitable T^2_ε. Denoting $T_\varepsilon = \max(T^1_\varepsilon, T^2_\varepsilon)$ we achieve that for each $\varepsilon > 0$,

$$\|\mathcal{D}u^\theta(t;u_0) - \mu^\theta\|^*_{L(L_1)} \le \varepsilon \quad \text{if } \theta\mathbf{N} \ni t \ge T_\varepsilon, \tag{10.2.19}$$

for all $0 < \theta \le 1/2$.

5°. *The theorem.* Now we can formulate and prove the main results of this section. We recall that $u^\theta(t;u_0) = u(t;u_0,\tilde{\xi}^\theta)$ and $u(t;u_0)$ are solution of respectively (10.2.1) and (1.1.12), equal u_0 at $t = 0$, and μ^θ is the stationary measure for the system (10.1.10) as in (10.1.12). For $0 \le \tau < \theta$ denote $\mu^\theta_\tau = \mathcal{S}_\tau \circ \mu^\theta$ (we recall that $\{\mathcal{S}_\tau\}$ are the flow-maps of the free Burgers equation). This measure equals $\mathcal{D}u^\theta_{stat}(k\theta + \tau)$ for any $k \in \mathbf{N}$, where $u^\theta_{stat}(t)$ is the solution of (10.2.1) as in (10.2.18). By μ and $u_{stat}(t)$ we denote the stationary measure and the stationary solution for the stochastic equation (1.1.12). Finally, for $t \in \mathbb{R}$ we denote by $\tau_\theta(t)$ "the phase of t", i.e. the real number $\tau \in [0,\theta)$ such that $t = k\theta + \tau$ for an integer k.

THEOREM 10.2.1. *If $B_m < \infty$ for all m and (10.2.2) holds, then for any $m \in \mathbf{N}$ the following properties are valid for solutions $u^\theta(t; u_0)$ with $u_0 \in H^m$ and for measures μ^θ_τ:*

1) for any $T \geq 1$,

$$\sup_{0 \leq t \leq T} \|\mathcal{D}u^\theta(t; u_0) - \mathcal{D}u(t; u_0)\|^*_{L(H^m)} \to 0 \quad \text{as} \quad \theta \to 0;$$

2) $\|\mathcal{D}u^\theta(t; u_0) - \mu^\theta_{\tau_\theta(t)}\|^*_{L(H^{m-1})} \to 0$ as $t \to \infty$, uniformly in $0 < \theta \leq 1/2$;

3) $\mu^\theta_\tau \rightharpoonup \mu$ in $\mathcal{P}(H^m)$ as $\theta \to 0$, uniformly in $0 \leq \tau < \theta$;

4) $\|\mathcal{D}u^\theta(t; u_0) - \mu\|^*_{L(H^m)} \to 0$ as $t \to \infty$ and $\theta \to 0$.

PROOF. The first assertion is proved in (10.2.14). To verify the second one we start with observing that due to Proposition 4.1.4, applied to eq. (B) with zero force η,

$$\begin{aligned}\|\mathcal{D}u^\theta(k\theta + \tau; u_0) - \mu^\theta_\tau\|^*_{L(L_1)} &= \|\mathcal{S}_\tau \circ \mathcal{D}u^\theta(k\theta; u_0) - \mathcal{S}_\tau \circ \mu^\theta\|^*_{L(L_1)} \\ &\leq \|\mathcal{D}u^\theta(k\theta; u_0) - \mu^\theta\|^*_{L(L_1)}.\end{aligned}$$

So by (10.2.19),

$$\|\mathcal{D}u^\theta(k\theta + \tau; u_0) - \mu^\theta_\tau\|^*_{L(L_1)} \to 0 \quad \text{as} \quad k \to \infty,$$

uniformly in $0 < \theta \leq 1/2$ and $0 \leq \tau < \theta$. Evoking estimate (10.2.4) and Proposition 11.4.2, similarly to Step 3, we see that the convergence above implies that of $\mathcal{D}u^\theta(k\theta + \tau; u_0)$ to μ^θ_τ in the Kantorovich distance (11.4.1) in the space $\mathcal{P}_1(L_1)$, uniformly in $0 \leq \tau < \theta \leq 1/2$. So by the Kantorovich–Rubinstein Theorem 11.4.3 for $k \geq 1$ there exist r.v.'s $U_{\tau,k}$, and $V_{\tau,k}$, valued in H^m (since the measures $\mathcal{D}u^\theta$ and μ^θ_τ are supported by H^m), such that $\mathcal{D}U_{\tau,k} = \mathcal{D}u^\theta(k\theta + \tau; u_0)$, $\mathcal{D}V_{\tau,k} = \mu^\theta_\tau$ and

$$\mathbb{E}|U_{\tau,k} - V_{\tau,k}|_1 \leq \kappa^7_{k^{-1}}(\|u_0\|_m). \tag{10.2.20}$$

Now for any $f \in \bar{B}^1_{L(H^{m-1})}$, using the Gagliardo–Nirenberg inequality (1.3.9), we have that

$$\begin{aligned}\langle f, \mathcal{D}u^\theta(k\theta + \tau; u_0)\rangle - \langle f, \mu^\theta_\tau\rangle &= \mathbb{E}\big(f(U_{\tau,k}) - f(V_{\tau,k})\big) \leq \mathbb{E}\|U_{\tau,k} - V_{\tau,k}\|_{m-1} \\ &\leq \mathbb{E}\Big[|U_{\tau,k} - V_{\tau,k}|_1^{\frac{2}{2m+1}} \|U_{\tau,k} - V_{\tau,k}\|_m^{\frac{2m-1}{2m+1}}\Big] \\ &\leq \big(\mathbb{E}|U_{\tau,k} - V_{\tau,k}|_1\big)^{\frac{2}{2m+1}} \big(\mathbb{E}\|U_{\tau,k} - V_{\tau,k}\|_m\big)^{\frac{2m-1}{2m+1}}.\end{aligned}$$

Estimate (10.2.6) with $u_0 = 0$ and convergence (10.1.12) immediately imply that $\langle \mu^\theta_\tau, \|u\|^2_m\rangle \leq C_m$. Since $\mathbb{E}\|U_{\tau,k} - V_{\tau,k}\|_m \leq \mathbb{E}(\|U_{\tau,k}\|_m + \|V_{\tau,k}\|_m)$, then in view of (10.2.6) and (10.2.20) the r.h.s. above is bounded by

$$C'_m(\|u_0\|_m)\big[\kappa^7_{k^{-1}}(\|u_0\|_m)\big]^{2/(2m+1)}.$$

Thus,

$$\|\mathcal{D}u^\theta(k\theta + \tau; u_0) - \mu^\theta_\tau\|^*_{L(H^{m-1})} \leq \kappa^8_{k^{-1}}(\|u_0\|_m).$$

Since the r.h.s. does not depend on $0 \leq \tau < \theta \leq 1/2$, then the second assertion is proved.

To establish the third we note that from (10.2.14) and (10.2.15) it folows that

$$\|\mathcal{D}u^\theta(t; u_0) - \mu\|^*_{L(H^m)} \leq \kappa^5_{t^{-1}}(m) + \kappa^4_\theta(t, \|u_0\|_m). \tag{10.2.21}$$

For a fixed $\tau \in [0,\theta)$, taking $t = k\theta + \tau$ with $k \in \mathbf{N}$, for any $\phi \in \bar{B}^1_{L(H^m)}$ we have

$$\langle \phi, \mu^\theta_\tau \rangle - \langle \phi, \mu \rangle = \int_{H^m} \mathbb{E}\big(\phi \circ u^\theta(t;v) - \phi \circ u_{stat}(t)\big)\mu^\theta(dv).$$

As at Step 5, splitting the integral over H^m to that over $\bar{B}^R_{H^m}$ and over the ball's complement we use (10.2.21) to bound the former by $\kappa^5_{t^{-1}}(m) + \kappa^4_\theta(t,R)$, and use (10.2.17) to bound the latter by $2C_m R^{-2}$. Choosing firstly R large and next k large, we get that $\|\mu^\theta_\tau - \mu\|^*_{L(H^m)} \le \kappa^9_\theta(m)$, uniformly in τ. This implies the third assertion.

The fourth claim of the theorem follows from the second and the third. □

The assertions of Theorem 10.2.1 may be summarised in a commutative diagram

$$\begin{array}{ccc} \mathcal{D}u^\theta(t;u_0) & \xrightarrow{t\to\infty} & \mu^\theta_{\tau(t)} \\ \Big\downarrow{\scriptstyle \theta\to 0} & & \Big\downarrow{\scriptstyle \theta\to 0} \\ \mathcal{D}u(t;u_0) & \xrightarrow{t\to\infty} & \mu \end{array}$$

where $u_0 \in H^m$, $m \in \mathbf{N}$. The top horizontal arrow signifies the weak convergence in $\mathcal{P}(H^{m-1})$, uniform in $\theta \in (0,1/2]$. The other three arrows stand for the weak convergence in $\mathcal{P}(H^m)$.

Corollary 10.2.2. *For any $u_0 \in H^m, m \in \mathbf{N}$,*

$$D(t,\theta) := \|\mathcal{D}u^\theta(t;u_0) - \mathcal{D}u(t;u_0)\|^*_{L(H^{m-1})} \to 0 \quad \text{as} \quad \theta \to 0,$$

uniformly in $t \ge 0$.

Proof. For any $T \ge 1$ and every $t \ge 0$, $0 < \theta \le 1/2$ we have that

$$\begin{aligned} D(t,\theta) \le & \sup_{0\le l\le T} D(l,\theta) \\ & + \sup_{l \ge T} \Big[\|\mathcal{D}u^\theta(l;u_0) - \mu^\theta_{\tau_\theta(l)}\|^*_{L(H^{m-1})} + \|\mu^\theta_{\tau_\theta(l)} - \mu\|^*_{L(H^{m-1})} \\ & \qquad + \|\mu - \mathcal{D}u(t;u_0)\|^*_{L(H^{m-1})} \Big]. \end{aligned}$$

By the item 1) of the theorem the first term in the r.h.s. is bounded from above by $\kappa^{10}_\theta(T, \|u_0\|_{m-1})$, while by the items 2), 3) and Theorem 3.3.4 the second term is bounded from above by $\kappa^{11}_{T^{-1}}(\|u_0\|_{m-1})$. This implies the assertion. □

The popular splitting up method, applied to the stochastic Burgers equation (1.1.12), suggests to construct its solutions on finite time intervals as limits when $\theta \to 0$ of solutions for eq. (10.1.8), where η^θ is a Gaussian kick-force (10.1.7) with the kicks

$$\eta_k = \xi(k\theta) - \xi\big((k-1)\theta\big) =: \eta^G_k, \quad k \ge 1$$

(e.g. see [**72, 123**] and references in the second paper). Our results show that the method still holds when the kicks η_k are bounded r.v.'s with zero mean and the same standard deviations as η^G_k. Moreover, solutions for the equations with bounded kick-forces approximate in distribution solutions for (1.1.12) uniformly in time.

Assertions 1) and 3) of Theorem 10.2.1 hold for a large class of nonlinear PDEs, perturbed by white forces and high-frequency kick forces. E.g. they hold for the 2d NSE (6.0.12), see in [**103**] Theorems 2.4.15 and 4.3.3. But the proofs of assertions 2), 4) of the theorem and of Corollary 10.2.2 use specific properties of the Burgers equation. Their validity for the 2d NSE is unknown, see the last paragraph of [**103**, Section 4.3.2].

10.3. Exponential mixing and its consequences.

Exponential mixing in equation (1.1.12). Exponential mixing in the "bigger brother" of eq. (1.1.12) – the stochastic 2d Navier–Stokes system (6.0.12) – has attracted much attention in the last 25 years, starting from the works [**63, 102**]; see [**46, 103**] for the statements and the history of their proofs. Without any doubt these results apply as well to the Burgers equation and allow to establish that under mild additional assumptions given below, convergence to zero in Theorem 3.3.4 is exponentially fast:

$$\|S_t^*\mu_0 - \mu\|_{L(H^m)}^* \le C_m e^{-\kappa_m t}, \qquad \forall t \ge 0, \tag{10.3.1}$$

with some $\kappa_m > 0$. In other words, the Markov process defined by the equation in H^m is *exponentially mixing*, provided that some non-degeneracy condition is imposed on the random force. Namely:

1) this is true if in (1.1.6) $b_s \neq 0$ for all $|s| \le N_\nu$, where N_ν is a suitable function of ν which goes to infinity as $\nu \to 0$ (in particular if $b_s \neq 0$ for all s). In literature there is a number of related proofs of the exponential mixing for the 2d Navier–Stokes system on a torus or on a bounded domain. To convert one of them into a proof for the Burgers equation is a useful but time-consuming exercise for the reader.

2) In [**75**] a stronger result was established: for the 2d NSE (6.0.12) let us decompose the random force η in the usual trigonometric basis $(e_s, s \in \mathbb{Z}^2 \setminus \{0\})$ of the space of zero-mean value divergence-free vector fields on $\mathbb{T}^2$ (see (2.29) in [**103**]),

$$\eta^\omega(t;x) = \partial_t\Big(\sum b_s\beta_s^\omega(t)e_s(x)\Big)$$

(here, as usual, $\beta_s(t)$ are independent standard Wiener processes). Assume that $b_s \neq 0$ if $s = (s_1, s_2)$ and $\max(|s_1|, |s_2|) \le 1$. Then eq. (6.0.12) is exponentially mixing. To prove a similar result for the Burgers equation is a task significantly more complicated than in 1) above. Needed results for the controllability of the Burgers equation may be found in [**138**] (there the equation is considered under the Dirichlet boundary conditions, so modifying the calculation for space-periodic solutions is a part of the job).

Note that the rate of convergence given by Theorem 4.2.4 is independent from ν with respect to the norm $\|\cdot\|_{L(L_1)}^*$, while the exponent of convergence κ_m in (10.3.1), given by the available proofs of the exponential mixing, tends to 0 with ν. On the other hand, it is also known that the limiting inviscid dynamics in the space L_1 is exponentially mixing, see the discussion after Theorem 8.5.2. It is unknown if for $\nu > 0$ the convergence $S_t^*\mu_0 \rightharpoonup \mu$ in $\mathcal{P}(L_1)$ is exponentially fast uniformly in $\nu \in (0, 1]$.

Exponential mixing in the Burgers equation with kick-forces and red forces.

Kick forces. If in (10.1.6) $b_s \neq 0$ for all $|s| \le N_\nu$ with a suitable $N_\nu \to \infty$, then the exponential convergence to a stationary measure μ, i.e. the relation

$$\|\mathcal{D}u(k) - \mu\|_L^* \le Ce^{-\kappa k}, \quad k \in \mathbf{N}, \tag{10.3.2}$$

with suitable $\kappa, C > 0$, for any solution $u(k) = u(k; u_0)$ follows from [**103**, Theorem 3.2.5.]. If only $b_s \neq 0$ for $|s| \leq N < \infty$, then the validity of the exponential mixing for small $\nu > 0$ is an open problem.

Red forces. If the force (10.1.19) is non-degenerate and $b_s \neq 0$ for all s, then the exponential mixing (10.3.2) follows from the result of [**105**] – all conditions required to apply the theorem in [**105**] easily follow from the second appendix in Section 1.8. If only $b_s \neq 0$ for $|s| \leq N_{red} < \infty$ with a suitable finite N_{red}, independent from ν, then the exponential mixing holds in view of an abstract theorem in [**101**]. The required analytic assumptions follow as before from the first appendix in Section 1.8. The approximate linearised controllability condition may be verified using the calculations in [**138**], similarly to how this is done in [**101**] for the more complicated 2d Navier–Stokes system.

Exponential mixing and ergodicity.

Markov processes with exponential mixing are ergodic in the sense that they satisfy the strong law of large numbers (SLLN), central limit theorem (CLT) (see [**55**]) and similar statements. In particular, this holds for solutions of (10.1.8), where the random force is such that the solutions $u(t)$ define an exponentially mixing process.

Let us start with the case when η is a kick-force or a red noise, assuming for simplicity that all numbers B_m are finite. In the former case (when η is a kick-force) we suppose that $b_s \neq 0$ for all s, while in the second – that $b_s \neq 0$ for $|s| \leq N_{red}$. Then a solution $u(t)$, evaluated at integer moments of time, defines an exponentially mixing Markov chain with a stationary measure μ.

THEOREM 10.3.1. *Let $f \in C_b(H^m)$ for some $m \in \mathbf{N}$. Then for any $u_0 \in H^1$,*

$$\bar{S}_N(f) := \frac{1}{N}\sum_{k=1}^{N} f(u(k; u_0)) \to \langle f, \mu\rangle \quad \text{as} \quad N \to \infty, \text{ a.s.} \tag{10.3.3}$$

This is the SLLN for solutions $u(k; u_0)$. Its proof may be obtained by a literal repetition of the argument in [**137**] (see also [**103**, Section 4.1]).

If $\langle f, \mu\rangle = 0$, then by the theorem above the time-average $\bar{S}_N(f)$ converges to 0 as $N \to \infty$, a.s. In fact the exponential mixing implies the stronger result

$$\lim_{N\to\infty} N^{1/2-\varepsilon}\bar{S}_N(f) = 0 \quad \text{a.s.},$$

for any $\epsilon > 0$, see [**103, 137**].

Applying to the Markov chain $u(k; u_0)$ Gordin's martingale approximation and literally repeating the argument in [**100, 103**] we get that the process satisfies the CLT:

THEOREM 10.3.2. *If under the assumptions of Theorem 10.3.1 $\langle f, \mu\rangle = 0$, then*

$$\mathcal{D}\left(\frac{1}{\sqrt{N}}\sum_{k=1}^{N} f(u(k; u_0))\right) \rightharpoonup N(0, \sigma_f^2) \quad \text{as} \quad N \to \infty,$$

for some $\sigma_f \geq 0$.

Moreover, it can be checked exactly as in [**103**] that $\sigma_f \neq 0$ if $f \not\equiv 0$.

Now let us turn to the stochastic Burgers equation (1.1.12). If there the force $\partial_t\xi$ is nondegenerate in the sense specified at the beginning of this subsection, then the Markov process $u(t;u_0), t \geq 0$, with $u_0 \in H^1$ is exponentially mixing, and continuous time versions of the two theorems above are true if the functional f has at most a polynomial growth at infinity (in the theorems' assertions we should replace the sums $\sum_{k=1}^{N} \cdots$ by integrals $\int_1^{N+1} \ldots dt$), cf. Problem 3.4.2. Again the proofs are the same as in [**100**, **103**, **137**].

EXAMPLE 10.3.3. Assume that $B_m < \infty$ for all m and the random force $\partial_t \xi$ is nondegenerate, so eq. (1.1.12) is exponentially mixing. Considering the functionals $e_k(u)$, defined in (7.3.2), and evoking the notation (7.5.6) we get that

$$\overline{e_k(u)} \to \langle e_k, \mu^\nu \rangle \quad \text{a.s. as} \quad \bar{\sigma}^\varepsilon \to \infty. \tag{10.3.4}$$

Indeed, let us in the definition of $e_k(u)$ replace $|\hat{u}_n|^2$ by $\min(M^2, |\hat{u}_n|^2)$, $M \in \mathbf{N}$, and denote the obtained functional $e_k^M(u)$. Then, firstly, for e_k replaced by e_k^M the convergence follows from the SLLN.

Secondly, the functionals e_k^M with large M approximate e_k well for solutions of (1.1.12). In view of the estimates used in the proof of Theorem 7.4.2, we have $|\langle e_k^M, \mu_\nu \rangle - \langle e_k, \mu_\nu \rangle| \leq CM^{-1}$, and $\mathbf{P}\{|e_k^M(u(t)) - e_k(u(t))| \geq \varepsilon\} \leq CM^{-1}\varepsilon^{-1}$, if $t \geq 1$. The two arguments jointly imply (10.3.4).

Note that convergence (10.3.4) agrees well with estimates (7.5.9) since the energy spectrum $\langle e_k, \mu_\nu \rangle$ satisfies (7.4.1).

10.4. Other equations.

10.4.1. Fractional Burgers equation. In [**30**] we consider the fractional Burgers equation

$$\frac{\partial u}{\partial t} + u\frac{\partial u}{\partial x} + \nu(-\Delta)^\gamma u = 0, \ t \geq 0, \ x \in S^1,$$

where $1/2 < \gamma < 1$ and $u_0 \in H^1$. Moreover, $\gamma > 1/2$ guarantees well-posedness of the equation and smoothness of solutions, and $\gamma \leq 1$ implies that the operator $(-\Delta)^\gamma$ has a positive kernel, so Oleinik's estimate holds; see the introduction to [**87**].

This equation was considered as a paradigm for turbulence models by Bardos, Frisch, Penel and Sulem [**5**], as well as Biler, Funaki and Woyczynski [**15**]. For other references, see the introduction to [**30**].

Using a nonlocal maximum principle from [**40**], we prove that the results in Theorem 10.1.1 still hold up to one modification: namely, ν needs to be replaced with ν^β, where $\beta = 1/(2\gamma - 1) \in (1, \infty)$. As before for solutions of the unforced equation 10.1.1, analogues of Theorem 7.2.3 and Theorem 7.3.2 hold with the caveats due to the purely dissipative nature of the system.

10.4.2. Multidimensional Burgers equation. Now we consider a multidimensional counterpart of the stochastic Burgers equation on a flat torus $\mathbf{T}^d$ of arbitrary dimension. The solution $\mathbf{u}$ is itself a d-dimensional vector.

$$\frac{\partial \mathbf{u}}{\partial t} + (\mathbf{u} \cdot \nabla)\mathbf{u} - \nu \Delta \mathbf{u} = \xi(t, \mathbf{x}),$$
$$t \geq 0, \quad \mathbf{x} \in \mathbf{T}^d = (\mathbb{R}/\mathbb{Z})^d.$$

The assumptions on the random forcing ξ are natural counterparts of the ones stated in Section 1.1; for simplicity, we assume that the initial condition and the forcing are C^∞-smooth in space.

Most importantly, we restrict ourselves to the case when $\mathbf{u}(0,\cdot)$ and ξ are gradients of scalar potentials. This ensures that $\mathbf{u}(t,\cdot)$ is also a gradient for all times. This assumption plays a crucial role, since it allows us to generalise Theorem 2.1.2 and its consequences, in particular for the upper estimates.

The statements of Theorem 2.3.4, Theorem 7.2.3 and Theorem 7.3.2 hold without any modification, excepted the change of notation due to the multidimensional setting. Defining Sobolev norms and the energy spectrum in the multidimensional setting is straightforward. The only definition for which one needs to be careful is the one for the structure function, where we now also need to average over all possible vectors of a given length.

Theorem 2.1.2 can be generalised without any difficulty in our potential setting. The counterparts of the quantity u_x^+ are the directional derivatives $\nabla_i u_i$. On the other hand, passing from this result to the upper estimates on Sobolev norms is more delicate. This is due to the fact that to get optimal estimates with respect to ν, we use the Gagliardo-Nirenberg inequality in a one-dimensional setting in order to take full advantage of the estimates on $\nabla_i u_i$. Passing from Sobolev norms on foliations of the torus consisting of circles with a rational slope to Sobolev norms on the torus itself requires elementary but tricky Hölder-type inequalities.

Finally, there is a major difficulty for the lower estimates. Namely, the energy balance is much more delicate to deal with than in dimension 1. Indeed, in the energy dissipation relation

$$\frac{d}{dt}\mathbb{E}|\mathbf{u}(t)|_2^2 = -2\nu\mathbb{E}\,\|\mathbf{u}(t)\|_1^2 - \mathbb{E}\,\langle \mathbf{u}(t),\ (\mathbf{u}\cdot\nabla)\mathbf{u}(t)\rangle_{L^2} + B_0,$$

the contribution of the nonlinearity does not vanish like in dimension 1. The idea of the proof is that with a strictly positive probability, during a long interval of time the noise mostly acts along one Fourier mode. Then the contribution of the nonlinearity can be neglected compared to that of the forcing.

The results above extend to the "kicked" setting, under some restrictions. Indeed, to prove the lower estimates we need some non-trivial assumptions on the support of the kick, again because of the additional trilinear term in the dissipation relation. For the same reason, the results in the unforced case are less general than in 1d.

For earlier results on the equation considered above, see the papers of Biryuk [**20**] (in the deterministic setting) and Brzezniak, Goldys and Neklyudov [**33**]. It was also studied by Gomes, Iturriaga, Khanin and Padilla [**70**] using Lagrangian techniques in the spirit of [**56**]. For the results on its inviscid version, see [**83**] and references therein.

10.4.3. Small viscosity complex Ginzburg–Landau equations and turbulence in the nonlinear Schrödinger equation. Since the Euler equation

$$\dot{u}(t,x) + (u\cdot\nabla)u + \nabla p = 0, \qquad \operatorname{div} u = 0, \tag{10.4.1}$$

is a Hamiltonian PDE with a quadratic nonlinearity, then the Navier–Stokes system (6.0.12), describing the hydrodynamical turbulence, is a homogeneous Hamiltonian PDE, damped by the small dissipation $\nu\Delta$, and driven by a random force, scaled in such a way that the solutions stay of order 1 when ν becomes very small. The Burgers equation (0.0.1) also is of this form since the Hopf equation $\dot{u} + uu_x = 0$ is Hamiltonian: it may be written as $\dot{u} + \frac{\partial}{\partial x}\nabla H(u) = 0$, where $H(u) = \frac{1}{6}\int u^3\,dx$.

Another popular equation from this class is obtained if we take for a Hamiltonian system the equation $\dot{u}(t,x) + i|u|^2u = 0$ (regarded as a zero-order PDE). Its damped/driven version is the following CGL (complex Ginzburg-Landau) equation

$$\dot{u} - \nu\Delta u + i|u|^2 u = \sqrt{\nu}\,\eta(t,x), \qquad u = u(t,x),\ \dim x = d \geq 1, \tag{10.4.2}$$

where the force $\eta(t,x)$ is smooth in x and white in time. Let us also consider a free version[2] of this equation:

$$\dot{u}(t,x) - \nu\Delta u + i|u|^2 u = 0. \tag{10.4.3}$$

Both equations and their variations have attracted significant attention in physical literature as models for hydrodynamical turbulence and as equations describing the "turbulence in the nonlinear Schrödinger equation", e.g. see [**6**, **25**]. A rigorous study of the small-viscosity equations (10.4.2) and (10.4.3), undertaken in late 1990's (see [**106**, **107**] for (10.4.3) and [**99**] for (10.4.2)) has motivated the study of the small-viscosity Burgers equation finalised in our book.

Now we present the above-mentioned results concerning the equations (10.4.2), (10.4.3) with small ν. Similarly to the rest of the book the solutions $u(t,x)$ are space-periodic. Moreover, now they are assumed to be odd in each coordinate x_j:

$$u(t,x) = u(t,\dots,x_j+2,\dots) = -u(t,\dots,-x_j,\dots), \quad j = 1,2,\dots d. \tag{10.4.4}$$

Such solutions vanish on the boundary of the cube of half-periods $[0,1]^d$, and have zero mean-value on the cube of periods $[0,2]^d$. [3]

The papers [**106**, **107**] established for eq. (10.4.3) some analogues of the results obtained above in Section 2 for the stochastic Burgers equation (in Section 10.1 we explain that similar estimates hold for solutions of the free Burgers equation). Namely, if $u(t,x)$ is a solution of (10.4.3), then the real-valued function $|u(t,x)|$ satisfies a parabolic inequality which admits the maximum principle. From here,

$$|u(t,\cdot)|_\infty \leq |u(0)|_\infty \min\left(1, C_1 e^{-\nu t K}\right), \qquad K = \pi^2 d/4.$$

This inequality, although weaker than Oleinik's maximum principle from Section 2.1, allows to estimate Sobolev norms of a solution u and get that

$$\|u(t)\|_m \leq C_m(u(0))\nu^{-m/2}e^{-\nu t K(m+1)}, \qquad m \in \mathbf{N} \tag{10.4.5}$$

(similar estimates were obtained in [**6**]). Next, an argument different from that in Section 2.3[4] allows to get in [**106**, **107**] lower bounds for the Sobolev and C^k–norms of a solution u, corresponding to a non-zero smooth initial data $u(0)$ of order 1. More precisely, it is proved in [**107**] that

$$\|u(t'_m)\|_{C^m} \geq c_m(u(0))\nu^{-m\gamma} \qquad m \in \mathbf{N},\ m \geq 3, \tag{10.4.6}$$

for any $\gamma < 1/3$ and a suitable $t'_m \in I_\nu := [0, \nu^{-1/3}]$ (estimates in [**106**] are a bit weaker). Similar lower bounds hold for the norms averaged along the time interval I_ν. Moreover, as it was later shown in [**21**], the lower bounds (10.4.6) remain true with $\gamma = 1/3$; also for smooth solutions u of equations (10.4.3) with complex "viscosity" ν (in this case in (10.4.6) ν should be replaced by $|\nu|$).

---

[2]Some results, presented below, apply to eq. (10.4.3), where ν is a small non-zero complex number such that $\Re\nu \geq 0$, $\Im\nu \leq 0$.

[3]Note that unlike for the Burgers and Navier–Stokes equations, the space mean-value is not preserved under the space-periodic dynamics of eq. (10.4.3).

[4]For solutions of (10.4.2) and (10.4.3) the balance of energy and the L_∞ bound do not allow to estimate $\|u(t)\|_1^2$ from below by a negative degree of ν, unlike for solutions to the Burgers equation.

Consider a solution u of (10.4.3) with a smooth initial data of order 1. Then as in Section 7.1 estimates (10.4.5) and (10.4.6) jointly imply that if it exists, the dissipative space scale $l_d(u)$ defined as in Section 7.1 (where $\langle\cdot\rangle$ stands for the time-averaging along intervals of length $\nu^{-1/3}$) satisfies [5]

$$l_d \in [\nu^{-1/3}, \nu^{-1/2}].$$

Since the exponents for ν^{-1} in estimates (10.4.5) and (10.4.6) are different, then more detailed small-scale analysis of solutions for (10.4.3), as in Sections 7.2-7.3, is not possible.

In [**99**] similar questions were studied for the stochastic equation (10.4.2), (10.4.4). Although the maximum principle cannot hold for solutions of the SPDE, it was shown that all moments of the L_∞ norms of solutions for (10.4.2), (10.4.4) are bounded by ν-independent constants. From here, similarly to the analysis of eq. (10.4.3), follow the estimates

$$\mathbb{E}\|u(t)\|_m^2 \leq C_m(u(0))\nu^{-m} \qquad m \in \mathbf{N}$$

(cf. (10.4.5)). The qualitative study of equation (10.4.2) with small ν was resumed in [**108**]. There the L_∞-estimates from [**99**] were sharpened and it was shown that suitable quadratic exponential moments of the L_∞-norms of solutions also are finite. From here, using an abstract theorem from [**103**], it was derived that equation (10.4.2) with a non-degenerate noise η is mixing. The research was continued in [**77**], where it is proved that for $m > 2$

$$\langle\langle \|u(t)\|_m^2 \rangle\rangle \geq C_m(\kappa_m)\nu^{-2m\kappa_m} \quad \text{for any} \quad \kappa_m > \kappa_m^0,$$

for a suitable $\kappa_m^0 > 0$. Here $\langle\langle\cdot\rangle\rangle$ stands for the averaging in ensemble and in time, along any time interval of length larger than $C\nu^{-1}$. For an integer $m \geq 3$ and any d a possible choice of κ_m^0 is an increasing function of m which is larger than 1/35 for all m and converges to 1/16 when m grows to infinity. This implies that for solutions of (10.4.2) the dissipative space scale satisfies

$$l_d \in [\nu^{-1/16}, \nu^{-1/2}].$$

[5] We mention that in [**6**] the upper bound $l_d \leq \nu^{-1/2}$ also was obtained and discussed.

CHAPTER 11

Appendices

11.1. Appendix A: Preliminaries from analysis.

Let $(X, \mathcal{F})$ be a measurable space, let μ be a σ-finite measure on it [1] and $(f_n)_{n\geq 0}$ be a sequence of integrable real-valued functions with finite integrals. The three theorems below are concerned with the integrability of $\lim_{n\to\infty} f_n$. They apply e.g. when $(X, \mathcal{F}, \mu)$ is a Banach space with the Borel sigma-algebra and a probability Borel measure on it, or when X is the set $\mathbf{N}$, given the counting measure.[2] In the latter case the theorems describe the passage to a limit for sums of infinite series. The proofs which we do not give may be found e.g. in [**51, 124, 139, 152**].

THEOREM 11.1.1. ***The Lebesgue dominated convergence theorem:*** *Assume that there exists a nonnegative μ-integrable function g such that μ-a.e. $|f_n| \leq g$ for all n and $\lim_{n\to+\infty} f_n = f$. Then f is μ-integrable and*

$$\lim_{n\to+\infty} \int_X f_n(x)d\mu(x) = \int_X f(x)d\mu(x).$$

THEOREM 11.1.2. ***Fatou's lemma:*** *Assume that $(f_n)_{n\geq 0}$ is a sequence of non-negative functions. Then we have:*

$$\int_X \liminf_{n\to+\infty} f_n(x)d\mu(x) \leq \liminf_{n\to+\infty} \int_X f_n(x)d\mu(x) \leq \infty.$$

THEOREM 11.1.3. ***The monotone convergence theorem:*** *Assume that the sequence of non-negative functions $(f_n(x))_{n\geq 0}$ is monotone increasing for μ- a.e. x. Then*

$$\lim_{n\to+\infty} \int_X f_n(x)d\mu(x) = \int_X \lim_{n\to+\infty} f_n(x)d\mu(x)$$

(where both sides of the equality may be infinite).

Theorems 11.1.4 and 11.1.6, given below, are corollaries of the monotone convergence theorem, frequently used in our book. There H^m and L_p stand for the Sobolev and Lebesgue spaces of functions on S^1, or for their subspaces formed by functions with zero mean-value.

THEOREM 11.1.4. *Let $m_1 \geq m_2$ and $\{\xi_n, n \geq 1\}$ be a sequence of r.v.'s in H^{m_1} such that $\mathbb{E}\|\xi_n\|_{m_1}^{\alpha} \leq K$ for all n and some $1 \leq \alpha < \infty$. Assume also that a.s. ξ_n converges in the space H^{m_2} to a random variable ξ. Then a.s. $\xi = \xi'$, where ξ' is a random variable in H^{m_1} and*

$$\mathbb{E}\|\xi'\|_{m_1}^{\alpha} \leq K. \tag{11.1.1}$$

[1] That is, $\mu(X) \leq \infty$ and $X = \cup X_n$, where $X_n \in \mathcal{F}$ for each n and $\mu(X_n) < \infty$.

[2] which assigns the measure one to each natural number.

PROOF. For $M \geq 1$ consider the r.v.'s $\|\xi_n\|^\alpha_{m_1,M}$, where

$$\|u\|^2_{m_1,M} = \sum_{|s|\leq M} (2\pi s)^{2m_1}|u_s|^2 \quad \text{for} \quad u = \sum u_s e_s.$$

By the first assumption $\mathbb{E}\|\xi_n\|^\alpha_{m_1,M} \leq K$ for all n, M. Now the second assumption and the Lebesgue dominated convergence theorem imply that for all M, $\mathbb{E}\|\xi\|^\alpha_{m_1,M}$ $\leq K$. Using the monotone convergence theorem we derive that $\mathbb{E}\|\xi\|^\alpha_{m_1} \leq K$. Consider the mapping $I : H^{m_2} \to H^{m_1}$ which is the identity on H^{m_1} and sends the complement to H^{m_1} to zero. It is measurable since H^{m_1} is a measurable subset of H^{m_2} (see (11.3.2)). Since $\mathbb{E}\|\xi\|^\alpha_{m_1} < \infty$, then $\xi \in H^{m_1}$ a.s., so the r.v. $\xi' := I(\xi)$ equals ξ a.s. This is a r.v. in H^{m_1} with the required properties. □

Applying this theorem we will not make a difference between the r.v. ξ and its modification ξ', and will regard ξ as a r.v. in H^{m_1}, taking infinite value on a null-set.

COROLLARY 11.1.5. *Let $m_1 \geq m_2$ and $\{\mu_n, n \geq 1\}$ be a sequence of measures in $\mathcal{P}(H^{m_1})$ such that $\langle \|u\|^p_{m_1}, \mu_n\rangle \leq K$ for all n and some $1 \leq p < \infty$. Assume also that $\mu_n \rightharpoonup \mu$ in $\mathcal{P}(H^{m_2})$. Then $\mu(H^{m_1}) = 1$, so μ may be regarded as a measure on H^{m_1}, and $\langle \|u\|^p_{m_1}, \mu\rangle \leq K$.*

PROOF. The result follows from the theorem above by applying Skorokhod's Theorem 1.7.2. □

THEOREM 11.1.6. *Let $1 \leq p, \alpha < \infty$ and $\{\xi_n, n \geq 1\}$ be a sequence of r.v.'s in L_p such that $\mathbb{E}|\xi_n|^\alpha_p \leq K$ for all n. Assume that a.s. $\xi^\omega_n \to \xi^\omega$ in L_1, where ξ is a r.v. in L_1. Then $\xi \in L_p$ a.s. and*

$$\mathbb{E}|\xi|^\alpha_p \leq K. \tag{11.1.2}$$

PROOF. For $M > 0$ consider the bounded continuous functional

$$\varphi^\alpha_M : L_1 \to \mathbb{R}, \quad u(x) \mapsto \big|\min(|u|, M)\big|^\alpha_p = \Big(\int \min(|u(x)|, M)^p dx\Big)^{\alpha/p} \tag{11.1.3}$$

(see Example 11.3.1 for a discussion of $\varphi^1_M =: \varphi_M$). Since the L_1-convergence $\xi^\omega_n \to \xi^\omega$ implies that $\xi^\omega_{n_j} \to \xi^\omega$ a.s. for a suitable sequence $n_j \to \infty$ and as $\varphi^\alpha_M \leq M^\alpha$ and $\mathbb{E}\varphi^\alpha_M(\xi_n) \leq \mathbb{E}|\xi_n|^\alpha_p \leq K$ for all n, then also $\mathbb{E}\varphi^\alpha_M(\xi) \leq K$ by the dominated convergence theorem, for any M. Sending in this inequality $M \to \infty$ and using the monotone convergence theorem we recover (11.1.2). By this estimate $\xi \in L_p$ a.s. □

Similarly to the above, using the Skorokhod representation theorem, we have the following corollary from this result:

COROLLARY 11.1.7. *Let $1 \leq p, \alpha < \infty$ and $\{\mu_n, n \geq 1\}$ be a sequence of measures in L_p such that $\langle |u|^\alpha_p, \mu_n\rangle \leq K$ for all n. Assume that $\mu_n \rightharpoonup \mu$ in $\mathcal{P}(L_1)$. Then $\mu(L_p) = 1$ and $\langle |u|^\alpha_p, \mu\rangle \leq K$.*

Since L_p is a Borel subset of L_1 (see Example 11.3.1), then in the corollary above μ may be regarded as a Borel measure on L_p. Similarly, in Theorem 11.1.6, ξ may be regarded as a r.v. in L_p.

Next we state a theorem about convergence to a limit for a family of uniformly integrable functions. The result below combines a sufficient condition for the uniform integrability and the corresponding limiting theorem:

THEOREM 11.1.8. ***(De La Vallée-Poussin):*** *Let $\mu(X) < \infty$. Assume that $f_n \to f$ μ–a.s. and*

$$\int_X G(|f_n(x)|)\, d\mu(x) \le C < \infty, \tag{11.1.4}$$

for some positive function G satisfying $G(t)/t \to \infty$ as $t \to \infty$. Then f is μ-integrable and

$$\lim_{n\to+\infty} \int_X f_n(x)d\mu(x) = \int_X f(x)d\mu(x).$$

A family of functions $f_n(x)$ satisfying (11.1.4) is *uniformly integrable.* See [**51**, **139**] for this notion.

Now we give a result from functional analysis, which may be found e.g. in [**152**]:

THEOREM 11.1.9. *Consider a sequence $(u_n)_{n\ge1}$ of elements of a Banach space X, converging weakly to u. Then we have:*

$$\|u\| \le \liminf_{n\to+\infty} \|u_n\|.$$

COROLLARY 11.1.10. *For any real numbers $m \le n$, if $u_j \to u$ as $j \to +\infty$ in H^m and $\|u_j\|_n \le C$ for all j, then $\|u\|_n \le C$. Therefore the closed H^n-ball $\overline{B_n^C}$ is a closed subset of H^m.*

Indeed, since the sequence $\{u_j\}$ is bounded in H^n, then its suitable subsequence weakly converges to a limit in H^n, which must equal u; so by the theorem above $\|u\|_n \le C$. The result can be easily proved directly:

PROBLEM 11.1.11. Derive the corollary from Fatou's lemma for series.

Now we recall Dini's theorem.

THEOREM 11.1.12. *If X is a metric compact and $\{f_n\} \subset C_b(X)$ is a monotonically increasing (or monotonically decreasing) sequence, pointwise converging to a function $f \in C_b(X)$, then the convergence is uniform. That is, $\|f_n - f\|_{C_b(X)} \to 0$.*

PROOF. For any $\varepsilon > 0$ consider the sets $U_n = \{x : |f_n(x) - f(x)| < \varepsilon\}$, $n \ge 1$. This is an increasing sequence of open domains, forming a cover of X. Since the latter is a compact set, there exists a finite subcover $\{U_{n_1}, \dots, U_{n_\varepsilon}\}$, $n_1 < \dots < n_\varepsilon$. But then $U_{n_\varepsilon} = X$, and $|f_n(x) - f(x)| < \varepsilon$ if $n \ge n_\varepsilon$. □

Finally we state the Riesz–Thorin inequality, also referred to as the log–convexity of Lebesgue norms.

THEOREM 11.1.13. *Consider the spaces $L_p := L_p(X, \mathcal{F}, \mu)$. Let $1 \le p_0 < p_1 \le \infty$, and for $0 < \theta < 1$ define p_θ by: $\frac{1}{p_\theta} = \frac{1-\theta}{p_0} + \frac{\theta}{p_1}$. Then*

$$|f|_{p_\theta} \le |f|_{p_0}^{1-\theta}\, |f|_{p_1}^{\theta}. \tag{11.1.5}$$

PROOF. Applying Hölder's inequality with $P = p_0/(p_\theta(1-\theta))$ and $Q = p_1/(p_\theta\theta)$ we get:

$$|f|_{p_\theta}^{p_\theta} = \int |f|^{p_\theta(1-\theta)}|f|^{p_\theta\theta} \le |f|_{p_0}^{p_\theta(1-\theta)}|f|_{p_1}^{p_\theta\theta}.$$

We have established (11.1.5). □

11.2. Appendix B: Spaces of functions.

Functions of x. In this book, unless otherwise stated, we work with functions of $x \in S^1$ with zero mean value. Accordingly, unless otherwise stated, we denote by L_p the subspace of $L_p(S^1)$ formed by functions of zero mean value, and denote by H^m, $m \in \mathbb{R}$, the L_2–Sobolev space of functions with zero mean value, given the homogeneous norms $\|\cdot\|_m$ as in Section 1.2. In particular, if $m \in \mathbf{N}$, then $\|u\|_m^2 = \int_{S^1} u^{(m)}(x)^2 dx$. Elements of spaces L_p and spaces $H^m(S^1)$ with $m \leq 1/2$ are equivalence classes, where functions coinciding a.e. are considered as the same object. If $m > 1/2$, then $H^m(S^1)$ is embedded in the space of continuous functions, and in this case a function in $H^m(S^1)$ is identified with the only continuous representative of the class of equivalence.

All the function spaces above are Hilbert or Banach; so they are metric spaces. A simple fact, given in the problem below, is useful to prove convergence in such spaces:

PROBLEM 11.2.1. Let X be a metric space. Show that a sequence $\{x_n\}$ in X converges to an $x \in X$ if and only if each subsequence $\{x_{n_j}\}$ contains itself a subsequence, converging to x.
Hint: Use that if $\{x_n\}$ does not converge to x, then there is $\delta > 0$ and a subsequence $\{x_{n_j}\}$ such that $\mathrm{dist}(x_{n_j}, x) \geq \delta$ for each j.

Functions of t and x. We usually regard functions of t, x as curves of t, valued in a certain space of functions of x, and we work with various functional spaces corresponding to the situations when these curves are continuous, or smooth, or only integrable with respect to t.

Continuous in t curves. For a Banach space B and $0 \leq T \leq \infty$ we denote by $C(0, T; B)$ the linear space of continuous mappings $[0, T] \to B$.[3] If $T < \infty$, we provide $C(0, T; B)$ with the supremum norm, which makes it a Banach space. On the other hand, for $T = \infty$, we endow $C(0, \infty; B)$ with the distance

$$\mathrm{dist}(u, v) = \sum_{n=1}^{+\infty} 2^{-n} \frac{\max_{0\leq t\leq n} \|u(t) - v(t)\|_B}{1 + \max_{0\leq t\leq n} \|u(t) - v(t)\|_B}, \tag{11.2.1}$$

which makes it a complete metric space. It is easy to see that a curve $u(t), t > 0$, belongs to $C(0, \infty; B)$ if and only if for each finite T its restriction to $[0, T]$ belongs to $C(0, T; B)$, and that a sequence of curves $(u_n(t), t \geq 0)$ converges to $(u(t), t > 0)$ if and only if for each finite T the restrictions of u_n's to $[0, T]$ converge to $u\,|_{[0,T]}$, in $C(0, T; B)$. So $C(0, \infty; B)$ may be regarded as a *Fréchet space* whose topology is defined by the countable family of seminorms

$$\sup_{0\leq t\leq k} \|u(t)\|_B, \qquad k = 1, 2, \dots, \tag{11.2.2}$$

see [**152**]. For any $0 < T < \infty$ the projection

$$C(0, \infty; B) \to C(0, T; B), \quad u \mapsto u\,|_{[0,T]}, \tag{11.2.3}$$

and the embedding

$$C(0, T; B) \to C(0, \infty; B),$$

[3] If $T = \infty$, then the interval $[0, T]$ becomes $[0, \infty)$.

which extends any curve $u(t), 0 \le t \le T$, by $u(T)$ for $t \ge T$, are continuous linear operators. Spaces $C(T_1, T_2; B)$, $T_1 < T_2 \le \infty$, are defined similarly.

If B is separable, the spaces $C(0,T;B)$, $T \le \infty$, are complete and separable.

In particular, if $T < \infty$ and $B = L_p$ or $B = H^m$, then $C(0,T;L_p)$ and $C(0,T;H^m)$ are Banach spaces whose elements are functions of t and x; they are separable with the exception of the spaces $C(0,T;L_\infty)$. For $0 < T \le \infty$ we abbreviate

$$C(0,T;H^m) = X_T^m\,, \quad X_T^0 = X_T\,.$$

The fundamental result concerning the spaces $C(0,T;B)$ is the *Arzelà–Ascoli theorem.* Below we give it in the form which may be found in [**115**].

THEOREM 11.2.2. ***The Arzelà-Ascoli theorem:*** *Let $T < \infty$, Z be a Banach space and let Y be a uniformly continuous subset of $C([0,T];Z)$, such that for every $t \in [0,T]$, the set $\{f(t),\ f \in Y\}$ is precompact in Z. Then Y is itself precompact in $C([0,T];Z)$.*

Measurable in t curves. We provide a Banach space B with the Borel sigma-algebra $\mathcal{B} = \mathcal{B}_B$, or with the sigma-algebra $C = C_B$ generated by cylindric sets $\{x \in B : x^*(x) \in [a,b]\}$, where x^* is an element of the dual space B^* and $a \le b$. Clearly $C \subset \mathcal{B}$. If the space B is separable, then $\mathcal{B}$ is generated by the functionals $x^* \in B^*$ (so is generated by cylindric sets as above), see the beginning of Appendix 11.3. In this case $C = \mathcal{B}$. For $0 < T \le \infty$ we provide the interval $[0,T]$ with the Borel sigma-algebra $\mathcal{B}$ and the Lebesgue measure dt. Then for $1 \le p \le \infty$ and a separable Banach space B we denote by $L_p(0,T;B)$ the L_p–space of measurable mappings $([0,T], \mathcal{L}, dt) \to (B, \mathcal{B}_B)$. This is a Banach space. When for a function $u(t,x)$, $t \ge 0$, $x \in S^1$, we consider $|u(t,\cdot)|_\infty$ (recall that the space $L_\infty(S^1)$ is not separable), we regard the latter as a function of t and not as an element of some space $L_p([0,T], L_\infty)$.

More generally, for a probability space $(\Omega, \mathcal{F}, \mathbf{P})$ and a separable Banach space B we use the L_p-spaces $L_p(\Omega; B)$, $1 \le p \le \infty$, formed by the classes of equivalence of measurable mappings $u : \Omega \to B$ with the norm $\left(\int \|u(\omega)\|_B^p \mathbf{P}(d\omega)\right)^{1/p}$ if $p < \infty$ and $\operatorname{ess\,sup} \|u(\omega)\|_B$ if $p = \infty$. These are Banach spaces; see [**24**, Section 6.10(vi)], [**52**, Section III.6] and [**117**, Section 2.1] for the spaces with $1 \le p < \infty$ and see the last two references for $p = \infty$.

We also use various spaces of maps $u : [0,T] \to B$ (B is a Banach space), defined in terms of $u(t)$ and its time-derivative $u_t(t)$, understood in the sense of generalised functions (e.g., see [**118, 119**]). Various embedding theorems relate these spaces. In particular, we use the two results below:

Let $B_1 \subset B_2 \subset B_3$ be separable Banach spaces (here $\subset$ indicates continuous embedding), such that the spaces B_1, B_3 are reflexive and the embedding of B_1 into B_2 is compact. For any $1 < p_1, p_3 < \infty$ and $0 < T < \infty$ consider the space

$$W = \{v \in L_{p_1}(0,T;B_1) : v_t \in L_{p_3}(0,T;B_3)\}\,.$$

Equipped with the norm $\|v\|_W = \|v\|_{L_{p_1}(0,T;B_1)} + \|v_t\|_{L_{p_3}(0,T;B_3)}$ it becomes a Banach space.

PROBLEM 11.2.3. Prove this assertion.

THEOREM 11.2.4. *The space W is compactly embedded in $L_{p_1}(0,T;B_2)$.*

For a proof of the theorem, see [**118**, Section 5.2].

Next for any real numbers $m_1 \le m_2$ consider the space

$$Z = \{u \in L_2(0,T;H^{m_2}) : u_t \in L_2(0,T;H^{m_1})\},$$

endowed with the Hilbert norm $\|u\|_Z^2 = \int_0^T \big(\|u\|_{m_2}^2 + \|u_t\|_{m_1}^2\big)dt$. Here H^m is the Sobolev space $H^m(S^1)$, or its subspace formed by functions with zero mean value.

PROBLEM 11.2.5. Prove that Z is a Hilbert space.

THEOREM 11.2.6. *The space Z is embedded in* $C(0,T;H^{(m_1+m_2)/2})$.

For a proof see [**119**, Theorem 3.1].

11.3. Appendix C: Measurable spaces, probability spaces and transition probabilities.

For definitions and basic facts concerning measurable spaces and probability spaces see [**51, 69, 124, 130, 139**]. In this Appendix we present some less known facts from this field, used in our book. Everywhere in the book, unless otherwise stated, a metric space X is assumed to be equipped with the Borel sigma-algebra $\mathcal{B}_X$, generated by all open subsets of X. Elements of $\mathcal{B}_X$ are called *Borel sets.* If X is a Polish space (i.e. a complete separable metric space), then $\mathcal{B}_X$ is generated by the countable family of open balls $B_X^{r_j}(v_l)$, $j,l \ge 1$, where the sequence $\{v_l, l \ge 1\}$, is dense in X and $\{r_j, j \ge 1\}$, is the set of positive rational numbers. Similarly, $\mathcal{B}_X$ is generated by the closed balls $\overline{B}_X^{r_j}(v_l)$; see [**22, 130**]. Each separable Banach space B is Polish. For such a space in view of the Hahn–Banach theorem every closed ball is an intersection of a countable family of closed affine half-spaces, so the Borel sigma-algebra of B is also generated by continuous linear functionals.

The extended real line $\mathbb{R} \cup \{\infty\}$ with the usual topology admits a metric which makes it a Polish space. Indeed, consider the mapping $A : \mathbb{R} \cup \{\infty\} \to (-\infty, \pi/2]$,

$$A(x) = \begin{cases} \arctan(x), & \text{if} \quad 0 \le x \le \infty, \\ x, \quad \text{if} \quad x \le 0, \end{cases}$$

and set $\mathrm{dist}(x,y) = |A(x) - A(y)|$. We equip $\mathbb{R} \cup \{\infty\}$ with the corresponding Borel sigma-algebra and understand accordingly the measurability of functions valued in $\mathbb{R} \cup \{\infty\}$. Cf. [**51**, Section 2.8], [**139**, Section II.2.2] and [**124**, Section II.2].

EXAMPLE 11.3.1. The separable Banach spaces $L_p = L_p(S^1)$, $1 \le p < \infty$ (where the norms are denoted $|\cdot|_p$) always are assumed to be equipped with the Borel sigma-algebras. For any $p < \infty$ and any $M > 0$ the bounded functional

$$\varphi_M : L_1 \to \mathbb{R}, \quad u(x) \mapsto \big|\min(|u|, M)\big|_p,$$

is continuous (and therefore measurable). Indeed, since the function $y \mapsto \min(|y|, M)$ is bounded and Lipschitz-continuous, then the functional $L_1 \ni u \mapsto \big|\min(|u|, M)\big|_p^p$ also is; so φ_M is continuous. By Theorem 11.1.3 for any $u \in L_1$, $|u|_p = \lim_{M\to\infty} \varphi_M(u) \le \infty$, so the functional

$$L_1 \to \mathbb{R} \cup \{\infty\}, \quad u \mapsto |u|_p,$$

is measurable (e.g., see [**51**, Theorem 4.2.2]). Accordingly, $L_p = \{u \in L_1 : |u|_p < \infty\}$, is a measurable subset of L_1, for any $p < \infty$.

If $m \le n$, the space H^n is continuously embedded in H^m:

$$i : H^n \mapsto H^m, \qquad m \le n. \tag{11.3.1}$$

So the pre-image mapping i^{-1} maps $\mathcal{B}_{H^m}$ to $\mathcal{B}_{H^n}$. But the direct map also sends measurable sets to measurable:

$$i(Q) \in \mathcal{B}(H^m) \quad \text{if} \quad Q \in \mathcal{B}(H^n). \tag{11.3.2}$$

Indeed, closed balls $\overline{B^r_{H^n}(v)}$, $v \in H^n, r > 0$, are closed in H^m by Corollary 11.1.10, so they belong to $\mathcal{B}(H^m)$. Therefore the sigma-algebra $i\Big(\mathcal{B}(H^n)\Big)$, generated by those balls, is a sub-algebra of $\mathcal{B}(H^m)$ and the result follows. By (11.3.2) the map of sigma-algebras naturally induced by the embedding i is surjective:

$$i^{-1}(\mathcal{B}_{H^m}) = \mathcal{B}_{H^n}. \tag{11.3.3}$$

Indeed, for any $Q \in \mathcal{B}(H^n)$, $i(Q) \in \mathcal{B}(H^m)$. But $i^{-1}(i(Q)) = Q$.

For a measurable space (M, C) we denote by $\mathcal{P}(M, C)$ the space of all probability measures on it, and for a metric space X we abbreviate $\mathcal{P}(X, \mathcal{B}_X)$ as $\mathcal{P}(X)$. A measurable mapping $F : (M, C_M) \to (Y, C_Y)$ induces a mapping of the space of measures

$$F\circ : \mathcal{P}(M, C_M) \to \mathcal{P}(Y, C_Y), \qquad m \longmapsto F \circ m,$$

where for $Q \in C_Y$, $(F \circ m)(Q) = m(F^{-1}(Q))$. The mapping $F\circ$ is linear in the sense that

$$F\circ(am_1 + (1-a)m_2) = aF\circ m_1 + (1-a)F\circ m_2 \quad \forall m_1, m_2 \in \mathcal{P}(M, C_M),\ \forall a \in [0, 1]. \tag{11.3.4}$$

If $(\Omega, \mathcal{F})$ is a measurable space and $\mathbb{P}$ is any element of $\mathcal{P}(\Omega, \mathcal{F})$, then the triplet $(\Omega, \mathcal{F}, \mathbb{P})$ is called a *probability space*. Elements of $\mathcal{F}$ are called *events*. Events of zero measure are called *null sets*.

A measurable mapping defined on a probability space $(\Omega, \mathcal{F}, \mathbb{P})$ is called a *random variable* (r.v.). The *law of a r.v.* $\xi : (\Omega, \mathcal{F}, \mathbf{P}) \to (X, \mathcal{B})$ is the measure

$$\mathcal{D}(\xi) = \xi \circ \mathbf{P} \in \mathcal{P}(X, \mathcal{B}).$$

So for a bounded continuous function f on X we have

$$\int_X f(x)\, \mathcal{D}(\xi)(dx) = \mathbb{E} f \circ \xi.$$

If X is a Polish space, then the space of measures $\mathcal{P}(X)$, given the Lipschitz-dual distance (1.7.4), again is a Polish space, see Theorem 1.7.7 and Problem 1.7.8. So for a measurable space (M, C) we can consider measurable maps

$$(M, C) \to (\mathcal{P}(X), \mathcal{B}), \quad m \mapsto \mu_m \in \mathcal{P}(X). \tag{11.3.5}$$

Such maps are called *random measures.*

THEOREM 11.3.2. *A mapping* (11.3.5) *is measurable (i.e. μ is a random measure) if and only if for each $Q \in \mathcal{B}$ the function $m \mapsto \mu_m(Q)$ is measurable, and if and only if for each $f \in C_b(X)$ the function $m \mapsto \langle \mu_m, f\rangle$ is measurable.*

For a proof see e.g. Corollary 5.1.9 and Theorem 5.8.5 in [**22**].

Consider again the embedding (11.3.1). It defines the mapping

$$i\circ : \mathcal{P}(H^n) \to \mathcal{P}(H^m). \tag{11.3.6}$$

PROPOSITION 11.3.3. *The image of the mapping* (11.3.6) *is exactly the collection M of measures in $\mathcal{P}(H^m)$ supported by the space H^n, $M = \{\mu \in \mathcal{P}(H^m) : \mu(H^n) = 1\}$.*

PROOF. Firstly note that for $\mu \in \mathcal{P}(H^m)$, by (11.3.2) the measure $\mu(H^n)$ is well-defined, and that obviously $i \circ (\mathcal{P}(H^n)) \subset M$. Next, for any $\mu \in M$ the function $\mathcal{B}_{H^n} \ni Q \mapsto \mu(i(Q))$ is well-defined by (11.3.2) and induces a sigma–additive measure. This measure is a probability one since $\mu(H^n) = 1$, and the assertion follows. □

In this book we often use the proposition above in the following setting. Let $\xi^\omega \in H^m$ be a r.v., denote $\mu = \mathcal{D}(\xi)$. Assume that $\mathbb{E}\|\xi\|_n^\alpha < \infty$ for some $n > m$ and $\alpha > 0$. Then $\mu(H^n) = 1$ and μ may be regarded as a Borel measure on H^n.

Denoting by $C_b(X)$ the space of bounded continuous functions on a Polish (i.e. a complete separable metric) space X, endowed with the uniform norm, we consider the natural pairing

$$\mathcal{P}(X) \times C_b(X) \to \mathbb{R}, \quad (\mu, f) \mapsto \int f(u)\, \mu(du) =: \langle \mu, f \rangle = \langle f, \mu \rangle.$$

It is continuous in the second variable, and also in the first one if the space of measures is given the weak topology (see Section 1.7).

This pairing is non-degenerate: if $\langle \mu, f \rangle = 0$ for all measures $\mu \in \mathcal{P}(X)$, then $f = 0$; if for some measures $\mu, \nu \in \mathcal{P}(X)$ we have $\langle \mu, f \rangle = \langle \nu, f \rangle$ for all functions $f \in C_b(X)$, then $\mu = \nu$.

The first assertion is obvious since $\mathcal{P}(X)$ contains all delta–measures on X. To prove the second let us take two different measures $\mu, \nu \in \mathcal{P}(X)$ and find for them a Borel set $Q \subset X$ such that $\mu(Q) > \nu(Q)$. Using Ulam's theorem (see in Section 1.7) we obtain a compact set $K \subset Q$ with the property that $\mu(K) > \nu(Q) \geq \nu(K)$. Next, for $n \in \mathbf{N}$ we define $f_n \in C_b(K)$ as

$$f_n(x) = \max\big(0, 1 - n\,\mathrm{dist}(x, K)\big), \quad 0 \leq f_n \leq 1.$$

The sequence $\{f_n\}$ monotonically decreases to χ_K. Applying the monotone convergence theorem to the functions $(1 - f_n)$ we get that $\lim\langle \mu, f_n \rangle = \mu(K)$, $\lim\langle \nu, f_n \rangle = \nu(K)$. So f_n's with $n \gg 1$ satisfy $\langle \mu, f_n \rangle > \langle \nu, f_n \rangle$: we have proved the second assertion.[4]

Every measure on a Polish space is a law of a suitable measurable mapping, defined on the unit interval:

THEOREM 11.3.4. *Consider a Polish space X and a measure $\mu \in \mathcal{P}(X)$. Let $(\Omega, \mathcal{F}, \mathbb{P})$ be the interval [0,1], given the Borel sigma-algebra and the Lebesgue measure. Then there exists a measurable mapping from Ω to X such that its law coincides with μ.*

This is a deep fact; for its proof see [**22**, Theorem 2.6.4], and [**130**, Assertion 26.6] (the imposed in [**130**] restriction that the measure μ is atomless may be easily removed). The assertion can be derived from a related result from the theory of measurable spaces. See in [**51**] Theorem 13.1.1 and a discussion of the long and involved history of its proof.

[4] Since the functions f_n are bounded and Lipschitz, then we have also established that the Lipschitz-dual distance (1.7.4) between the measures μ and ν is positive. Note that the latter fact follows from the much stronger assertion of Theorem 1.7.7, and again implies the property of measures $\mu \neq \nu$ which we have just proved.

Theorem 11.3.4 allows "to think about measures in terms of random variables", cf. the Skorokhod Theorem 1.7.2.

Let $(X_1, \mathcal{B}_1)$ and $(X_2, \mathcal{B}_2)$ be measurable spaces. Consider also the space $(X_1 \times X_2, \mathcal{B}_1 \otimes \mathcal{B}_2)$ (if X_1, X_2 are Polish spaces, endowed with their Borel sigma-algebras $\mathcal{B}_{X_1}, \mathcal{B}_{X_2}$, then $\mathcal{B}_1 \otimes \mathcal{B}_2 = \mathcal{B}_{X_1 \times X_2}$; see [**51**, Proposition 4.1.7]). Then for any measurable function f on $(X_1 \times X_2, \mathcal{B}_1 \otimes \mathcal{B}_2)$ and each $x_2 \in X_2$, the function $x_1 \mapsto f(x_1, x_2)$ is measurable on $(X_1, \mathcal{B}_1)$, and if in addition f is integrable, then *Fubini's theorem* holds (see [**51, 122, 130, 139**]):

THEOREM 11.3.5. *Let μ_1 and μ_2 be probability measures on $(X_1, \mathcal{B}_1)$ and $(X_2, \mathcal{B}_2)$. Then a measurable function f on $(X_1 \times X_2, \mathcal{B}_1 \otimes \mathcal{B}_2)$ is integrable with respect to $\mu_1 \times \mu_2$ if and only if*

i) $\int |f(x_1, x_2)| \, \mu_2(dx_2) < \infty, \quad \mu_1$*–a.s;*

ii) $\int \Big(\int |f(x_1, x_2)| \, \mu_2(dx_2) \Big) \mu_1(dx_1) < \infty$.

If i) and ii) hold, then

$$\int f(x_1, x_2)(\mu_1 \times \mu_2)(dx_1 \, dx_2) = \int \Big(\int f(x_1, x_2) \, \mu_2(dx_2) \Big) \mu_1(dx_1) .$$

For measurable spaces $(X_1, \mathcal{B}_1)$ and $(X_2, \mathcal{B}_2)$ as in the theorem above, a mapping

$$\lambda : X_1 \times \mathcal{B}_2 \to [0, 1] \tag{11.3.7}$$

is called a *transition probability* (see [**124, 130, 138**]) if

i) for each $B \in \mathcal{B}_2$ the function $x_1 \mapsto \lambda(x_1, B)$ is measurable on $(X_1, \mathcal{B}_1)$;

ii) for each fixed $x_1 \in X_1$, $\lambda(x_1, \cdot)$ is a probability measure on $(X_2, \mathcal{B}_2)$.

The transition probabilities as in (11.3.7) appear naturally when one studies stochastic systems and equations. The property *ii)* usually holds trivially. To verify *i)* the lemma below is useful:

LEMMA 11.3.6. *Let $E \in \mathcal{B}_1 \otimes \mathcal{B}_2$ and $\nu \in \mathcal{P}(X_2, \mathcal{B}_2)$. Then the relation*

$$X_1 \ni x_1 \mapsto \nu\{x_2 \in X_2 : (x_1, x_2) \in E\}$$

defines a measurable function on $(X_1, \mathcal{B}_1)$.

The result follows e.g. from [**130**, Assertion 35.9], where f is the characteristic function of E. The importance of the transition probabilities is due to the fact that they naturally define a linear (in the sense of relation (11.3.4)) mapping from $\mathcal{P}(X_1, \mathcal{B}_1)$ to $\mathcal{P}(X_2, \mathcal{B}_2)$:

LEMMA 11.3.7. *Let λ be a transition probability and μ be a probability measure on $(X_1, \mathcal{B}_1)$. Then the relation*

$$\nu(E) = \int \lambda(x_1, E) \, \mu(dx_1), \qquad E \in \mathcal{B}_2 , \tag{11.3.8}$$

defines a probability measure ν on $(X_2, \mathcal{B}_2)$.

PROOF. Among the properties which have to be checked in order to verify that ν is a probability measure, only the σ–additivity is not obvious. So, let $A_1, A_2, \dots$

be a collection of pairwise non-intersecting sets from $\mathcal{B}_2$. Then

$$\nu(\cup A_j) = \int \lambda(x_1, \cup A_j)\, \mu(dx_1) = \int \sum_j \lambda(x_1, A_j)\, \mu(dx_1)$$
$$= \sum_j \int \lambda(x_1, A_j)\, \mu(dx_1) = \sum_j \nu(A_j),$$

where the third equality follows from the monotone convergence theorem since

$$\int \sum_{j=1}^{\infty} \lambda(x_1, A_j)\, \mu(dx_1) = \int \lim_{M\to\infty} \sum_{j=1}^{M} \lambda(x_1, A_j)\, \mu(dx_1)$$
$$= \lim_{M\to\infty} \int \sum_{j=1}^{M} \lambda(x_1, A_j)\, \mu(dx_1) = \lim_{M\to\infty} \sum_{j=1}^{M} \nu(A_j).$$

So the σ–additivity follows. □

We denote the measure ν defined by (11.3.8), as

$$\nu = \int \lambda(x_1, \cdot)\, \mu(dx_1)\,.$$

See [**124**, Section III.2].

11.4. Appendix D: Kantorovich distance.

Let M be a Polish space and $\mathcal{P}_1(M)$ be the space of probability measures on it with finite first moment:

$$\mathcal{P}_1(M) = \{\mu \in \mathcal{P}(M) : \int_M \mathrm{dist}_M(x, x_0)\mu(dx) < \infty\},$$

where x_0 is a fixed point in M (the space will not change if we replace x_0 by another point in M). In his celebrated study of the optimal transport problem in the 1930's–1940's[5] Leonid Kantorovich introduced in $\mathcal{P}_1(M)$ the distance

$$\|\mu - \nu\|_{Kant} = \|\mu - \nu\|_{Kant,M} := \sup_{\mathrm{Lip} f \le 1} \big|\langle f, \mu\rangle - \langle f, \nu\rangle\big| \le \infty \tag{11.4.1}$$

(obviously, we may assume that $f(x_0) = 0$). This distance is greater or equal than the Lipschitz–dual distance (1.7.4), and it is easy to see that if M is bounded then the two are equivalent. However if M is unbounded, then the convergence in the Kantorovich distance is stronger than the weak convergence in $\mathcal{P}(M)$ (equivalent to the convergence in the Lipschitz–dual distance by Theorem 1.7.7).

EXAMPLE 11.4.1. In $\mathcal{P}(\mathbb{R})$ consider the sequence of measures $\mu_n = (1-n^{-1/2})\delta_0 + n^{-3/2}\chi_{[0,n]}dx$. Obviously $\mu_n \rightharpoonup \delta_0$ as $n \to \infty$, but choosing in (11.4.1) $f(x) = x$ we see that $\|\mu_n - \delta_0\|_{Kant} \ge \sqrt{n}/2$.

To bound (11.4.1) in terms of the Lipschitz-dual distance between μ and ν one has to control "the behaviour of the measures at infinity", see [**147**, Theorem 7.12]. In particular, the following is true:

[5]Which was a part of the research for which later Kantorovich received the Nobel price in economics.

PROPOSITION 11.4.2. *Let μ and ν be two measures in $\mathcal{P}_1(M)$ such that*

$$\langle dist_M(x,x_0)^\gamma, \mu\rangle,\ \langle dist_M(x,x_0)^\gamma, \nu\rangle \le K, \quad K \ge 1, \tag{11.4.2}$$

for some $\gamma > 1$. Then

$$\|\mu-\nu\|_{Kant} \le C_\gamma K^{1/\gamma}\big(\|\mu-\nu\|_L^*\big)^{1-1/\gamma}. \tag{11.4.3}$$

PROOF. For any 1-Lipschitz function f on M, vanishing at x_0, and any $R > 0$ denote $f^R = \min(|f|, R)\,\mathrm{sgn} f$. Then

$$|\langle f,\mu\rangle - \langle f,\nu\rangle| \le |\langle f^R,\mu\rangle - \langle f^R,\nu\rangle| + |\langle f - f^R,\mu\rangle| + |\langle f - f^R,\nu\rangle|.$$

Since $f^R = f$ if $|f| < R$ and $|f(x) - f^R(x)| \le |f(x)| \le \mathrm{dist}_M(x,x_0)$ for all $x \in M$, then by (11.4.2) the second term in the r.h.s. is bounded by

$$\langle \mathrm{dist}_M(x,x_0)\chi_{\mathrm{dist}_M(x,x_0)\ge R},\ \mu\rangle \le R^{1-\gamma}\langle \mathrm{dist}_M(x,x_0)^\gamma,\mu\rangle \le R^{1-\gamma}K,$$

and the third term is bounded by the same quantity. As f^R is a 1-Lipschitz function whose modulus is bounded by R, then the first term in the r.h.s. is at most $(1+R)\|\mu-\nu\|_L^*$. So

$$\|\mu-\nu\|_{Kant} \le (1+R)\|\mu-\nu\|_L^* + 2R^{1-\gamma}K \qquad \forall R > 0.$$

Minimising the r.h.s. in $R > 0$ we see that $\|\mu-\nu\|_{Kant} \le \|\mu-\nu\|_L^* + C_\gamma K^{1/\gamma}\big(\|\mu-\nu\|_L^*\big)^{1-1/\gamma}$. Since $\|\mu-\nu\|_L^* \le 2$ and $K \ge 1$, then (11.4.3) follows. □

In particular, if $\mu_n \rightharpoonup \mu$ is a weakly converging sequence in $\mathcal{P}(M)$ such that each μ_n satisfies (11.4.2) for a fixed $\gamma > 0$ and $K > 1$ (so μ_n belongs to $\mathcal{P}_1(M)$), then it is easy to see that μ also meets (11.4.2), and by (11.4.3) $\|\mu_n - \mu\|_{Kant} \to 0$.

If measures $\mu,\nu \in \mathcal{P}_1(M)$ are the laws of some r.v.'s ξ^μ and ξ^ν and f is a 1-Lipschitz function, then

$$|\langle f,\mu\rangle - \langle f,\nu\rangle| = \big|\mathbb{E}\big(f(\xi^\mu) - f(\xi^\nu)\big)\big| \le \mathbb{E}\,\mathrm{dist}_M(\xi^\mu,\xi^\nu).$$

So

$$\|\mu-\nu\|_{Kant} \le \mathbb{E}\,\mathrm{dist}_M(\xi^\mu,\xi^\nu). \tag{11.4.4}$$

A remarkable feature of the Kantorovich distance is that for given measures μ and ν the r.v.'s ξ^μ and ξ^ν may be chosen in such a way that the inequality above becomes an equality. This is the assertion of the celebrated *Kantorovich–Rubinstein theorem*:

THEOREM 11.4.3. *For any $\mu,\nu \in \mathcal{P}_1(M)$ there exist r.v.'s $\xi^\mu,\xi^\nu \in M$ such that $\mathcal{D}\xi^\mu = \mu$, $\mathcal{D}\xi^\nu = \nu$ and*

$$\|\mu-\nu\|_{Kant} = \mathbb{E}\ dist_M(\xi^\mu,\xi^\nu). \tag{11.4.5}$$

For a proof see [**51**, Theorem 11.8.2] and [**147**, Theorem 1.14].

11.5. Appendix E: Random processes and the Wiener process.

A *random process* is a random variable, valued in a space of curves in a certain measurable space. In this book we only deal with random processes with continuous trajectories in Banach spaces, equipped with the Borel sigma-algebras, which are defined on finite or infinite time intervals $[T_0,T]$, $T_0 < T \le \infty$ (we recall that if

$T = \infty$, then the interval becomes $[T_0, \infty)$). Accordingly, a *random process in a Banach space* B is a random variable

$$\xi : (\Omega, \mathcal{F}, \mathbb{P}) \to (C([T_0, T], B), \mathcal{B}),$$

where $\mathcal{B}$ is the Borel sigma-algebra in $C([T_0, T], B)$. See Section 11.2 for the spaces $C([T_0, T], B)$ with $T \leq \infty$. We usually write a process ξ as $\xi^\omega(t)$ or $\xi(t)$, where $(\xi^\omega(t), T_0 \leq t \leq T \leq \infty)$, are r.v.'s in B.

If the space B is formed by functions of $x \in S^1$, then a random process $\xi^\omega \in C([T_0, T_1], B)$ defines random variables $\xi^\omega(t, x)$. In this case we may call ξ a *random field* on $[T_0, T_1] \times S^1$.

Consider a random process $w^\omega(t) \in \mathbb{R}$, $0 \leq t < \infty$, such that:

(1) $w^\omega(0) = 0$ for every ω.
(2) For all $0 \leq t_1 \leq \ldots \leq t_N$ the random variable $\omega \longmapsto (w^\omega(t_1), \ldots, w^\omega(t_N)) \in \mathbb{R}^N$ is Gaussian, and for any $1 \leq j_1 < j_2 \leq j_3 < j_4 \leq N$ the Gaussian r.v.'s $w(t_{j_4}) - w(t_{j_3})$ and $w(t_{j_2}) - w(t_{j_1})$ are independent.
(3) For all $t \geq 0$, $\mathbb{E}[w^\omega(t)] = 0$, and for $0 \leq t_1 \leq t_2$, $\mathbb{E}(w^\omega(t_2) - w^\omega(t_1))^2 = t_2 - t_1$.

A random process $w(t)$ as above is called the *(standard) Wiener process*. Its law is the (standard) *Wiener measure*; this is a Borel measure on the Polish space $C(\mathbb{R}^+)$.

PROBLEM 11.5.1. Show that if w is a Wiener process and $\tilde{t} \geq 0, a > 0$, then $\widetilde{w}(t) = w(\tilde{t} + t) - w(\tilde{t})$ and $\hat{w}(t) = a^{-1/2} w(at)$ also are Wiener processes.

The restriction $w^\omega \mid_{[0,T]} =: w^{T,\omega}$, $0 < T < \infty$, is called the *Wiener process on* $[0, T]$; its law is called *the Wiener measure on* $[0, T]$. Denote by $\dot{X}$ the space of continuous functions on $[0, T]$ vanishing at zero. Consider any continuous functional $f \in (\dot{X})^*$ and the r.v. $f(w^\omega)$. Since the functionals $\phi \in (\dot{X})^*$ of the form

$$\phi(w(\cdot)) = \sum_{j=1}^{M} c_j w(t_j), \qquad 0 \leq t_1 < \cdots < t_M \leq T,$$

with any $M \in \mathbf{N}$ and any real numbers c_j are dense in $(\dot{X})^*$, then

$$f(w^\omega) = \lim_{n \to \infty} \phi_n(w^\omega) \quad a.s.,$$

for a suitable sequence of functionals as above. By the property (2), each $\phi_n(w^\omega)$ is a Gaussian random variable. So $f(w^\omega)$ also is Gaussian as an a.s.–limit of Gaussian random variables (e.g. see [**139**, II.13.5]).

DEFINITION 11.5.2. A Borel measure μ on a Banach space B is called Gaussian if for any $f \in B^*$ the measure $f \circ \mu \in \mathcal{P}(\mathbb{R})$ is Gaussian.

We have seen that for any $0 < T < \infty$ the law of w^T is a Gaussian measure on $C([0, T])$. Gaussian measures on a Banach space B possess many deep properties, e.g. see in [**23, 109**]. In particular, if μ is such a measure, then by Fernique's theorem [**109**, Theorem II.3.1] there exists $\kappa > 0$ such that

$$\int_B e^{\kappa \|u\|_B^2} \mu(du) < \infty. \tag{11.5.1}$$

It is well-known that one can build a countable family $w_1, w_2, \ldots$ of independent standard Wiener processes, taking for $(\Omega, \mathcal{F}, \mathbb{P})$ the segement $[0, 1]$, equipped with the Borel sigma-algebra and the Lebesgue measure; e.g. see [**80**].

Let $w_1,\dots,w_N$ be independent standard Wiener processes and $c_1,\dots,c_N$ be real numbers. Doob's moment inequality [6] applied to the process $|\mathbf{w}(t)|=(c_1^2w_1(t)^2+\dots+c_N^2w_N(t)^2)^{1/2}$ tells us that

$$\mathbb{E}\Big(\sup_{0\le t\le T}|\mathbf{w}(t)|\Big)^p\le\Big(\frac{p}{p-1}\Big)^p\,\mathbb{E}[|\mathbf{w}(T)|^p],\quad\forall p>1. \tag{11.5.2}$$

Next, if $\sigma>0$ is such that $\mathbb{E}e^{\sigma|\mathbf{w}(T)|^2}<\infty$, then denoting $X_T=\sup_{0\le t\le T}|\mathbf{w}(t)|$ and using (11.5.2) we have:

$$\mathbb{E}e^{\sigma X_T^2}=\sum_{p=0}^{\infty}\frac{1}{p!}\sigma^p\mathbb{E}X_T^{2p}\le1+\sum_{p=1}^{\infty}\frac{1}{p!}\sigma^p\Big(\frac{2p}{2p-1}\Big)^{2p}\,\mathbb{E}\,|\mathbf{w}(T)|^{2p}.$$

Since $\max_{x\ge2}(x/(x-1))^x=4$, then

$$\mathbb{E}e^{\sigma\sup_{0\le t\le T}|\mathbf{w}(t)|^2}=\mathbb{E}\sup_{0\le t\le T}e^{\sigma|\mathbf{w}(t)|^2}\le4\mathbb{E}e^{\sigma|\mathbf{w}(T)|^2}-3. \tag{11.5.3}$$

11.6. Appendix F: Filtered probability spaces and Markov processes.

Consider a probability space $(\Omega,\mathcal{F},\mathbf{P})$ and assume that it is endowed with a family of σ–algebras $\{\mathcal{F}_t\subset\mathcal{F},t\ge0\}$, which is non-decreasing with time. The quadruple $(\Omega,\mathcal{F},\mathcal{F}_t,\mathbf{P})$ is called a *filtered probability space*, and the triple $(\Omega,\mathcal{F},\mathcal{F}_t)$ – a *filtered measurable space*. A random process $x(t)\in X$, $t\ge0$, where X is a complete metric space, provided with the Borel σ–algebra $\mathcal{B}_X$ (as always in this book, all trajectories $x^\omega(t)$ are continuous) is called *adapted to* $\mathcal{F}_t$ if $x(t)$ is $\mathcal{F}_t$–measurable for each $t\ge0$.

EXAMPLE 11.6.1. Let $B_{m_*}<\infty$ for some $m_*>m\in\mathbf{N}$, and $\xi(t)\in H$ be the process (1.1.6) as in Theorem 1.2.3. Let $\{\mathcal{F}_t,t\ge0\}$ be the filtration generated by the process ξ. That is, $\mathcal{F}_t$, $t\ge0$, is the σ–algebra generated by the r.v. $\xi^\omega\mid_{[0,t]}\in\dot{X}_t^{m_*}$. Now assume that u_0 is a non-random vector, and let $u^\omega(t)=u(t;u_0)$ be the (strong) solution of (1.1.13) constructed in Theorem 1.4.2. By Corollary 1.4.3 the process $u^\omega(t)$ is adapted to $\mathcal{F}_t$.

Let $u^\omega(t)=u(t;u_0)$ be the solution as in the example above. Recalling the notation (1.4.7) we have the following fundamental result concerning the conditional expectation of $u^\omega(t)$ with respect to $\mathcal{F}_t$ (the classical notion of conditional expectation can be found for example in [**69**, **80**]; it is not used in our book apart from this appendix):

LEMMA 11.6.2. *If $t\le s$ and $f\in C_b(H^m,\mathbb{R})$, then $\mathbb{E}\big[f(u(t))\mid\mathcal{F}_s\big]=f(u(t))$, while for $t\ge s$ we have*

$$\mathbb{E}\big[f(u(t))\mid\mathcal{F}_s\big]=\langle\Sigma_{t-s}u(s),f\rangle=(S_{t-s}f)(u(s)). \tag{11.6.1}$$

PROOF. The first assertion is obvious since $u(t)$ is $\mathcal{F}_t$–measurable and therefore $\mathcal{F}_s$–measurable; see the example above. Now let $t\ge s$. Since $\mathcal{F}_s$ is generated by the r.v. $\xi\mid_{[0,s]}$, then each $\mathcal{F}_s$–measurable function is a Borel function of $\xi\mid_{[0,s]}$ (this is the Doob–Dynkin lemma, see [**124**, Section I, Theorem T18]). So we should check that for any bounded Borel function Φ on $\dot{X}_s$, we have

$$\mathbb{E}\big[f(u(t))\Phi(\xi\mid_{[0,s]})\big]=\mathbb{E}\big[S_{t-s}f(u(s))\Phi(\xi\mid_{[0,s]})\big].$$

[6] See in [**80**] Theorem 1.3.8, Proposition 1.3.6 and Problem 1.3.7

Since the solution $u(t)$ is given by (1.4.3), then the expectation in the l.h.s. depends on the law of the process ξ rather than on the process itself. So, as in Proposition 1.6.5 we can replace ξ by another process with the same law. Choosing for ξ the process of the form (1.1.6) with $\beta_s(t)$ defined by (1.6.11) and repeating the calculation (1.6.12) with $\Phi(u\,|_{[0,s]})$ replaced by $\Phi(\xi\,|_{[0,s]})$, we get the assertion. □

Let X be a Polish space. A *Markov family of random processes* in X (with continuous trajectories, as usual in this book), or simply a *Markov process* in X, is the following collection of objects (see [**80**]):[7]

– a filtered measurable space $(\Omega, \mathcal{F}, \mathcal{F}_t)$;

– a family of probability measures $\{\mathbf{P}_v, v \in X\}$ on $(\Omega, \mathcal{F})$ such that the function $v \mapsto \mathbf{P}_v(Q)$ is Borel-measurable for each $Q \in \mathcal{F}$;

– a random process $u(t) \in X$, $t \geq 0$, adapted to the filtration and such that

$$\begin{aligned} &\mathbf{P}_v\{u(0) = v\} = 1 \quad \forall v, \\ &\mathbf{P}_v\big[u(t+s) \in \Gamma \mid \mathcal{F}_s\big] = P_t(u(s), \Gamma), \quad \mathbf{P}_v\text{–a.s.}, \end{aligned} \tag{11.6.2}$$

where P_t is a transition probability on X, $P_t(v, \cdot) = \mathcal{D}^{\mathbf{P}_v} u(t)$.

Under the assumptions of Example 11.6.1 the solutions $u(t; u_0)$ define a Markov family of processes in H^m. To see this, let us take the following filtered measurable space:

$$(\tilde{\Omega}, \tilde{\mathcal{F}}, \tilde{\mathcal{F}}_t), \quad \tilde{\Omega} = H^m \times \Omega, \quad \tilde{\mathcal{F}} = \mathcal{B}_{H^m} \otimes \mathcal{F}, \quad \tilde{\mathcal{F}}_t = \mathcal{B}_{H^m} \otimes \mathcal{F}_t.$$

For the family of measures we take $\{\mathbf{P}_v = \delta_v \times \mathbf{P}, v \in H^m\}$, and define the random process

$$t \mapsto u^{\tilde{\omega}}(t) = u^{\omega}(t; v) \quad \text{for} \quad \tilde{\omega} = (v, \omega). \tag{11.6.3}$$

The property (11.6.2) follows from (11.6.1) and the rest is obvious.

Similarly, if $B_{1_*} < \infty$ for some $1_* > 1$, then the generalised solutions $u(t, x)$ of the problem (1.1.13) with $u_0 \in L_1$ define a Markov process in L_1. Indeed, let us take for a filtered measurable space $(\tilde{\Omega}, \tilde{\mathcal{F}}, \tilde{\mathcal{F}}_t)$ and a family of measures $\mathbf{P}_v$, $v \in L_1$, the same objects as above, but with the space H^m replaced by L_1, and define $u^{\tilde{\omega}}(t)$ as in (11.6.3). To prove (11.6.2), for any $u_0 \in L_1$ let us choose an H^1–sequence converging to u_0 in L_1, and denote $u^{n\omega}(t) = u(t; u_0^n)$. Let $f \in C_b(L_1)$. For u replaced by u^n relation (11.6.1) holds by Lemma 11.6.2. Passing to the limit as $n \to \infty$ using the dominated convergence theorem we recover (11.6.1) for u. This implies (11.6.2), so the $u^{\tilde{\omega}}(t)$ which we have just defined is a Markov family of processes in L_1.

11.7. Appendix G: Stopping and hitting times and the strong Markov property.

For $m \geq 1$ such that $B_{m_*} < \infty$ for some $m_* > m$ consider a solution $u(t) \in H^m$ of eq. (B). Let Q be a closed subset of H^m and T a positive constant. Denote by τ the random time

$$\tau^{\omega}_{Q,T} = \inf\{0 \leq t \leq T : u^{\omega}(t) \in Q\}.$$

Here $\tau = T$ if $u^{\omega}(t) \notin Q$ for $t \leq T$. We call τ the *hitting time* for the closed set Q; see Figure 11.7.1. This is a special case of a *stopping time*, see [**80, 128**]. Note that $\tau = 0$ if $Q = H^m$, and $\tau = T$ if $Q = \varnothing$.

[7] our definition is a bit less general than that in [**80**], where the measurability of the function $P_v(Q)$ is understood in a wider sense.

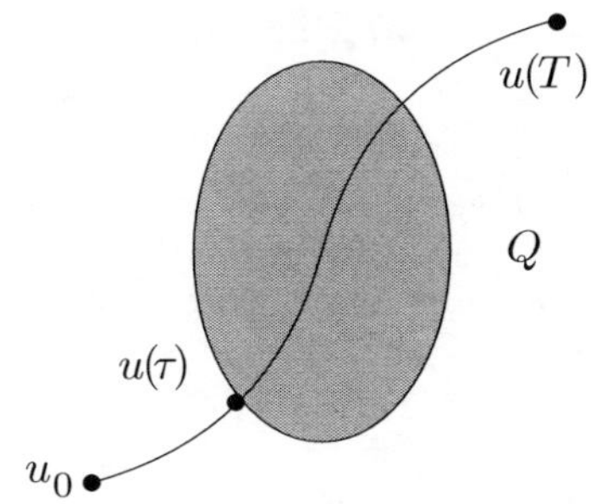

FIGURE 11.7.1. Hitting time τ corresponding to a set $Q \in H^m$.

The importance of the two concepts above is due to the *strong Markov property*, which we define now for a solution $u(t) = u(t; u_0)$, where u_0 is a constant vector in H^m, or a random variable in H^m (independent from the force η):

(11.7.1)

Let $T_1 \geq T$ and $f \in C_b(H^m)$. Then $\mathbb{E}[f(u(T_1))] = \mathbb{E}[(S_{T_1-\tau^\omega_{Q,T}} f)(u(\tau^\omega_{Q,T}))]$.

Here $S_{T_1-\tau^\omega_{Q,T}} f(v)$ is defined by the relation (1.6.4), where we replaced t by the r.v. $T_1 - \tau^\omega_{Q,T}$. See [**69**, **80**].

Note that if $\tau = T$ and $u(0)$ is a non-random vector in H^m, then (11.7.1) coincides with the Kolmogorov-Chapman property (1.6.9).

11.8. Appendix H: Stochastic differential equations and Ito's formula.

In this appendix we present a very brief account of stochastic differential equations (SDEs) in $\mathbb{R}^n$ and of Ito's formula for their solutions, sufficient for the purposes of our book. For a good introduction to the non-abridged theory of SDEs and to Ito's formula we refer the reader to [**69**, **84**, **98**], while for a sufficiently complete account of the modern state of affairs in the two fields – to [**80**].

In this book we only use SDEs with diagonal additive random forces. That is, equations of the form

$$\dot{x}_j(t) - f_j(t, x(t)) = \sum_{j=1}^n b_j \dot{\beta}_j(t), \quad 1 \leq j \leq n, \quad x = (x_1, \ldots, x_n) \in \mathbb{R}^n, \tag{11.8.1}$$

and the initial-value problems for them with prescribed initial data

$$x_j(0) = x_{0j}, \qquad 1 \leq j \leq n. \tag{11.8.2}$$

Here $t \geq 0$, $f_j : \mathbb{R}_+ \times \mathbb{R}^n \to \mathbb{R}^n$, $1 \leq j \leq n$, are continuous locally Lipschitz mappings, b_j's are any real numbers (we do not exclude that all of them vanish), $\beta_j(t)$'s – standard independent Wiener processes, and x_0 is a r.v. in $\mathbb{R}^n$, independent from those processes. The derivatives $\dot{\beta}_j(t)$ are defined only in the sense of distributions, so as in the Section 1.1 we adopt the convention that a random process $x^\omega(t)$, $t \in [0, T]$, $0 < T \leq \infty$, is a strong solution of the problem (11.8.1), (11.8.2) if

$$x^\omega_j(t) = x^\omega_{0j} + \int_0^t f_j(s, x^\omega(s))\, ds + \sum_{j=1}^n b_j \beta^\omega_j(t), \qquad 0 \leq t \leq T, \tag{11.8.3}$$

for a.e. ω. Below in this appendix we will drop the epithet "strong" and only use the term "solutions".

A solution of (11.8.1), (11.8.2), if it exists, is unique in the sense that any two solutions $x^1(t)$ and $x^2(t)$ must coincide a.s. Indeed, for a.e. ω the difference $x^1 - x^2$ satisfies a deterministic equation with zero r.h.s. and zero initial data, so it vanishes identically for the same reason as in the theory of ordinary differential equations. As in the deterministic case, a solution for (11.8.1), (11.8.2) may not exist, but it does exist if the vector field f is globally Lipschitz:

THEOREM 11.8.1. *If for any $T > 0$ the mapping $x \mapsto f(t,x)$ is Lipschitz on $\mathbb{R}^n$ with a Lipschitz constant, independent from $0 \le t \le T$, then the problem has a unique solution $x(t)$.*

PROOF. Let us denote $\beta(t) = (b_j\beta_j(t), 1 \le j \le n) \in \mathbb{R}^n$ and, as in Section 1.3, write a solution $x(t)$ as $x(t) = \beta(t) + y(t)$. Then

$$\dot{y}(t) = f(t, \beta(t) + y(t)), \qquad y(0) = x_0. \tag{11.8.4}$$

This is an ODE corresponding to a vector field which is continuous in time and Lipschitz in space. So for every ω it has a unique solution $y^\omega(t)$, and the corresponding $x^\omega(t)$ solves (11.8.1), (11.8.2). Since (for a fixed ω) $y(\cdot)$ continuously depends on u_0 and $\beta(\cdot)$, then $x(\cdot)$ is a measurable function of ω, i.e. a random process. □

The Lipschitz property restriction, imposed in this theorem on the equation, is strong and is violated too often. But if f is only locally Lipschitz in x, then still the problem (11.8.1), (11.8.2) is soluble if eq. (11.8.1) has a Lyapunov function. The result bellow is taken from [**84**, Theorem 3.5]:

THEOREM 11.8.2. *Assume that there exist a C^2–smooth function $V(x)$ and a constant $c > 0$ such that*

$$\frac{1}{2}\sum_j b_j^2 \frac{\partial^2}{\partial x_j^2} V(x) + \sum_j f_j(t,x)\frac{\partial}{\partial x_j} V(x) \le cV(x) \quad \forall x, \ \forall t \ge 0, \tag{11.8.5}$$

and

$$\inf_{|x|\ge R} V(x) \to \infty \quad \textit{as} \quad R \to \infty. \tag{11.8.6}$$

Then the problem (11.8.1), (11.8.2) *has a unique solution $x(t)$, and*

$$\mathbb{E}V(x(t)) \le e^{ct}\mathbb{E}V(x_0) \tag{11.8.7}$$

(the r.h.s. may be infinite).

The result remains true if all b_j vanish. Then the theorem presents the classical method of Lyapunov's function for the ODE $(11.8.1)|_{b_j=0\,\forall j}$.

EXAMPLE 11.8.3. Assume that the vector field f satisfies

$$f(t,x)\cdot x \le -\alpha\|x\|^2 \qquad \forall t \ge 0, \, \forall x, \tag{11.8.8}$$

with some $\alpha > 0$. Let us take $V(x) = e^{\sigma\|x\|^2}$, where $0 < \sigma \le \alpha/b_{max}^2$, where $b_{max} = \max_{1\le j\le n}|b_j|$. Then

$$\sum_j f_j(t,x)\frac{\partial}{\partial x_j}V(x) = e^{\sigma\|x\|^2}2\sigma x\cdot f(x) \le -V(x)2\sigma\alpha\|x\|^2,$$

and

$$\frac{1}{2}\sum_j b_j^2 \frac{\partial^2}{\partial x_j^2} V(x) \le \sigma V(x)\big(\sum b_j^2 + 2\sigma (b_{max})^2 \|x\|^2\big).$$

So the l.h.s. of (11.8.5) is bounded by

$$\sigma V(x)\big(- 2\alpha \|x\|^2 + 2\sigma (b_{max})^2 \|x\|^2 + \sum b_j^2\big).$$

Since $\sigma(b_{max})^2 \le \alpha$, then (11.8.5) holds with $c = \sigma \sum b_j^2$. Relation (11.8.6) obviously is valid, so the last theorem implies

(11.8.9)
$$\mathbb{E} e^{\sigma \|x(t)\|^2} \le e^{t\sigma \sum b_j^2}\, \mathbb{E} e^{\sigma \|x_0\|^2}.$$

A powerful tool to study equation (11.8.1) is given by Ito's formula. Below we give a relation, obtained by taking its expectation. Sometimes it is called *the weak Ito formula*, but we prefer to drop the epithet "weak". We state the result in application to a solution of (11.8.1):

THEOREM 11.8.4. *Let $x^\omega(t)$ be a solution of (11.8.1) for $0 \le t \le T$ and $F(x)$ be a C^2-smooth function. Assume that*
(11.8.10)
$$\int_0^T \mathbb{E}\left[|\nabla F(x^\omega(t)) \cdot f(t, x^\omega(t))| + \|\nabla F(x^\omega(t))\| + |\sum_j b_j^2 \frac{\partial^2}{\partial x_j^2} F(x^\omega(t))| \right]^2 dt < \infty.$$

Then for $0 \le t \le T$ we have
(11.8.11)
$$\frac{\partial}{\partial t} \mathbb{E} F(x(t)) = \mathbb{E}\left(\nabla F(x(t)) \cdot f(t, x(t))\right) + \tfrac{1}{2}\mathbb{E} \sum_j b_j^2 \frac{\partial^2}{\partial x_j^2} F(x(t)), \quad \textit{a.s. in } t.$$

As in similar situations above, (11.8.11) should be understood in the integrated sense: $\mathbb{E}F(x(t)) - \mathbb{E}F(x(0))$ equals to the integral of the term in the r.h.s., for all $t \ge 0$. Since in view of (11.8.10) the expectation $\mathbb{E}F(x(t))$ is an absolutely continuous function of t, then by the latter relation (11.8.11) holds for a.e. t. As before, the assertion remains true if (11.8.1) is an ODE, i.e. if all b_j vanish, and $u(0)$ is a constant vector or a random variable (in the former case the averaging sign may be dropped).

Two important comments have to be made:

i) as it was said, the relation (11.8.11) is just the expected value of the complete Ito formula; the latter contains a stochastic term, vanishing when taking the expectation.

ii) The formula applies to solutions of equations with much more general stochastic terms than in (11.8.1). Moreover, it applies to a class of random processes which is significantly larger than the class of solutions for various SPDEs, see [**80**].

PROBLEM* 11.8.5. Prove the theorem above by applying the complete Ito formula and verifying that relation (11.8.10) implies that the expectation of the stochastic term in the latter vanishes. In a convenient form the complete Ito formula may be found in [**69, 98**].

PROPOSITION 11.8.6. *Consider a vector field* $f(t,x)$ *satisfying* (11.8.8) *and a* C^2*–function* $F(x)$. *Assume that* $\mathbb{E}e^{\sigma\|x_0\|^2} < \infty$ *for some* $0 < \sigma \leq \alpha/b^2_{max}$ *and* (11.8.12)

$$|\nabla F(x) \cdot f(t,x)| + \|\nabla F(x)\| + |\sum_j b_j^2 \frac{\partial^2}{\partial x_j^2} F(x)| \leq C_T e^{\sigma\|x\|^2} \quad \forall\, x,\ \forall\, t \in [0,T],$$

with a suitable constant $C_T > 0$. *Then a solution* $x(t)$ *of the problem* (11.8.1), (11.8.2) *satisfies* (11.8.10) *for* $0 \leq t \leq T$. *So* $x(t)$ *also meets* (11.8.11).

PROBLEM 11.8.7. Prove the proposition above, using the result of Example 11.8.3.

CHAPTER 12

Solutions for selected exercises

SOLUTION 1.3.2. *Assume firstly that* $u_0 = 0$. *Following the proof of Theorem 1.2.1, consider the random curves* $\xi_N(t) \in H^{(N)}$ *and denote by* $v^{N,\omega}$ *a solution of* (1.3.1) *with* $u_0 = 0$ *and* $\xi = \xi^{N,\omega}$. *By Lemma 1.3.1 with* $m := m+1$ *this* $v^{N,\omega}$ *continuously depends on* $\xi^{N,\omega}$, *so the mapping* $\Sigma^N : \omega \mapsto v^{N,\omega} \in X_T^m$ *is measurable. Using the complete Ito formula we find that* $\Sigma^N \in \mathbb{L}_2^{T,m} = L_2(\Omega; \dot{X}_T^m)$ *and that the r.v.'s* $\Sigma^N, N \geq 1$, *form a fundamental system in* $\mathbb{L}_2^{T,m}$. *Let us denote the limiting element by* $\Sigma(0,\omega)$.

By linearity, for any $u_0 \in H^m$ *it remains to define* $\Sigma(u_0,\omega)$ *as* $\Sigma(0,\omega) + e^{-\nu\Delta t}u_0$, *and define the full-measure set* $\Omega \setminus Q_m$ *as the set of convergence of any sub-sequence* $\{\Sigma^{N_j}\}$, *which converges in* X_T^m *a.s. For details of this argument see* [**103**, Section 2.4.1] *and* [**43**, Section 5].

SOLUTION 1.3.5. *Using the boundedness of the derivatives* $f^{(k)}$ *with* $k \geq 2$ *and Hölder's inequality we get:*

$$N_m^{0f}(h)^2 \leq C(m)\sum_{k=1}^{m} \sum_{\beta_1+\dots+\beta_k=m} \int \left(f^{(k)}(h)h^{(\beta_1)}\dots h^{(\beta_k)}\right)^2 dx$$

$$\leq C(m)\int \left(hh^{(m)}\right)^2 dx + C(m)\int \sum_{k=2}^{m} \sum_{\beta_1+\dots+\beta_k=m} \left(h^{(\beta_1)}\dots h^{(\beta_k)}\right)^2 dx$$

$$\leq C(m)|h|_\infty^2 \|h\|_m^2 + C(m)\sum_{k=2}^{m} \sum_{\beta_1+\dots+\beta_k=m} \prod_{i=1}^{k} |h|^2_{\beta_i,2m/\beta_i}.$$

The first term in the r.h.s. and the sum in β_j *with* $k=2$ *may be estimated exactly as when proving Lemma 1.3.4, and they are bounded by the r.h.s.'s of* (1.3.10) *or of* (1.3.11). *In particular, this fact establishes the assertions* (1.3.17) *and* (1.3.18) *if* $m=1$ *or* $m=2$. *Consider now the sum in* β_j *with* $k \geq 3$. *Due to* (1.3.11),

$$|h|_{\beta_i,2m/\beta_i} \leq C\|h\|_{m+1}^{\theta_i}|h|_{1,1}^{1-\theta_i}, \quad \theta_i = \frac{\beta_i}{m}\frac{2m-1}{2m+1}, \quad 1 \leq i \leq k.$$

Since $\sum_1^k \beta_i = m$, *then* $\sum_1^k \theta_i = \frac{2m-1}{2m+1}$ *and so*

$$\prod_{i=1}^{k} |h|_{\beta_i,2m/\beta_i} \leq C\|h\|_{m+1}^{\frac{2m-1}{2m+1}} |h|_{1,1}^{\frac{2m+3}{2m+1}} |h|_{1,1}^{k-2}.$$

Since this holds true for any $3 \leq k \leq m$, *then we got* (1.3.17) *with* $m \geq 3$.

Similarly by (1.3.10), *for* $k \geq 3$

$$|h|_{\beta_i,2m/\beta_i} \leq C\|h\|_{m+1}^{\gamma_i}\|h\|^{1-\gamma_i}, \quad \gamma_i = \frac{\beta_i(2m-1)+m}{2m(m+1)}, \quad 1 \leq i \leq k.$$

Now $\sum_1^k \gamma_i = \frac{2m+k-1}{2m+2}$, *so* $\prod_{i=1}^k |h|_{\beta_i,2m/\beta_i}$ *is bounded by*

$$C\|h\|_{m+1}^{\frac{2m+k-1}{2m+2}} \|h\|^{k-\frac{2m+k-1}{2m+2}} = C\|h\|_{m+1}^{\frac{2m+1}{2m+2}} \|h\|^{\frac{2m+3}{2m+2}} \big(\|h\|_{m+1}^{\frac{1}{2m+2}} \|h\|^{\frac{2m+1}{2m+2}}\big)^{k-2}.$$

This implies (1.3.18) *for* $m \geq 3$.

SOLUTION 1.3.7. *Consider the mapping* $S^1 \to S^1$, $y \mapsto x = jy$. *For a function* $u(x)$ *denote* $U(y) = u(jy)$. *Then*

i) a function $U(y)$ *may be obtained as* $u(jy)$ *for a function* u *on* S^1 *if and only if in the decomposition of* U *in Fourier series in* y *only modes* $e^{ijky}, k = \pm 1, \pm 2, \dots$ *are present.*

ii) Since $\partial^k U/\partial y^k = j^k \partial^k u/\partial x^k$ *for* $k \in \mathbf{N}$, *then* $u(t,x)$ *solves the problem*

$$u_t + j^2 \nu u_{xx} + ju u_x = \partial_t \xi(t,x); \quad u(0,x) = u_0(x), \tag{12.0.1}$$

if and only if $U(t,y) = u(t,jy)$ *solves* (1.1.13) *with* u_0 *and* ξ *replaced by* $U_0(y)$ *and* $\Xi(y)$, *obtained from* u_0 *and* ξ *as above.*

Now, for any given $U_0(y)$ *and* $\Xi(t,y)$ *as in i) we consider the corresponding* $u_0(x)$ *and* $\xi(t,x)$, *solve for them problem* (12.0.1), *and for the solution* $u(t,x)$ *construct the corresponding* $U(t,y)$. *This is a solution of* (1.1.13) *as required; it is unique by Theorem* 1.3.6.

SOLUTION 1.3.8. *We restrict ourselves to deriving estimate* (1.3.22). *As in the proof of Theorem* 1.3.6 *we write a solution* u *of* (0.0.12) *as* $u = v + w$, *where* v *solves* (1.3.1). *Then*

$$w_t - \nu w_{xx} = -\frac{\partial}{\partial x} f(w+v). \tag{12.0.2}$$

As before, we firstly multiply the equation by $w(t)$ *in* L_2. *Since* $\langle \frac{\partial}{\partial x} f(w+v), w\rangle = -\langle \frac{\partial}{\partial x} f(w+v), v\rangle$, *then we get*

$$\frac{1}{2}\frac{d}{dt}\|w\|^2 + \nu \|w\|_1^2 \leq |\langle f(w+v), \frac{\partial}{\partial x} v\rangle| \leq \|f(w+v)\|\,\|v\|_1.$$

But as $|f(w+v)| \leq C(|w|^2 + |v|^2)$, *then* $\|f(w+v)\| \leq C(|w|_4^2 + |v|_4^2) \leq C_1(\|w\|\,\|w\|_1 + \|v\|_1^2)$, *and we again arrive at estimate* (1.3.20). *Next multiplying* (12.0.2) *in* L_2 *by* w_{xx} *and using* (1.3.18) *we find that the l.h.s. of* (1.3.21) *with* $m = 1$ *is bounded by*

$$\Big|\langle \frac{\partial}{\partial x} f(w+v), w_{xx}\rangle\Big| \leq N_1^{0f}(w+v)\|w\|_2 \leq C\|w\|_2 \|w+v\|_2^{3/4}\|w+v\|^{5/4}.$$

Arguing as at Step 2 of the proof and using (1.3.20) *we derive from here that* w *satisfies* (1.3.22) *with* $m = 1$. *So*

$$|w(t)|_{1,1} \leq C\|w(t)\|_1 \leq C'. \tag{12.0.3}$$

Finally, multiplying (12.0.2) *in* L_2 *by* $w^{(2m)}$ *and using* (1.3.17) *we bound the l.h.s. of* (1.3.21) *by*

$$|\langle \partial_x^m f(w+v), \partial_x^{m+1} w\rangle| \leq C_1 \|w\|_{m+1} \|w+v\|_{m+1}^{(2m-1)/(2m+1)} (1 + |w+v|_{1,1})^{c_m}.$$

Jointly with (12.0.3) *this relation implies for* w *the final estimate* (1.3.22) *for the same reason as when proving Theorem* 1.3.6.

SOLUTION 1.3.10. *Firstly we check that relation* (1.3.33) *holds for the Galerkin approximations* $u^N(t,x)$. *Then the balance relation still is valid for the solution* u *since by Proposition* 1.3.9 *with* $m = 1$ *and* $T \geq \sigma$ *the sequence* $\{u^N\}$ *converges to* u *strongly in* $L_2(0,T;H^1)$ *and in* X_T^0.

SOLUTION 1.4.5. *Step 1) of the proof of Theorem 1.4.4 and Step 2) for $m = 1$ go without changes compared to the case of the Burgers equation, since for $m = 1$ inequality (1.3.18) coincides with (1.3.15). To repeat Step 2) for $m \geq 2$ we start with an auxiliary construction. Namely, for an integer $k \geq 2$ we consider $G_k(u^N)=\|u^N\|_1^{2k}$ and apply to it Ito's formula (its applicability follows from the same argument as at Step 1)). Denoting by C, C_1 various N-independent positive constants (which may depend on ν, k and u_0) and using (1.3.18) with $m = 1$, followed by Young's inequality, we get:*

$$\begin{aligned}
&\frac{\partial}{\partial t}\mathbb{E}G_k(u^N)\\
&\quad\leq -2k\nu\mathbb{E}\Big[\|u^N\|_2^2 G_{k-1}(u^N)\Big] - 2k\mathbb{E}\left[\langle(-\Delta)u^N, \partial_x f(u^N)\rangle_1 G_{k-1}(u^N)\right]\\
&\qquad + k(B_1^N + 2(k-1)\max_{|s|\geq 1}(2\pi|s||b_s|)^2)\ \ \mathbb{E}\ G_{k-1}(u^N)\\
&\quad\leq -\nu\mathbb{E}\ \|u^N\|_2^2 G_{k-1}(u^N)\\
&\qquad + \mathbb{E}\ \Big[(-(2k-1)\nu\|u^N\|_2^2 + (2k-1)\nu\|u^N\|_2^2 + C\|u^N\|^a + C)G_{k-1}(u^N)\Big]\\
&\quad\leq -\nu\mathbb{E}\ G_k(u^N) + C\mathbb{E}\ \Big[(1+\|u^N\|^{ak})^{1/k}(G_k(u^N))^{(k-1)/k}\Big], \qquad a > 0.
\end{aligned}$$

Now applying (1.4.9) jointly with Hölder's inequality we find that

$$\frac{\partial}{\partial t}\mathbb{E}G_k(u^N)\leq -\nu\mathbb{E}\ G_k(u^N)+C_1\Big[\mathbb{E}\ G_k(u^N)\Big]^{(k-1)/k}\leq C_2(-\mathbb{E}\ G_k(u^N)+1), \qquad \textit{a.s. in } t.$$

Therefore, by Gronwall's lemma, $\mathbb{E}G_k(u^N(t))$ is bounded uniformly in t and N, for all $k \geq 2$. Thus we have obtained upper bounds for all moments of $\|u^N(t)\|_1$, and therefore also of $|u^N(t)|_{1,1}$. Then we use them in combination with (1.3.17) and Young's inequality in order to generalise estimate (1.4.16) for $m \geq 2$, which completes Step 2).

Step 3) is the same as previously.

SOLUTION 1.7.4. *It is obvious that (1.7.3) is a norm. If $\{f_j\}$ is a sequence in $L(O)$, fundamental with respect to this norm, then $f_j \to f$ in $C_b(O)$ for some $f \in C_b(O)$. It is easy to check that f is a Lipschitz function on O and that $Lip(f_j - f) \to 0$, so $\|f_j - f\|_{L(O)} \to 0$.*

To prove the second assertion it suffices to note that $\{f_\lambda(x) = \min(\lambda x, 1),\quad 1 \leq \lambda \leq 2\}$ is an uncountable family in $L([0,1])$ such that for $\lambda_1 \neq \lambda_2$

$$\|f_{\lambda_1} - f_{\lambda_2}\|_L \geq Lip(f_{\lambda_1} - f_{\lambda_2})\geq 1.$$

SOLUTION 1.7.6. *If $\phi \in L(Y)$, then $\phi \circ F \in L(X)$ and $Lip\,\phi \circ F \leq C\, Lip\,\phi$. So $\|\phi \circ F\|_{L(X)} \leq C'\|\phi\|_{L(Y)}$. Then*

$$\begin{aligned}
\|\mu - \nu\|_{L(X)}^* &= \sup_{\|f\|_{L(X)}\leq 1}\ \big(\langle f, \mu\rangle - \langle f, \nu\rangle\big)\\
&\geq \sup_{\|\phi\|_{L(Y)}\leq 1}\ \big(\langle {C'}^{-1}\phi \circ F, \mu\rangle - \langle {C'}^{-1}\phi \circ F, \nu\rangle\big)\\
&= {C'}^{-1}\|F \circ \mu - F \circ \nu\|_{L(Y)}^*
\end{aligned}$$

which proves (1.7.5).

SOLUTION 1.8.7. *The constants below depend on $r, T, \nu, \|u_0\|_{r+1}$ and $\|\eta\|_{U^r}$.*

A solution of the problem is unique for the same reason as at Step 1 of the proof of Theorem 1.3.6. *To derive a priori estimates we argue as when proving Theorem* 1.3.6*: firstly we multiply the equation in* L_2 *by* $u(t)$ *and get that* $\|u(t)\| \le C$ *for* $0 \le t \le T$. *Next we denote* $m = r+1$ *and multiply the equation in* L_2 *by* $u^{(2m)}$. *Arguing as at Step 2 of the proof with* $v = 0$ *we find that*

$$\frac{1}{2}\frac{d}{dt}\|u(t)\|_m^2 + \frac{\nu}{2}\|u(t)\|_{m+1}^2 \le C_1 + C_2\|\eta(t)\|_{m-1}^2.$$

So u *meets estimate* (1.3.22). *After the a priori estimates are obtained, Galerkin's method implies existence of a solution* $u(t,x)$ *of the smoothness prescribed in the theorem, as at Steps 3-4 of the proof of Theorem* 1.3.6. *See Section 2.1 of* [**103**], *where solutions of the 2d Navier–Stokes system with a regular force are constructed similarly.*

The mapping Ψ *is locally Lipschitz for the same reason as in Problem* 1.3.12*; so it is bounded on bounded sets since* $\Psi(0,0) = 0$. *Alternatively, the local Lipschitz property follows from the result of Problem* 1.8.8.

SOLUTION 2.1.6. *The function* $v = u - \xi$ *satisfies*

$$v_t - \nu v_{xx} = \nu\xi_{xx} - uu_x =: g, \qquad v(0) = u_0. \tag{12.0.4}$$

Due to Theorem 1.3.6, g *lies in* X_T^0 *and continuously depends on* $\xi \in \dot{X}_T^4$. *Let us write* v *as* $v = v^1 + v^2$, *where* v^1 *solves* (12.0.4) *with* $u_0 := u_0$, $g := 0$ *and* v^2 *solves it for* $u_0 := 0$, $g := g$. *Decomposing in Fourier series, it immediately follows that* $v^1(\tau) \in H^s$ *continuously depends on* $\tau > 0$, *for any* $s > 0$. *Now note that* $v_t^2 = w$ *satisfies*

$$w_t - \nu w_{xx} = \partial g(t,x)/\partial t, \quad w(0) = g(0),$$

where $g(0) \in L_2$ *continuously depends on* ξ. *By Lemma* 1.3.1, $v_t^2 = w \in X_T^{-\varepsilon}$ *continuously depends on* $g \in X_T^0$, *for any* $\varepsilon > 0$. *In view of the equation* (12.0.4) *this implies that* $v \in X_T^{2-\varepsilon}$ *continuously depends on* $g \in X_T^0$, *which proves the problem's assertion.*

SOLUTION 2.3.5. *See in* [**18**].

SOLUTION 3.2.3. *Consider the function*

$$v(t,x) = \exp(\tfrac{1}{2}(t-1)(\pi/2\varepsilon')^2)\cos(x\pi/2\varepsilon').$$

It satisfies the equation in (3.2.3), *vanishes for* $x = \pm\varepsilon'$ *and* $0 \le v \le 1$ *for* $t = 1$. *The difference* $w = u - v$ *solves the equation, vanishes for* $x = \pm\varepsilon'$ *and is non-negative for* $t = 1$. *So* $w \ge 0$ *in* Π *by the maximum principle, and* $u(0,0) \ge v(0,0) = \exp\left(\frac{-\pi^2}{8\varepsilon'^2}\right)$.

SOLUTION 3.4.3. *Case* $a < 3/2$. *Let* $t = \nu^b\tau$ *(the parameter* b *will be fixed later), and consider* $\tilde{\xi}(\tau) = \nu^{-b/2}\xi(\nu^b\tau)$ *This is also a Wiener process in* H *with the same law as* ξ *(since it may be written in the form* (1.1.6) *with other standard independent Wiener processes* β_s*; see Problem* 11.5.1*). Let us substitute* $v = \nu^b u$. *Then eq.* (3.4.10) *can be rewritten as* $v_\tau + vv_x - \nu^{b+1}v_{xx} = \nu^{a+3b/2}\tilde{\xi}_\tau$. *Choose* $b = -2a/3$. *Then* $u = \nu^{2a/3}v$, *and*

$$v_\tau + vv_x - \tilde{\nu}v_{xx} = \tilde{\xi}_\tau, \quad \tilde{\nu} = \nu^{1-2a/3}.$$

The assumption $a < 3/2$ ensures that $\tilde{\nu}$ goes to zero with ν. By Theorem 3.4.1 the energy of the solution v is of order one, so the energy of u for small ν is large if $a < 0$ and is small if $a > 0$.

Case $a \geq 3/2$. Applying Ito's formula to $h(t) = \mathbb{E}|u|_2^2$ (rigorously speaking, as in the proof of Theorem 1.4.4 we should apply the formula to the Galerkin approximations and next pass to the limit), we obtain that

$$dh(t)/dt = -2\nu\|u(t)\|_1^2 + \nu^{2a}B_0 \leq -2\nu h(t) + \nu^{2a}B_0.$$

Consider the solution to (3.4.10) with $u_0 = 0$. By Gronwall's lemma we get

$$h(t) \leq B_0\nu^{2a}\frac{1}{2\nu}(1 - e^{-2\nu t}) \leq \frac{B_0}{2}\nu^{2a-1}, \quad t \geq 0.$$

Arguing as when proving Theorem 3.4.1 we derive from here that $\int |u|_2^2 d\mu_\nu(u) \leq C\nu^{2a-1}$, which completes the proof.

SOLUTION 7.2.4. *By the result of Problem 1.3.7, if the force ξ and the initial data u_0 are periodic with period $1/j$, where $j \geq 2$ is an integer, then the solution u also is $1/j$-periodic, so $S_{p,1/j} = 0$, for any p.*

SOLUTION 8.2.5. *For any $(u_0, \xi) \in H^2 \times \dot{X}_T^4$ the set $\{u^\nu(t; u_0, \xi), 0 < \nu \leq 1\}$ is bounded in L_∞. So by Alaoglu's theorem* [**52**] *each sequence $\nu_j \to 0$ has a subsequence $\nu_j' \to 0$ such that $\{u^{\nu_j'}(t; u_0, \xi)\}$ converges to a limit in L_∞^*. By (8.2.4) the limit equals $u^e(t; u_0, \xi)$, so $u^\nu(t; u_0, \xi) \to u^e(t; u_0, \xi)$ in L_∞^*. Accordingly, for each $\phi \in L_1$, $\int u^\nu(t; u_0, \xi)\phi\, dx$ converges to $\int u^e(t; u_0, \xi)\phi\, dx$, for every (u_0, ξ). So the latter is a measurable function on $H^2 \times \dot{X}_T^4$.*

SOLUTION 10.1.2. *See in* [**27**].

SOLUTION 10.1.3. *See in* [**27**].

SOLUTION 11.1.11. *Let us decompose the functions u_j and u in the base $\{e_s\}$: $u_j = \sum u_j^s e_s$, $u = \sum u^s e_s$. Then $\sum_s |2\pi s|^{2n}|u_j^s|^2 \leq C^2$ for each j, and $u_j^s \to u^s$ as $j \to \infty$. So by Fatou's lemma $\sum_s |2\pi s|^{2n}|u^s|^2 \leq C^2$.*

SOLUTION 11.2.3. *It is obvious that $\|\cdot\|_W$ is a norm. So it remains to show that W is complete with respect to it. Let $v_1, v_2, \dots$ be a Cauchy sequence of elements in W with respect to $\|\cdot\|_W$. Then*

$$v_n \to v \text{ in } L_{p_1}(0,T;B_1), \qquad v_{nt} \to w \text{ in } L_{p_3}(0,T;B_3).$$

From here we get that $(\partial/\partial t)v = w$ in the sense of generalised functions, so $v \in W$ and $\|v_n - v\|_W \to 0$ as $n \to \infty$.

SOLUTION 11.2.5. *See* [**118**, Chapter 5.2].

Acknowledgements

AB would like to acknowledge the support of the grants ANR-12-BS01-0020 WKBHJ and ANR-16-CE40-0013 ISDEEC. Both authors have also been supported by the grant ANR-17-CE40-0006 NONSTOPS, and SK was supported by the grant 18-11-00032 of Russian Science Foundation.

We would also like to thank Vladimir Bogachev, Vahagn Nersesyan and Armen Shirikyan for discussions during the preparation of the manuscript, as well as the anonymous referee for the remarks which allowed us to improve the quality of presentation in our text.

Bibliography

[1] E. Aurell, U. Frisch, J. Lutsko, and M. Vergassola, *On the multifractal properties of the energy dissipation derived from turbulence data*, J. Fluid Mech. **238** (1992), 467–486, DOI 10.1017/S0022112092001782. MR1166090 ↑1, 6, 96, 133

[2] Y. Bakhtin, E. Cator, and K. Khanin, *Space-time stationary solutions for the Burgers equation*, J. Amer. Math. Soc. **27** (2014), no. 1, 193–238, DOI 10.1090/S0894-0347-2013-00773-0. MR3110798 ↑99

[3] Y. Bakhtin and K. Khanin, *On global solutions of the random Hamilton-Jacobi equations and the KPZ problem*, Nonlinearity **31** (2018), no. 4, R93–R121, DOI 10.1088/1361-6544/aa99a6. MR3816628 ↑99

[4] Y. Bakhtin and L. Li, *Thermodynamic limit for directed polymers and stationary solutions of the Burgers equation*, Comm. Pure Appl. Math. **72** (2019), no. 3, 536–619, DOI 10.1002/cpa.21779. MR3911894 ↑99

[5] C. Bardos, P. Penel, U. Frisch, and P.-L. Sulem, *Modified dissipativity for a nonlinear evolution equation arising in turbulence*, Arch. Rational Mech. Anal. **71** (1979), no. 3, 237–256, DOI 10.1007/BF00280598. MR531061 ↑153

[6] M. V. Bartuccelli, J. D. Gibbon, and M. Oliver, *Length scales in solutions of the complex Ginzburg-Landau equation*, Phys. D **89** (1996), no. 3-4, 267–286, DOI 10.1016/0167-2789(95)00275-8. MR1369236 ↑85, 133, 155, 156

[7] G. K. Batchelor, *The theory of homogeneous turbulence*, Cambridge Monographs on Mechanics and Applied Mathematics, Cambridge at the University Press, 1953. MR0052268 ↑7, 94, 98

[8] H. Bateman, *Some recent researches on the motion of fluids*, Monthly Weather Review **43** (1915), 163–170. 95

[9] C. Bauzet, G. Vallet, and P. Wittbold, *The Cauchy problem for conservation laws with a multiplicative stochastic perturbation*, J. Hyperbolic Differ. Equ. **9** (2012), no. 4, 661–709, DOI 10.1142/S0219891612500221. MR3021756 ↑119

[10] U. Frisch and J. Bec, *"Burgulence"*, Turbulence: nouveaux aspects/New trends in turbulence (Les Houches, 2000), EDP Sci., Les Ulis, 2001, pp. 341–383, DOI 10.1007/3-540-45674-0_7. MR1999766 ↑1, 96, 133, 138

[11] J. Bec and K. Khanin, *Burgers turbulence*, Phys. Rep. **447** (2007), no. 1-2, 1–66, DOI 10.1016/j.physrep.2007.04.002. MR2318582 ↑97

[12] E. Ben-Naim, S. Y. Chen, G. D. Doolen, and S. Redner, *Shocklike dynamics of inelastic gases*, Phys. Rev. Lett. **83** (1999), 4069–4072. 2

[13] P. Biler, A. Boritchev, G. Karch, and P. Laurençot, *Concentration phenomena in a diffusive aggregation model*, J. Differential Equations **271** (2021), 1092–1108, DOI 10.1016/j.jde.2020.09.035. MR4154938 ↑8, 133

[14] ———, *Concentration phenomena in a diffusive aggregation model*, J. Differential Equations **271** (2021), 1092–1108, DOI 10.1016/j.jde.2020.09.035. MR4154938 ↑8, 133

[15] P. Biler, T. Funaki, and W. A. Woyczynski, *Fractal Burgers equations*, J. Differential Equations **148** (1998), no. 1, 9–46, DOI 10.1006/jdeq.1998.3458. MR1637513 ↑153

[16] P. Billingsley, *Convergence of probability measures*, 2nd ed., Wiley Series in Probability and Statistics: Probability and Statistics, John Wiley & Sons, Inc., New York, 1999. A Wiley-Interscience Publication, DOI 10.1002/9780470316962. MR1700749 ↑42, 44, 142, 146

[17] A. Biryuk, *Estimates for Spatial Derivatives of Solutions for Quasilinear Parabolic Equations with Small Viscosity*, Ph.D. thesis, Heriot-Watt University, Edinburgh, Scotland, United Kingdom, 2001. 85, 133

[18] ———, *Spectral properties of solutions of the Burgers equation with small dissipation*, Functional Analysis and its Applications **35:1** (2001), 1–12. 85, 119, 133, 137, 138, 178

[19] ———, *Note on the transformation that reduces the Burgers equation to the heat equation*, 2003, Mathematical Physics Preprint Archive, mp arc: 03-370. 2

[20] ———, *On multidimensional Burgers type equations with small viscosity*, Contributions to current challenges in mathematical fluid mechanics (G.Galdi, J.Heywood, and R.Rannacher, eds.), Advances in Mathematical Fluid Mechanics, Birkhäuser, 2004, pp. 1–30. 154

[21] A. Biryuk, *Lower bounds for derivatives of solution for nonlinear Schrödinger equations*, Proc. Roy. Soc. Edinburgh Sect. A **139** (2009), no. 2, 237–251, DOI 10.1017/S0308210507000133. MR2496962 ↑155

[22] V. I. Bogachev, *Weak convergence of measures*, Mathematical Surveys and Monographs, vol. 234, American Mathematical Society, Providence, RI, 2018, DOI 10.1090/surv/234. MR3837546 ↑42, 44, 162, 163, 164

[23] ———, *Gaussian measures*, Mathematical Surveys and Monographs, vol. 62, American Mathematical Society, Providence, RI, 1998, DOI 10.1090/surv/062. MR1642391 ↑20, 67, 168

[24] V. I. Bogachev and O. G Smolyanov, *Real and functional analysis*, Springer, 2020. 161

[25] T. Bohr, M. H. Jensen, G. Paladin, and A. Vulpiani, *Dynamical systems approach to turbulence*, Cambridge Nonlinear Science Series, vol. 8, Cambridge University Press, Cambridge, 1998, DOI 10.1017/CBO9780511599972. MR1643112 ↑155

[26] A. Boritchev, *Generalised Burgers equation with random force and small viscosity*, Ph.D. thesis, Ecole Polytechnique, 2012. 85, 133

[27] A. Boritchev, *Estimates for solutions of a low-viscosity kick-forced generalized Burgers equation*, Proc. Roy. Soc. Edinburgh Sect. A **143** (2013), no. 2, 253–268, DOI 10.1017/S0308210511000989. MR3039811 ↑179

[28] ———, *Sharp estimates for turbulence in white-forced generalised Burgers equation*, Geom. Funct. Anal. **23** (2013), no. 6, 1730–1771, DOI 10.1007/s00039-013-0245-4. MR3132902 ↑8, 85, 133

[29] ———, *Decaying turbulence in the generalised Burgers equation*, Arch. Ration. Mech. Anal. **214** (2014), no. 1, 331–357, DOI 10.1007/s00205-014-0766-5. MR3237888 ↑137, 138

[30] ———, *Decaying turbulence for the fractional subcritical Burgers equation*, Discrete Contin. Dyn. Syst. **38** (2018), no. 5, 2229–2249, DOI 10.3934/dcds.2018092. MR3809034 ↑153

[31] ———, *Exponential convergence to the stationary measure for a class of 1D Lagrangian systems with random forcing*, Stoch. Partial Differ. Equ. Anal. Comput. **6** (2018), no. 1, 109–123, DOI 10.1007/s40072-017-0104-7. MR3768996 ↑129, 134

[32] H. Brezis and P. Mironescu, *Where Sobolev interacts with Gagliardo-Nirenberg*, J. Funct. Anal. **277** (2019), no. 8, 2839–2864, DOI 10.1016/j.jfa.2019.02.019. MR3990737 ↑22

[33] Z. Brzeźniak, B. Goldys, and M. Neklyudov, *Multidimensional stochastic Burgers equation*, SIAM J. Math. Anal. **46** (2014), no. 1, 871–889, DOI 10.1137/120866117. MR3166959 ↑154

[34] J. M. Burgers, *A mathematical model illustrating the theory of turbulence*, Advances in Applied Mechanics, Academic Press, Inc., New York, N. Y., 1948, pp. 171–199. edited by Richard von Mises and Theodore von Kármán,. MR0027195 ↑3, 6, 95, 96

[35] ———, *The nonlinear diffusion equation: asymptotic solutions and statistical problems*, Reidel, 1974. 2, 95

[36] H. Cartan, *Calcul différentiel* (French), Hermann, Paris, 1967. MR0223194 ↑9, 44, 45

[37] G.-Q. G. Chen and P. H. C. Pang, *Invariant measures for nonlinear conservation laws driven by stochastic forcing*, Chin. Ann. Math. Ser. B **40** (2019), no. 6, 967–1004, DOI 10.1007/s11401-019-0169-x. MR4032936 ↑119

[38] D. Chowdhury, L. Santen, and A. Schadschneider, *Statistical physics of vehicular traffic and some related systems*, Phys. Rep. **329** (2000), no. 4-6, 199–329, DOI 10.1016/S0370-1573(99)00117-9. MR1755758 ↑2

[39] J. D. Cole, *On a quasi-linear parabolic equation occurring in aerodynamics*, Quart. Appl. Math. **9** (1951), 225–236, DOI 10.1090/qam/42889. MR42889 ↑2

[40] P. Constantin and V. Vicol, *Nonlinear maximum principles for dissipative linear nonlocal operators and applications*, Geom. Funct. Anal. **22** (2012), no. 5, 1289–1321, DOI 10.1007/s00039-012-0172-9. MR2989434 ↑153

[41] G. Da Prato, A. Debussche, and R. Temam, *Stochastic Burgers' equation*, NoDEA Nonlinear Differential Equations Appl. **1** (1994), no. 4, 389–402, DOI 10.1007/BF01194987. MR1300149 ↑8

[42] G. Da Prato and D. Gatarek, *Stochastic Burgers equation with correlated noise*, Stochastics Stochastics Rep. **52** (1995), no. 1-2, 29–41, DOI 10.1080/17442509508833962. MR1380259 ↑85

[43] G. Da Prato and J. Zabczyk, *Stochastic equations in infinite dimensions*, Encyclopedia of Mathematics and its Applications, vol. 44, Cambridge University Press, Cambridge, 1992, DOI 10.1017/CBO9780511666223. MR1207136 ↑1, 175

[44] G. Da Prato and J. Zabczyk, *Ergodicity for infinite-dimensional systems*, London Mathematical Society Lecture Note Series, vol. 229, Cambridge University Press, Cambridge, 1996, DOI 10.1017/CBO9780511662829. MR1417491 ↑1, 3, 85

[45] C. M. Dafermos, *Hyperbolic conservation laws in continuum physics*, 3rd ed., Grundlehren der Mathematischen Wissenschaften [Fundamental Principles of Mathematical Sciences], vol. 325, Springer-Verlag, Berlin, 2010, DOI 10.1007/978-3-642-04048-1. MR2574377 ↑124

[46] A. Debussche, *Ergodicity results for the stochastic Navier-Stokes equations: an introduction*, Topics in mathematical fluid mechanics, Lecture Notes in Math., vol. 2073, Springer, Heidelberg, 2013, pp. 23–108, DOI 10.1007/978-3-642-36297-2_2. MR3076070 ↑94, 151

[47] A. Debussche and J. Vovelle, *Scalar conservation laws with stochastic forcing*, J. Funct. Anal. **259** (2010), no. 4, 1014–1042, DOI 10.1016/j.jfa.2010.02.016. MR2652180 ↑119

[48] ———, *Invariant measure of scalar first-order conservation laws with stochastic forcing*, Probability Theory and Related Fields **163** (2013), 575–611. 119

[49] N. Dirr and P. E. Souganidis, *Large-time behavior for viscous and nonviscous Hamilton-Jacobi equations forced by additive noise*, SIAM J. Math. Anal. **37** (2005), no. 3, 777–796, DOI 10.1137/040611896. MR2191776 ↑119

[50] B. Dubrulle, *Beyond Kolmogorov cascades*, J. Fluid Mech. **867** (2019), P1, 63, DOI 10.1017/jfm.2019.98. MR3925888 ↑93

[51] R. M. Dudley, *Real analysis and probability*, Cambridge Studies in Advanced Mathematics, vol. 74, Cambridge University Press, Cambridge, 2002. Revised reprint of the 1989 original, DOI 10.1017/CBO9780511755347. MR1932358 ↑42, 43, 44, 77, 157, 159, 162, 164, 165, 167

[52] N. Dunford and J. T. Schwartz, *Linear operators. Part I*, Wiley Classics Library, John Wiley & Sons, Inc., New York, 1988. General theory; With the assistance of William G. Bade and Robert G. Bartle; Reprint of the 1958 original; A Wiley-Interscience Publication. MR1009162 ↑122, 123, 161, 179

[53] A. Dunlap, *Existence of stationary stochastic Burgers evolutions on* $\mathbf{R}^2$ *and* $\mathbf{R}^3$, Nonlinearity **33** (2020), no. 12, 6480–6501, DOI 10.1088/1361-6544/aba50a. MR4173562 ↑99

[54] A. Dunlap, C. Graham, and L. Ryzhik, *Stationary solutions to the stochastic Burgers equation on the line*, Comm. Math. Phys. **382** (2021), no. 2, 875–949, DOI 10.1007/s00220-021-04025-x. MR4227165 ↑99

[55] R. Durrett, *Probability. Theory and examples* , third ed., Thomson, Brooks Cole, 2005. 143, 152

[56] W. E, K. Khanin, A. Mazel, and Ya. Sinai, *Invariant measures for Burgers equation with stochastic forcing*, Ann. of Math. (2) **151** (2000), no. 3, 877–960, DOI 10.2307/121126. MR1779561 ↑3, 96, 97, 119, 129, 134, 154

[57] S. F. Edwards, *The statistical dynamics of homogeneous turbulence*, J. Fluid Mech. **18** (1964), 239–273, DOI 10.1017/S0022112064000180. MR178673 ↑92

[58] W. C. Van Etten, *Introduction to Random Signals and Noise*, John Wiley and Sons, 2006. 142

[59] L. C. Evans, *Partial differential equations*, Graduate Studies in Mathematics, vol. 19, American Mathematical Society, Providence, RI, 1998. MR1625845 ↑65, 69, 124

[60] ———, *An introduction to stochastic differential equations*, American Mathematical Society, Providence, RI, 2013, DOI 10.1090/mbk/082. MR3154922 ↑1

[61] W. Feller, *An introduction to probability theory and its applications. Vol. II.*, Second edition, John Wiley & Sons, Inc., New York-London-Sydney, 1971. MR0270403 ↑67

[62] J. Feng and D. Nualart, *Stochastic scalar conservation laws*, J. Funct. Anal. **255** (2008), no. 2, 313–373, DOI 10.1016/j.jfa.2008.02.004. MR2419964 ↑119

[63] F. Flandoli and B. Maslowski, *Ergodicity of the 2-D Navier-Stokes equation under random perturbations*, Comm. Math. Phys. **172** (1995), no. 1, 119–141. MR1346374 ↑85, 151

[64] V. A. Florin, *Some of the simplest nonlinear problems arising in the consolidation of wet soil* (Russian), Izvestiya Akad. Nauk SSSR. Otd. Tehn. Nauk **1948** (1948), 1389–1402. For English translation, see ref. [17] at `www.mathnet.ru/eng/person161417%7D`. MR0029605 ↑2

[65] ———, *Fundamentals of Soil Mechanics (in Russian)*, vol. 1,2, Gosstroiizdat, Leningrad–Moscou, 1959-1961, Vol. 1 translated into English by "National Technical Information Service" (1961). 2

[66] C. Foias, O. Manley, R. Rosa, and R. Temam, *Navier-Stokes equations and turbulence*, Encyclopedia of Mathematics and its Applications, vol. 83, Cambridge University Press, Cambridge, 2001, DOI 10.1017/CBO9780511546754. MR1855030 ↑92

[67] A. R. Forsyth, *Theory of differential equations. Part 4. Partial differential equations*, vol. 5-6, Cambridge University Press, 1906. 95

[68] U. Frisch, *Turbulence*, Cambridge University Press, Cambridge, 1995. The legacy of A. N. Kolmogorov. MR1428905 ↑89, 92, 93, 133

[69] I. I. Gikhman and A. V. Skorokhod, *Introduction to the theory of random processes*, Dover Publications, Inc., Mineola, NY, 1996. Translated from the 1965 Russian original; Reprint of the 1969 English translation; With a preface by Warren M. Hirsch. MR1435501 ↑67, 78, 162, 169, 171, 173

[70] D. Gomes, R. Iturriaga, K. Khanin, and P. Padilla, *Viscosity limit of stationary distributions for the random forced Burgers equation* (English, with English and Russian summaries), Mosc. Math. J. **5** (2005), no. 3, 613–631, 743, DOI 10.17323/1609-4514-2005-5-3-613-631. MR2241814 ↑97, 119, 154

[71] M. Gubinelli and N. Perkowski, *KPZ reloaded*, Comm. Math. Phys. **349** (2017), no. 1, 165–269, DOI 10.1007/s00220-016-2788-3. MR3592748 ↑8

[72] I. Gyöngy and N. Krylov, *On the splitting-up method and stochastic partial differential equations*, Ann. Probab. **31** (2003), no. 2, 564–591, DOI 10.1214/aop/1048516528. MR1964941 ↑150

[73] I. Gyöngy and C. Rovira, *On stochastic partial differential equations with polynomial nonlinearities*, Stochastics Stochastics Rep. **67** (1999), no. 1-2, 123–146, DOI 10.1080/17442509908834205. MR1717799 ↑119

[74] M. Hairer, *Solving the KPZ equation*, Ann. of Math. (2) **178** (2013), no. 2, 559–664, DOI 10.4007/annals.2013.178.2.4. MR3071506 ↑8

[75] M. Hairer and J. C. Mattingly, *Ergodicity of the 2D Navier-Stokes equations with degenerate stochastic forcing*, Ann. of Math. (2) **164** (2006), no. 3, 993–1032, DOI 10.4007/annals.2006.164.993. MR2259251 ↑151

[76] E. Hopf, *The partial differential equation $u_t + uu_x = \mu u_{xx}$*, Comm. Pure Appl. Math. **3** (1950), 201–230, DOI 10.1002/cpa.3160030302. MR47234 ↑2

[77] G. Huang and S. Kuksin, *On the energy transfer to high frequencies in damped/driven nonlinear Schrödinger equation*, Stoch. PDE Anal. Appl., to appear (2021). 156

[78] R. Iturriaga and K. Khanin, *Burgers turbulence and random Lagrangian systems*, Comm. Math. Phys. **232** (2003), no. 3, 377–428, DOI 10.1007/s00220-002-0748-6. MR1952472 ↑97, 119

[79] R. Iturriaga, K. Khanin, and K. Zhang, *Exponential convergence of solutions for random Hamilton-Jacobi equations*, Stoch. Partial Differ. Equ. Anal. Comput. **8** (2020), no. 3, 544–579, DOI 10.1007/s40072-019-00153-7. MR4147697 ↑129, 134

[80] I. Karatzas and S. E. Shreve, *Brownian motion and stochastic calculus*, 2nd ed., Graduate Texts in Mathematics, vol. 113, Springer-Verlag, New York, 1991, DOI 10.1007/978-1-4612-0949-2. MR1121940 ↑36, 78, 144, 168, 169, 170, 171, 173

[81] M. Kardar, G. Parisi, and Y.-C. Zhang, *Dynamic scaling of growing interfaces*, Phys. Rev. Lett. **56** (1986), 889–892. 3

[82] T. Kelaï and S. Kuksin, *Introduction à l'équation de Burgers stochastique et à la Burgulence* (French, with French summary), Bull. Sci. Math. **140** (2016), no. 2, 140–187, DOI 10.1016/j.bulsci.2015.11.004. MR3456186 ↑8, 133

[83] K. Khanin and K. Zhang, *Hyperbolicity of minimizers and regularity of viscosity solutions for a random Hamilton-Jacobi equation*, Comm. Math. Phys. **355** (2017), no. 2, 803–837, DOI 10.1007/s00220-017-2919-5. MR3681391 ↑119, 154

[84] R. Khasminskii, *Stochastic stability of differential equations*, 2nd ed., Stochastic Modelling and Applied Probability, vol. 66, Springer, Heidelberg, 2012, DOI 10.1007/978-3-642-23280-0. MR2894052 ↑67, 171, 172

[85] S. Kida, *Asymptotic properties of Burgers turbulence*, J. Fluid Mech. **93** (1979), no. 2, 337–377, DOI 10.1017/S0022112079001932. MR544560 ↑95

[86] J. U. Kim, *On a stochastic scalar conservation law*, Indiana Univ. Math. J. **52** (2003), no. 1, 227–256, DOI 10.1512/iumj.2003.52.2310. MR1970028 ↑119

[87] A. Kiselev, F. Nazarov, and R. Shterenberg, *Blow up and regularity for fractal Burgers equation*, Dyn. Partial Differ. Equ. **5** (2008), no. 3, 211–240, DOI 10.4310/DPDE.2008.v5.n3.a2. MR2455893 ↑153

[88] A. N. Kolmogorov, *Dissipation of energy in the locally isotropic turbulence*, Proc. Roy. Soc. London Ser. A **434** (1991), no. 1890, 15–17, DOI 10.1098/rspa.1991.0076. Translated from the Russian by V. Levin. MR1124923 ↑89

[89] A. N. Kolmogoroff, *On degeneration of isotropic turbulence in an incompressible viscous liquid*, C. R. (Doklady) Acad. Sci. URSS (N. S.) **31** (1941), 538–540. MR0004568 ↑89

[90] A. N. Kolmogorov, *The local structure of turbulence in incompressible viscous fluid for very large Reynolds numbers*, Proc. Roy. Soc. London Ser. A **434** (1991), no. 1890, 9–13, DOI 10.1098/rspa.1991.0075. Translated from the Russian by V. Levin. MR1124922 ↑89

[91] A. Kolmogorov and S. Fomin, *Elements of the Theory of Functions and Functional Analysis*, Dover, 1999. 77, 122

[92] A. N. Kolmogorov, *Remarks on statistical solutions of Navier–Stokes equations (in Russian)*, Russian Math. Surveys **33** (1978), no. 3, 124. 92

[93] R. H. Kraichnan, *Inertial ranges in two-dimensional turbulence*, Phys. Fluids **10** (1967), 1417–1423. 95

[94] J. Krug and H. Spohn, *Universality classes for deterministic surface growth*, Phys. Rev. A (3) **38** (1988), no. 8, 4271–4283, DOI 10.1103/PhysRevA.38.4271. MR966369 ↑3

[95] S. N. Kružkov, *The Cauchy problem in the large for non-linear equations and for certain first-order quasilinear systems with several variables* (Russian), Dokl. Akad. Nauk SSSR **155** (1964), 743–746. MR0164137 ↑2, 49, 119, 124, 133

[96] ———, *First order quasilinear equations in several independent variables*, Mathematics of the USSR Sbornik **10** (1970), no. 2, 217–243. 2, 119, 133

[97] N. V. Krylov, *An analytic approach to SPDEs*, Stochastic partial differential equations: six perspectives, Math. Surveys Monogr., vol. 64, Amer. Math. Soc., Providence, RI, 1999, pp. 185–242, DOI 10.1090/surv/064/05. MR1661766 ↑1

[98] ———, *Introduction to the theory of random processes*, Graduate Studies in Mathematics, vol. 43, American Mathematical Society, Providence, RI, 2002, DOI 10.1090/gsm/043. MR1885884 ↑171, 173

[99] S. B. Kuksin, *A stochastic nonlinear Schrödinger equation. I. A priori estimates* (Russian, with Russian summary), Tr. Mat. Inst. Steklova **225** (1999), no. Solitony Geom. Topol. na Perekrest., 232–256; English transl., Proc. Steklov Inst. Math. **2(225)** (1999), 219–242. MR1725943 ↑85, 155, 156

[100] S. B. Kuksin, *Randomly forced nonlinear PDEs and statistical hydrodynamics in 2 space dimensions*, Zurich Lectures in Advanced Mathematics, European Mathematical Society (EMS), Zürich, 2006, DOI 10.4171/021. MR2225710 ↑94, 95, 152, 153

[101] S. Kuksin, V. Nersesyan, and A. Shirikyan, *Exponential mixing for a class of dissipative PDEs with bounded degenerate noise*, Geom. Funct. Anal. **30** (2020), no. 1, 126–187, DOI 10.1007/s00039-020-00525-5. MR4081057 ↑152

[102] S. Kuksin and A. Shirikyan, *Stochastic dissipative PDEs and Gibbs measures*, Comm. Math. Phys. **213** (2000), no. 2, 291–330, DOI 10.1007/s002200000237. MR1785459 ↑94, 151

[103] ———, *Mathematics of two-dimensional turbulence*, Cambridge Tracts in Mathematics, vol. 194, Cambridge University Press, Cambridge, 2012, DOI 10.1017/CBO9781139137119. MR3443633 ↑35, 43, 70, 94, 95, 133, 138, 139, 140, 146, 150, 151, 152, 153, 156, 175, 178

[104] ———, *Rigorous results in space-periodic two-dimensional turbulence*, Physics of Fluids **29** (2017), 125106. 94, 95

[105] S. Kuksin and H. Zhang, *Exponential mixing for dissipative PDEs with bounded non-degenerate noise*, Stochastic Process. Appl. **130** (2020), no. 8, 4721–4745, DOI 10.1016/j.spa.2020.01.014. MR4108469 ↑152

[106] S. B. Kuksin, *Oscillations in space-periodic nonlinear Schrödinger equations*, Geom. Funct. Anal. **7** (1997), no. 2, 338–363, DOI 10.1007/PL00001622. MR1445390 ↑155

[107] ———, *Spectral properties of solutions for nonlinear PDEs in the turbulent regime*, Geom. Funct. Anal. **9** (1999), no. 1, 141–184, DOI 10.1007/s000390050083. MR1675893 ↑85, 133, 155

[108] S. Kuksin and V. Nersesyan, *Stochastic CGL equations without linear dispersion in any space dimension*, Stoch. Partial Differ. Equ. Anal. Comput. **1** (2013), no. 3, 389–423, DOI 10.1007/s40072-013-0010-6. MR3327512 ↑156

[109] H. H. Kuo, *Gaussian measures in Banach spaces*, Lecture Notes in Mathematics, Vol. 463, Springer-Verlag, Berlin-New York, 1975. MR0461643 ↑168

[110] A. Kupiainen and M. Marcozzi, *Renormalization of generalized KPZ equation*, J. Stat. Phys. **166** (2017), no. 3-4, 876–902, DOI 10.1007/s10955-016-1636-3. MR3607594 ↑8

[111] L. Bertini, N. Cancrini, and G. Jona-Lasinio, *The stochastic Burgers equation*, Comm. Math. Phys. **165** (1994), no. 2, 211–232. MR1301846 ↑8

[112] O. A. Ladyženskaja, V. A. Solonnikov, and N. N. Ural′ceva, *Linear and quasilinear equations of parabolic type* (Russian), Translated from the Russian by S. Smith. Translations of Mathematical Monographs, Vol. 23, American Mathematical Society, Providence, R.I., 1968. MR0241822 ↑69

[113] J. W. Lamperti, *Probability*, 2nd ed., Wiley Series in Probability and Statistics: Probability and Statistics, John Wiley & Sons, Inc., New York, 1996. A survey of the mathematical theory; A Wiley-Interscience Publication, DOI 10.1002/9781118150443. MR1406796 ↑142

[114] L. D. Landau and E. M. Lifschitz, *Fluid mechanics*, 2nd ed., Pergamon Press, 1987. 89, 93, 133

[115] S. Lang, *Real analysis*, 2nd ed., Addison-Wesley Publishing Company, Advanced Book Program, Reading, MA, 1983. MR783635 ↑161

[116] P. D. Lax, *Hyperbolic systems of conservation laws. II*, Comm. Pure Appl. Math. **10** (1957), 537–566, DOI 10.1002/cpa.3160100406. MR93653 ↑2, 119, 124

[117] M. Ledoux and M. Talagrand, *Probability in Banach spaces*, Ergebnisse der Mathematik und ihrer Grenzgebiete (3) [Results in Mathematics and Related Areas (3)], vol. 23, Springer-Verlag, Berlin, 1991. Isoperimetry and processes, DOI 10.1007/978-3-642-20212-4. MR1102015 ↑161

[118] J.-L. Lions, *Quelques méthodes de résolution des problèmes aux limites non linéaires* (French), Dunod; Gauthier-Villars, Paris, 1969. MR0259693 ↑26, 139, 161, 162, 179

[119] J.-L. Lions and E. Magenes, *Non-homogeneous boundary value problems and applications. Vol. I*, Springer-Verlag, New York-Heidelberg, 1972. Translated from the French by P. Kenneth; Die Grundlehren der mathematischen Wissenschaften, Band 181. MR0350177 ↑9, 26, 161, 162

[120] P.-L. Lions, B. Perthame, and E. Tadmor, *A kinetic formulation of multidimensional scalar conservation laws and related equations*, J. Amer. Math. Soc. **7** (1994), no. 1, 169–191, DOI 10.2307/2152725. MR1201239 ↑119

[121] W. Liu and M. Röckner, *Stochastic partial differential equations: an introduction*, Universitext, Springer, Cham, 2015, DOI 10.1007/978-3-319-22354-4. MR3410409 ↑1

[122] P.-A. Meyer, *Probability and potentials*, Blaisdell Publishing Co. Ginn and Co., Waltham, Mass.-Toronto, Ont.-London, 1966. MR0205288 ↑165

[123] N. Nagase, *Remarks on nonlinear stochastic partial differential equations: an application of the splitting-up method*, SIAM J. Control Optim. **33** (1995), no. 6, 1716–1730, DOI 10.1137/S036301299324618X. MR1358091 ↑150

[124] J. Neveu, *Mathematical foundations of the calculus of probability*, Translated by Amiel Feinstein, Holden-Day, Inc., San Francisco, Calif.-London-Amsterdam, 1965. MR0198505 ↑157, 162, 165, 166, 169

[125] L. Nirenberg, *On elliptic partial differential equations*, Ann. Scuola Norm. Sup. Pisa Cl. Sci. (3) **13** (1959), 115–162. MR109940 ↑22

[126] A. Obukhoff, *On the energy distribution in the spectrum of a turbulent flow*, C. R. (Doklady) Acad. Sci. URSS (N.S.) **32** (1941), 19–21. MR0005852 ↑92

[127] ———, *Spectral energy distribution in a turbulent flow*, Izvestiya Akademii Nauk SSSR, Seriya Geografii i Geofiziki **5** (1941), no. 4-5, 453–466. 92

[128] B. Øksendal, *Stochastic differential equations*, 6th ed., Universitext, Springer-Verlag, Berlin, 2003. An introduction with applications, DOI 10.1007/978-3-642-14394-6. MR2001996 ↑170

[129] O. A. Oleĭnik, *Discontinuous solutions of non-linear differential equations* (Russian), Uspehi Mat. Nauk (N.S.) **12** (1957), no. 3(75), 3–73. MR0094541 ↑2, 49, 119, 124

[130] K. R. Parthasarathy, *Introduction to probability and measure*, The Macmillan Co. of India, Ltd., Delhi, 1977. MR0651012 ↑42, 81, 162, 164, 165

[131] J. Pöschel and E. Trubowitz, *Inverse spectral theory*, Pure and Applied Mathematics, vol. 130, Academic Press, Inc., Boston, MA, 1987. MR894477 ↑44, 45, 46

[132] C. Prévôt and M. Röckner, *A concise course on stochastic partial differential equations*, Lecture Notes in Mathematics, vol. 1905, Springer, Berlin, 2007. MR2329435 ↑1

[133] J. Quastel and H. Spohn, *The one-dimensional KPZ equation and its universality class*, J. Stat. Phys. **160** (2015), no. 4, 965–984, DOI 10.1007/s10955-015-1250-9. MR3373647 ↑3, 8

[134] B. L. Rozovskiĭ, *Stochastic evolution systems*, Mathematics and its Applications (Soviet Series), vol. 35, Kluwer Academic Publishers Group, Dordrecht, 1990. Linear theory and applications to nonlinear filtering; Translated from the Russian by A. Yarkho, DOI 10.1007/978-94-011-3830-7. MR1135324 ↑1

[135] B. Saussereau and I. L. Stoica, *Scalar conservation laws with fractional stochastic forcing: existence, uniqueness and invariant measure*, Stochastic Process. Appl. **122** (2012), no. 4, 1456–1486, DOI 10.1016/j.spa.2012.01.005. MR2914759 ↑119

[136] S. F. Shandarin and Ya. B. Zel′dovich, *The large-scale structure of the universe: turbulence, intermittency, structures in a self-gravitating medium*, Rev. Modern Phys. **61** (1989), no. 2, 185–220, DOI 10.1103/RevModPhys.61.185. MR989562 ↑2

[137] A. Shirikyan, *A version of the law of large numbers and applications*, Probabilistic methods in fluids, World Sci. Publ., River Edge, NJ, 2003, pp. 263–271, DOI 10.1142/9789812703989_0017. MR2083377 ↑152, 153

[138] ———, *Control theory for the Burgers equation: Agrachev-Sarychev approach*, Pure Appl. Funct. Anal. **3** (2018), no. 1, 219–240. MR3777005 ↑151, 152, 165

[139] A. N. Shiryaev, *Probability*, 2nd ed., Graduate Texts in Mathematics, vol. 95, Springer-Verlag, New York, 1996. Translated from the first (1980) Russian edition by R. P. Boas, DOI 10.1007/978-1-4757-2539-1. MR1368405 ↑15, 17, 42, 44, 78, 144, 157, 159, 162, 165, 168

[140] Ya. G. Sinaĭ, *Two results concerning asymptotic behavior of solutions of the Burgers equation with force*, J. Statist. Phys. **64** (1991), no. 1-2, 1–12, DOI 10.1007/BF01057866. MR1117645 ↑3, 119

[141] Ya. G. Sinai, *Burgers system driven by a periodic stochastic flow*, Itô's stochastic calculus and probability theory, Springer, Tokyo, 1996, pp. 347–353. MR1439535 ↑3, 119

[142] D. W. Stroock and S. R. S. Varadhan, *Multidimensional diffusion processes*, Grundlehren der Mathematischen Wissenschaften [Fundamental Principles of Mathematical Sciences], vol. 233, Springer-Verlag, Berlin-New York, 1979. MR532498 ↑36

[143] G. I. Taylor, *Statistical theory of turbulence, I-IV*, Proc. Roy. Soc. London **A151** (1935), 421–478. 89, 94

[144] M. E. Taylor, *Partial differential equations I. Basic theory*, 2nd ed., Applied Mathematical Sciences, vol. 115, Springer, New York, 2011, DOI 10.1007/978-1-4419-7055-8. MR2744150 ↑9

[145] ———, *Partial differential equations III. Nonlinear equations*, 2nd ed., Applied Mathematical Sciences, vol. 117, Springer, New York, 2011, DOI 10.1007/978-1-4419-7049-7. MR2744149 ↑28

[146] A. Tsinober, *An informal conceptual introduction to turbulence*, Fluid Mechanics and its Applications, vol. 92, Springer, Dordrecht, 2009. Second edition of *An informal introduction to turbulence*, DOI 10.1007/978-90-481-3174-7. MR2723330 ↑93

[147] C. Villani, *Topics in optimal transportation*, Graduate Studies in Mathematics, vol. 58, American Mathematical Society, Providence, RI, 2003, DOI 10.1090/gsm/058. MR1964483 ↑42, 166, 167

[148] M. J. Vishik and A. V. Fursikov, *Mathematical problems of statistical hydromechanics*, Mathematics and its Applications (Soviet Series), vol. 9, Kluwer Academic Publishers Group, Dordrecht, 1988. Translated from the 1980 Russian original [MR0591678] by D. A. Leites, DOI 10.1007/978-94-009-1423-0. MR3444271 ↑1, 92, 99

[149] Th. von Karman and L. Howard, *On the statistical theory of isotropic turbulence*, Proc. Roy. Soc. London **A164** (1938), 192–215. 89, 94

[150] J. von Neumann, *Collected works (1949-63)*, vol. 6, ch. "Recent theories of turbulence", pp. 437–472, Pergamon Press, 1963. 95

[151] G. B. Whitham, *Linear and nonlinear waves*, Wiley-Interscience [John Wiley & Sons], New York-London-Sydney, 1974. Pure and Applied Mathematics. MR0483954 ↑2

[152] K. Yosida, *Functional analysis*, Classics in Mathematics, Springer-Verlag, Berlin, 1995. Reprint of the sixth (1980) edition, DOI 10.1007/978-3-642-61859-8. MR1336382 ↑123, 157, 159, 160

Index

Selected Published Titles in This Series

255 **Alexandre Boritchev and Sergei Kuksin,** One-Dimensional Turbulence and the Stochastic Burgers Equation, 2021
252 **Julie Déserti,** The Cremona Group and Its Subgroups, 2021
251 **David Hoff,** Linear and Quasilinear Parabolic Systems, 2020
250 **Bachir Bekka and Pierre de la Harpe,** Unitary Representations of Groups, Duals, and Characters, 2020
249 **Nikolai M. Adrianov, Fedor Pakovich, and Alexander K. Zvonkin,** Davenport–Zannier Polynomials and Dessins d'Enfants, 2020
248 **Paul B. Larson and Jindrich Zapletal,** Geometric Set Theory, 2020
247 **István Heckenberger and Hans-Jürgen Schneider,** Hopf Algebras and Root Systems, 2020
246 **Matheus C. Bortolan, Alexandre N. Carvalho, and José A. Langa,** Attractors Under Autonomous and Non-autonomous Perturbations, 2020
245 **Aiping Wang and Anton Zettl,** Ordinary Differential Operators, 2019
244 **Nabile Boussaïd and Andrew Comech,** Nonlinear Dirac Equation, 2019
243 **José M. Isidro,** Jordan Triple Systems in Complex and Functional Analysis, 2019
242 **Bhargav Bhatt, Ana Caraiani, Kiran S. Kedlaya, Peter Scholze, and Jared Weinstein,** Perfectoid Spaces, 2019
241 **Dana P. Williams,** A Tool Kit for Groupoid C^*-Algebras, 2019
240 **Antonio Fernández López,** Jordan Structures in Lie Algebras, 2019
239 **Nicola Arcozzi, Richard Rochberg, Eric T. Sawyer, and Brett D. Wick,** The Dirichlet Space and Related Function Spaces, 2019
238 **Michael Tsfasman, Serge Vlăduţ, and Dmitry Nogin,** Algebraic Geometry Codes: Advanced Chapters, 2019
237 **Dusa McDuff, Mohammad Tehrani, Kenji Fukaya, and Dominic Joyce,** Virtual Fundamental Cycles in Symplectic Topology, 2019
236 **Bernard Host and Bryna Kra,** Nilpotent Structures in Ergodic Theory, 2018
235 **Habib Ammari, Brian Fitzpatrick, Hyeonbae Kang, Matias Ruiz, Sanghyeon Yu, and Hai Zhang,** Mathematical and Computational Methods in Photonics and Phononics, 2018
234 **Vladimir I. Bogachev,** Weak Convergence of Measures, 2018
233 **N. V. Krylov,** Sobolev and Viscosity Solutions for Fully Nonlinear Elliptic and Parabolic Equations, 2018
232 **Dmitry Khavinson and Erik Lundberg,** Linear Holomorphic Partial Differential Equations and Classical Potential Theory, 2018
231 **Eberhard Kaniuth and Anthony To-Ming Lau,** Fourier and Fourier-Stieltjes Algebras on Locally Compact Groups, 2018
230 **Stephen D. Smith,** Applying the Classification of Finite Simple Groups, 2018
229 **Alexander Molev,** Sugawara Operators for Classical Lie Algebras, 2018
228 **Zhenbo Qin,** Hilbert Schemes of Points and Infinite Dimensional Lie Algebras, 2018
227 **Roberto Frigerio,** Bounded Cohomology of Discrete Groups, 2017
226 **Marcelo Aguiar and Swapneel Mahajan,** Topics in Hyperplane Arrangements, 2017
225 **Mario Bonk and Daniel Meyer,** Expanding Thurston Maps, 2017
224 **Ruy Exel,** Partial Dynamical Systems, Fell Bundles and Applications, 2017
223 **Guillaume Aubrun and Stanisław J. Szarek,** Alice and Bob Meet Banach, 2017
222 **Alexandru Buium,** Foundations of Arithmetic Differential Geometry, 2017

For a complete list of titles in this series, visit the
AMS Bookstore at **www.ams.org/bookstore/survseries/**.